AF718765

Methods in Cell Biology

Cilia: From Mechanisms to Disease - Part A

Volume 175

Series Editor

Lorenzo Galluzzi
Weill Cornell Medical College,
New York, NY, United States

Methods in Cell Biology

Cilia: From Mechanisms to Disease - Part A

Volume 175

Edited by

José Manuel Bravo-San Pedro
Universidad Complutense de Madrid, Madrid, Spain
Centro de Investigación Biomédica en Red en Enfermedades Neurodegenerativas (CIBERNED), Madrid, Spain

Lorenzo Galluzzi
Weill Cornell Medicine,
New York, NY, United States

ELSEVIER

ACADEMIC PRESS
An imprint of Elsevier

Academic Press is an imprint of Elsevier
50 Hampshire Street, 5th Floor, Cambridge, MA 02139, United States
525 B Street, Suite 1650, San Diego, CA 92101, United States
The Boulevard, Langford Lane, Kidlington, Oxford OX5 1GB, United Kingdom
125 London Wall, London, EC2Y 5AS, United Kingdom

First edition 2023

Copyright © 2023 Elsevier Inc. All rights reserved.

No part of this publication may be reproduced or transmitted in any form or by any means, electronic or mechanical, including photocopying, recording, or any information storage and retrieval system, without permission in writing from the publisher. Details on how to seek permission, further information about the Publisher's permissions policies and our arrangements with organizations such as the Copyright Clearance Center and the Copyright Licensing Agency, can be found at our website: www.elsevier.com/permissions.

This book and the individual contributions contained in it are protected under copyright by the Publisher (other than as may be noted herein).

Notices
Knowledge and best practice in this field are constantly changing. As new research and experience broaden our understanding, changes in research methods, professional practices, or medical treatment may become necessary.

Practitioners and researchers must always rely on their own experience and knowledge in evaluating and using any information, methods, compounds, or experiments described herein. In using such information or methods they should be mindful of their own safety and the safety of others, including parties for whom they have a professional responsibility.

To the fullest extent of the law, neither the Publisher nor the authors, contributors, or editors, assume any liability for any injury and/or damage to persons or property as a matter of products liability, negligence or otherwise, or from any use or operation of any methods, products, instructions, or ideas contained in the material herein.

ISBN: 978-0-443-18586-1
ISSN: 0091-679X

For information on all Academic Press publications
visit our website at https://www.elsevier.com/books-and-journals

Publisher: Zoe Kruze
Developmental Editor: Naiza Ermin Mendoza
Production Project Manager: Vijayaraj Purushothaman
Cover Designer: Matthew Limbert

Typeset by STRAIVE, India

Working together to grow libraries in developing countries
www.elsevier.com • www.bookaid.org

Contents

Contributors

Ruxandra Bachmann-Gagescu
Institute of Medical Genetics; Department of Molecular Life Sciences, University of Zurich, Zurich, Switzerland

Katja Baur
Department of Neurobiology, Universität Heidelberg, Heidelberg, Germany

José Manuel Bravo-San Pedro
Departamento de Fisiología, Facultad de Medicina, Universidad Complutense de Madrid; Centro de Investigación Biomédica en Red en Enfermedades Neurodegenerativas (CIBERNED), Madrid, Spain

Chia-Hsiang Chang
Institute of Biomedical Sciences, Academia Sinica, Taipei, Taiwan

Ting-Yu Chen
Institute of Biomedical Sciences, Academia Sinica, Taipei, Taiwan

Francesca Ciccolini
Department of Neurobiology, Universität Heidelberg, Heidelberg, Germany

Yaiza Corral Nieto
Departamento de Fisiología, Facultad de Medicina, Universidad Complutense de Madrid, Madrid, Spain

Bin Deng
Vermont Biomedical Research Network, University of Vermont, Burlington, VT, United States

Aurore M.M. Dupuy
Nantes Université, Inserm, CNRS, Université d'Angers, CRCI^{2}NA, Nantes, France

Marco Garbelli
Department of Molecular Life Sciences, University of Zurich, Zurich, Switzerland

Allison E. Gatz
Department of Biological Sciences, Center for Stem Cells and Regenerative Medicine, Center for Zebrafish Research, Boler-Parseghian Center for Rare and Neglected Diseases, University of Notre Dame, Notre Dame, IN, United States

Vincent J. Guen
Nantes Université, Inserm, CNRS, Université d'Angers, CRCI^{2}NA, Nantes, France

Gabriele Hölzl-Wenig
Department of Neurobiology, Universität Heidelberg, Heidelberg, Germany

Kosuke Hosoba
Program of Biomedical Science; Program of Mathematical and Life Science, Graduate School of Integrated Sciences for Life, Hiroshima University, Hiroshima, Japan

Faryal Ijaz
Department of Anatomy and Developmental Biology, Graduate School of Biomedical and Health Sciences, Hiroshima University, Hiroshima, Japan

Koji Ikegami
Department of Anatomy and Developmental Biology, Graduate School of Biomedical and Health Sciences, Hiroshima University, Hiroshima, Japan

Philippe P. Juin
Nantes Université, Inserm, CNRS, Université d'Angers, CRCI^{2}NA, Nantes, France

Tatsuya Katsuno
Kyoto University Graduate School of Medicine, Kyoto, Japan

Stephen M. King
Department of Molecular Biology and Biophysics, University of Connecticut Health Center, Farmington, CT, United States

Hiroko Kishi
Department of Molecular and Cellular Physiology, Graduate School of Medicine, Yamaguchi University, Hiroshima, Japan

Yo Kishimoto
Kyoto University Graduate School of Medicine, Kyoto, Japan

Yuki Kobayashi
Graduate School of Integrated Sciences for Life, Hiroshima University, Hiroshima, Japan

Sukanya Lodh
Marquette University, Milwaukee, WI, United States

Raj Luxmi
Department of Molecular Biology and Biophysics, University of Connecticut Health Center, Farmington, CT, United States

Laura Manrique Garcia
Departamento de Fisiología, Facultad de Medicina, Universidad Complutense de Madrid, Madrid, Spain

Markus Masek
Institute of Medical Genetics; Department of Molecular Life Sciences, University of Zurich, Zurich, Switzerland

José M. Mateos
Center for Microscopy and Image Analysis, University of Zurich, Zurich, Switzerland

Tatsuo Miyamoto
Department of Molecular and Cellular Physiology, Graduate School of Medicine, Yamaguchi University, Hiroshima, Japan

Paula Moreno-Cruz
Departamento de Fisiología, Facultad de Medicina, Universidad Complutense de Madrid, Madrid, Spain

Tomoka Morita
Department of Molecular and Cellular Physiology, Graduate School of Medicine, Yamaguchi University, Hiroshima, Japan

Ashikun Nabi
Luminex, Austin, TX, United States

Ryosuke Nakamura
New York University Grossman School of Medicine, New York, NY, United States

Ryota Nakazato
Department of Anatomy and Developmental Biology, Graduate School of Biomedical and Health Sciences, Hiroshima University, Hiroshima, Japan

Stephan C.F. Neuhauss
Department of Molecular Life Sciences, University of Zurich, Zurich, Switzerland

Koichi Omori
Kyoto University Graduate School of Medicine, Kyoto, Japan

Hiroshi Otani
Department of Anatomy and Developmental Biology, Graduate School of Biomedical and Health Sciences, Hiroshima University, Hiroshima, Japan

Seiji Oyagi
Kyoto University Graduate School of Medicine, Kyoto, Japan

Amanda Gabrielly Pereira
Departamento de Fisiología, Facultad de Medicina, Universidad Complutense de Madrid, Madrid, Spain

Yumiko Saito
Graduate School of Integrated Sciences for Life, Hiroshima University, Hiroshima, Japan

Tang K. Tang
Institute of Biomedical Sciences, Academia Sinica, Taipei, Taiwan

Megan Valentine
State University of New York, Plattsburgh, NY, United States

Judith Van Houten
University of Vermont, Burlington, VT, United States

Hannah M. Wesselman
Department of Biological Sciences, Center for Stem Cells and Regenerative Medicine, Center for Zebrafish Research, Boler-Parseghian Center for Rare and Neglected Diseases, University of Notre Dame, Notre Dame, IN, United States

Rebecca A. Wingert
Department of Biological Sciences, Center for Stem Cells and Regenerative Medicine, Center for Zebrafish Research, Boler-Parseghian Center for Rare and Neglected Diseases, University of Notre Dame, Notre Dame, IN, United States

Takashi Yamamoto
Program of Biomedical Science; Program of Mathematical and Life Science, Graduate School of Integrated Sciences for Life, Hiroshima University, Hiroshima, Japan

Junji Yano
University of Vermont, Burlington, VT, United States

Jingjing Zang
Department of Molecular Life Sciences, University of Zurich, Zurich, Switzerland

Ying Zhang
Department of Molecular and Cellular Physiology, Graduate School of Medicine, Yamaguchi University, Hiroshima, Japan

Urs Ziegler
Center for Microscopy and Image Analysis, University of Zurich, Zurich, Switzerland

Cilia: From Mechanisms to Disease - Part A

José Manuel Bravo-San Pedro

Departamento de Fisiología, Facultad de Medicina, Universidad Complutense de Madrid, Madrid, Spain

Centro de Investigación Biomédica en Red en Enfermedades Neurodegenerativas (CIBERNED), Madrid, Spain

In the late 17th century, Antonie van Leeuwenhoek described cilia for the first time as "nimbly moving feet" (Mitchell, 2017). Currently, this cellular organelle is classified into two general classes: motile cilia (MC) and primary cilia (PC). MC, found in epithelial cells of the respiratory tract, brain ventricles, fallopian tubes, and middle ear, consist of multiple microtubules that generate directional fluid flow (Ringers, Olstad, & Jurisch-Yaksi, 2020). Conversely, PC are nonmotile sensory organelles, which stem from the basal body of most human cells and are formed by doublet microtubules. PC play a crucial role in pre- and postembryonic development and in adult tissue homeostasis through transducing and coordination of extracellular signals into intracellular pathways (Satir & Christensen, 2007). PC are cylindrical organelles surrounded by a continuous lipid bilayer with the plasma membrane (PM) and comprise an axoneme, a basal body, and a transition zone (TZ). The axoneme comprises a ring of 9+2 doublet microtubules coupled to several proteins responsible for anterograde and retrograde intraflagellar transport (Kozminski, Beech, & Rosenbaum, 1995; Kozminski, Johnson, Forscher, & Rosenbaum, 1993) essential for the formation (ciliogenesis) and maintenance of most eukaryotic cilia and flagella (Ishikawa & Marshall, 2017). The axoneme extends from the basal body, which is a modified mother centriole that contains specialized appendixes such as transition fibers for fitting the PM or vesicles during the first stages of ciliogenesis (Sorokin, 1962) and basal feet that interact with the actine cytoskeleton and mediate the alignment in multiciliated cells (Meunier & Azimzadeh, 2016). Last, the TZ is located at the base of the PC and contains specific gating structures for soluble, transmembrane, and membrane-associated proteins and lipids (Gonçalves & Pelletier, 2017).

PC have great functional importance for the cell. In fact, a wide variety of signaling pathways have been linked to the cilium, including Hedgehog, Wnt, Notch, Hippo, GPCR, PDGF (and other RTKs, including FGF), mTOR (Medina, 2021), and TGF-β, indicating the participation of this cellular organelle in multiple physiological processes (Wheway, Nazlamova, & Hancock, 2018). Numerous studies have indicated its importance in the control of DNA damage (Chao et al., 2022; Li, Zimmerman, & Weyemi, 2021), the cytoskeleton (Fewins, Whitehouse, & Leinster, 1988; Mul, Mitra, & Peterman, 2022), and different types of cellular stresses (Renaudin, 2021) and have demonstrated that various types of cancer cells fail to express cilia (Higgins, Obaidi, & McMorrow, 2019). Moreover, due to the importance of cilia in development and homeostasis and abnormalities in their structure and function, PC are directly responsible for a lot of severe human disorders, commonly referred to as ciliopathies (Waters & Beales, 2011). More than 400 ciliopathy-associated genes triggering over 35 types of ciliopathies have been described so far (Reiter & Leroux, 2017), including Bardet–Biedl syndrome (BBS) (Zhao & Rahmouni, 2022), Alström syndrome (Bea-Mascato, Neira-Goyanes, Iglesias-Rodríguez, & Valverde, 2022), Joubert syndrome (JBTS) (Tsai, Hsu, Liu, Chang, & Tang, 2019), polycystic kidney disease, retinal degeneration, abnormal situs, obesity (Engle, Bansal, Antonellis, & Berbari, 2021), diabetes, etc. (Hildebrandt, Benzing, & Katsanis, 2011).

In *Cilia: From Mechanisms to Disease* (an issue of the successful *Methods in Cell Biology* series), the world's leading experts in the cilia field offer a range of techniques and protocols that can be used to study aspects of this interesting cellular organelle both in vitro and in vivo. Each chapter of this book provides one or more protocols detailing different techniques such as flow cytometry, live imaging, analysis of motility, left–right organization analysis, morphology studies, electroporation, CRISPR/Cas9-mediated gene targeting (Gillyard & Davis, 2021), siRNA delivery, isolation of ciliary ectosomes applied in different in vitro and in vivo models, including *Mus musculus*, *Danio rerio*, *Chlamydomonas reinhardtii*, or *Paramecium caudatum*, and using specific tissues such as the human airway, tracheal epithelia, retina, or cerebellum. Topics such as autophagy, motility, and ciliogenesis and specific disorders—cardiac injury, BBS, and JBTS, for instance—are also addressed. This book comprises two volumes aimed at both expert and beginner researchers in this area of research, offering an interesting guide to broaden their methodological skills.

Acknowledgments

J.M.B.-S.P. is funded by the "Ramon y Cajal Program" (RYC-2018-025099-I) and supported by Spain's Ministerio de Ciencia e Innovación (PID2019-108827RA-I00 and PDC2022-133398-I00).

References

Bea-Mascato, B., Neira-Goyanes, E., Iglesias-Rodríguez, A., & Valverde, D. (2022). Depletion of *ALMS1* affects TGF-β signalling pathway and downstream processes such as cell migration and adhesion capacity. *Frontiers in Molecular Biosciences*, *13*(9), 992313. https://doi.org/10.3389/fmolb.2022.992313 (PMID: 36325276).

Chao, Y. Y., Huang, B. M., Peng, I. C., Lee, P. R., Lai, Y. S., Chiu, W. T., et al. (2022). ATM- and ATR-induced primary ciliogenesis promotes cisplatin resistance in pancreatic ductal adenocarcinoma. *Journal of Cellular Physiology*. https://doi.org/10.1002/jcp.30898 (Epub ahead of print. PMID: 36251015).

Engle, S. E., Bansal, R., Antonellis, P. J., & Berbari, N. F. (2021). Cilia signaling and obesity. *Seminars in Cell & Developmental Biology*, *110*, 43–50. https://doi.org/10.1016/j.semcdb.2020.05.006 (Epub 2020 May 25. PMID: 32466971).

Fewins, H. E., Whitehouse, G. H., & Leinster, S. J. (1988). The spontaneous disappearance of breast calcification. *Clinical Radiology*, *39*(3), 257–261. https://doi.org/10.1016/s0009-9260(88)80522-1 (PMID: 3396275).

Gillyard, T., & Davis, J. (2021). DNA double-strand break repair in cancer: A path to achieving precision medicine. *International Review of Cell and Molecular Biology*, *364*, 111–137. https://doi.org/10.1016/bs.ircmb.2021.06.003 (Epub 2021 Aug 3. PMID: 34507781).

Gonçalves, J., & Pelletier, L. (2017). The ciliary transition zone: Finding the pieces and assembling the gate. *Molecules and Cells*, *40*(4), 243–253. https://doi.org/10.14348/molcells.2017.0054 (PMID: 28401750).

Higgins, M., Obaidi, I., & McMorrow, T. (2019). Primary cilia and their role in cancer. *Oncology Letters*, *17*(3), 3041–3047. https://doi.org/10.3892/ol.2019.9942 (Epub 2019 Jan 17. PMID: 30867732).

Hildebrandt, F., Benzing, T., & Katsanis, N. (2011). Ciliopathies. *The New England Journal of Medicine*, *364*(16), 1533–1543. https://doi.org/10.1056/NEJMra1010172 (PMID: 21506742).

Ishikawa, H., & Marshall, W. F. (2017). Intraflagellar transport and ciliary dynamics. *Cold Spring Harbor Perspectives in Biology*, *9*(3), a021998. https://doi.org/10.1101/cshperspect.a021998 (PMID: 28249960).

Kozminski, K. G., Beech, P. L., & Rosenbaum, J. L. (1995). The Chlamydomonas kinesin-like protein FLA10 is involved in motility associated with the flagellar membrane. *The Journal of Cell Biology*, *131*(6 Pt. 1), 1517–1527. https://doi.org/10.1083/jcb.131.6.1517 (PMID: 8522608).

Kozminski, K. G., Johnson, K. A., Forscher, P., & Rosenbaum, J. L. (1993). A motility in the eukaryotic flagellum unrelated to flagellar beating. *Proceedings of the National Academy of Sciences of the United States of America*, *90*(12), 5519–5523. https://doi.org/10.1073/pnas.90.12.5519 (PMID: 8516294).

Li, H., Zimmerman, S. E., & Weyemi, U. (2021). Genomic instability and metabolism in cancer. *International Review of Cell and Molecular Biology*, *364*, 241–265. https://doi.org/10.1016/bs.ircmb.2021.05.004 (Epub 2021 Jul 1. PMID: 34507785).

Medina, D. L. (2021). Lysosomal calcium and autophagy. *International Review of Cell and Molecular Biology*, *362*, 141–170. https://doi.org/10.1016/bs.ircmb.2021.03.002. Epub 2021 Apr 20. PMID: 34253294.

Meunier, A., & Azimzadeh, J. (2016). Multiciliated cells in animals. *Cold Spring Harbor Perspectives in Biology*, *8*(12), a028233. https://doi.org/10.1101/cshperspect.a028233 (PMID: 27580630).

Mitchell, D. R. (2017). Evolution of cilia. *Cold Spring Harbor Perspectives in Biology*, *9*(1), a028290. https://doi.org/10.1101/cshperspect.a028290 (PMID: 27663773).

Mul, W., Mitra, A., & Peterman, E. J. G. (2022). Mechanisms of regulation in intraflagellar transport. *Cells*, *11*(17), 2737. https://doi.org/10.3390/cells11172737 (PMID: 36078145).

Reiter, J. F., & Leroux, M. R. (2017). Genes and molecular pathways underpinning ciliopathies. *Nature Reviews. Molecular Cell Biology*, *18*(9), 533–547. https://doi.org/10.1038/nrm.2017.60 (Epub 2017 Jul 12. PMID: 28698599).

Renaudin, X. (2021). Reactive oxygen species and DNA damage response in cancer. *International Review of Cell and Molecular Biology*, *364*, 139–161. https://doi.org/10.1016/bs.ircmb.2021.04.001 (Epub 2021 May 3. PMID: 34507782).

Ringers, C., Olstad, E. W., & Jurisch-Yaksi, N. (2020). The role of motile cilia in the development and physiology of the nervous system. *Philosophical Transactions of the Royal Society of London. Series B, Biological Sciences*, *375*(1792), 20190156. https://doi.org/10.1098/rstb.2019.0156 (PMID: 31884916).

Satir, P., & Christensen, S. T. (2007). Overview of structure and function of mammalian cilia. *Annual Review of Physiology*, *69*, 377–400. https://doi.org/10.1146/annurev.physiol.69.040705.141236 (PMID: 17009929).

Sorokin, S. (1962). Centrioles and the formation of rudimentary cilia by fibroblasts and smooth muscle cells. *The Journal of Cell Biology*, *15*(2), 363–377. https://doi.org/10.1083/jcb.15.2.363 (PMID: 13978319).

Tsai, J. J., Hsu, W. B., Liu, J. H., Chang, C. W., & Tang, T. K. (2019). CEP120 interacts with C2CD3 and Talpid3 and is required for centriole appendage assembly and ciliogenesis. *Scientific Reports*, *9*(1), 6037. https://doi.org/10.1038/s41598-019-42577-0 (Erratum in: Sci Rep. 2020 Jan 22;10(1):1265. PMID: 30988386).

Waters, A. M., & Beales, P. L. (2011). Ciliopathies: An expanding disease spectrum. *Pediatric Nephrology*, *26*(7), 1039–1056. https://doi.org/10.1007/s00467-010-1731-7 (Epub 2011 Jan 6. PMID: 21210154).

Wheway, G., Nazlamova, L., & Hancock, J. T. (2018). Signaling through the primary cilium. *Frontiers in Cell and Development Biology*, *8*(6), 8. https://doi.org/10.3389/fcell.2018.00008 (PMID: 29473038).

Zhao, Y., & Rahmouni, K. (2022). BBSome: A new player in hypertension and other cardiovascular risks. *Hypertension*, *79*(2), 303–313. https://doi.org/10.1161/HYPERTENSIONAHA.121.17946. Epub 2021 Dec 6. PMID: 34865504.

CHAPTER

Protocols to induce and study ciliogenesis

1

Paula Moreno-Cruz[a,†], Yaiza Corral Nieto[a,†], Laura Manrique Garcia[a], Amanda Gabrielly Pereira[a], and José Manuel Bravo-San Pedro[a,b,*]

[a]*Departamento de Fisiología, Facultad de Medicina, Universidad Complutense de Madrid, Madrid, Spain*

[b]*Centro de Investigación Biomédica en Red en Enfermedades Neurodegenerativas (CIBERNED), Madrid, Spain*

[*]*Corresponding author: e-mail address: josemabr@ucm.es*

Chapter outline

[†]P.M.-C and Y.C.N. contributed equally to this work.

Methods in Cell Biology, Volume 175, ISSN 0091-679X, https://doi.org/10.1016/bs.mcb.2022.10.002
Copyright © 2023 Elsevier Inc. All rights reserved.

Abstract

Primary cilia (PC) are sensory organelles that function as cellular antennas, transmitting signals between the extracellular and intracellular spaces in many vertebrate tissues. The cell generates and assembles PC through a highly regulated process called ciliogenesis. This complex process is involved in several physiological functions, including embryonic development, locomotion, cell cycle regulation or energetic homeostasis control. In general, when a cell finishes its cell division, the oldest centriole usually migrates to the plasma membrane and becomes a basal body that gives rise to the formation of a cilium. For this reason, the presence of cilia is incompatible with cell division, so when a cell is going to divide, the cilium and the basal body disappear.

Ciliogenesis is triggered by various stimuli, all of them related to cell cycle blockade. This cell cycle, and ciliogenesis induction, can be observed by: (1) the influence of growth factors (lack of serum and consequent inability to promote cell cycle exit and increase the proportion of cells in G_0); (2) pharmacological cell cycle inhibitors (staurosporine or etoposide); or (3) physiological cell cycle inhibition (excessive contact between neighboring cells).

Evaluation of ciliogenesis induction is vitally important for the study of diseases related to ciliary dysfunction, called ciliopathies. That is why the use of correct protocols for inducing cilia formation and an accurate posterior visualization of the cilia after performing said protocols are essential parts in the study of these diseases. To facilitate this task, here we described detailed protocols to induce ciliogenesis *in vitro* and visualize PC by immunofluorescence microscopy in cultured cells.

1 Introduction

Primary cilia (PC) are organelles extended from the membrane of most mammalian cells and formed by microtubular axoneme, matrix, transition zone (TZ), and basal body (Sanchez & Dynlacht, 2016). The first consists of 9 parallel pairs of microtubules associated with proteins such as the intraflagellar transport proteins (essentials for the correct development of ciliogenesis, regulating anterograde and retrograde transport along the axoneme) (Ishikawa & Marshall, 2017). This axoneme is covered by the ciliary membrane, which is continuous with the cell plasma membrane, and has a large number of ion channels (transient receptor potential (TRP) ion channels or various transporter proteins) and receptors such as (G-protein-coupled receptors (GPCRs), Receptor tyrosine kinases (RTKs), Notch receptors or extracellular matrix proteins receptors (ECM) (Kim & Dynlacht, 2013; Satir & Christensen, 2007). The TZ at the base of the cilium acts as a molecular gate to regulate protein and lipid transport and has vital importance for the functionality of the cilium (Goncalves & Pelletier, 2017).

The PC have multiple functions, not only playing an essential role in locomotion or embryonic development, but also acting as sensory organelle by capturing many kinds of signals (Corbit et al., 2005; Francis, Sfakianos, Lo, & Mellman, 2011; Kukic, Rivera-Molina, & Toomre, 2016; Rohatgi & Snell, 2010; Singla & Reiter, 2006). Due to the extensive number of PC functions, defects in its structure, or in

PC-related signaling pathways, affect multiple tissues and developing a group of diseases called ciliopathies. The spectrum of these ciliary diseases is broad and often includes neurodevelopmental defects, renal disease, retinal degeneration or metabolic problems (Fliegauf, Benzing, & Omran, 2007). For this reason, it is crucial to know well both their functions and their structures (Pan, Wang, & Snell, 2005; Roberson et al., 2015; Valente et al., 2010).

There are two pathways of ciliogenesis induction: (1) some cells assemble their primary cilia within the cytoplasm before coupling to the plasma membrane (intracellular ciliogenesis) and conversely; (2) other cells use an extracellular pathway (extracellular ciliogenesis) to assemble their motile cilia, which begin with the coupling of basal bodies in the membrane (Shakya & Westlake, 2021). It is known that cell cycle exit is strongly associated with primary ciliogenesis in cultured cells, hence serum starvation has become a very common method of inducing ciliogenesis in order to stop the cell cycle. On the one hand, pharmacological cell cycle inhibitors such as staurosporine, a protein kinase inhibitor like protein kinase C (PKC) and cyclin-dependent kinase (CDKs) (Antonsson & Persson, 2009); and etoposide, a topoisomerase II inhibitor (Teng et al., 2021), are also used to arrest cell cycle and consequently induce ciliogenesis. On the other hand, cell confluence is equally important because it causes cell cycle arrest by contact inhibition of cell proliferation, so that it facilitates ciliogenesis in cultured cells (Alieva, Gorgidze, Komarova, Chernobelskaya, & Vorobjev, 1999; Pitaval, Tseng, Bornens, & Thery, 2010; Wheatley, Feilen, Yin, & Wheatley, 1994). Besides that, cilia assembly and disassembly are modulated by other processes, such as cell polarization (Fan et al., 2004; Lutz & Burk, 2006) and cell migration (Albrecht-Buehler, 1977; Schneider et al., 2009). Generally, cilia assemble in G_1 phase, they abound in G_0 phase, and disassembled with entry into mitosis, but depending on cell line, there may be some differences in the relation between cilia and cell cycle phases (Rieder, Jensen, & Jensen, 1979; Sanchez & Dynlacht, 2016; Santos & Reiter, 2008; Tucker, Pardee, & Fujiwara, 1979).

Cilia have been shown to have important functions in the metabolism of Hepatic cells (Morleo & Franco, 2019; Nechipurenko, 2020). HepG2 cell line is a human hepatocarcinoma that is commonly used in metabolism studies. They are adherent cells growing in small aggregates. Furthermore, HepG2 cells exhibit ciliogenesis when the cell cycle is inhibited either by contact or by serum starvation for 24–72h (Liu et al., 2019).

Since the PC are small organelles, composed of many different proteins with specific locations, the best technique to visualize them is the fluorescence microscopy. In order to understand that we are properly labeling PCs, it is necessary to use antibodies against two different proteins; acetylated tubulin, that is accumulated in stable microtubules of the ciliary axoneme as part of heterodimers formed by α and β tubulin subunits (Wloga, Joachimiak, Louka, & Gaertig, 2017) and γ-tubulin as part of the ciliary basal body (Liang et al., 1996). Therefore, the use of HepG2 cells can serve as a model to study ciliogenesis by immunofluorescence and to understand the importance of PC in metabolism.

2 Material

2.1 Disposables materials

- 24-well plates for cell culture (142475, Thermo Scientific)
- Cell counting slides (87144E, Kova International)
- Conical polypropylene centrifuge tubes (15 and 50 mL) (11889640, Fisherbrand) and microtubes (1.5 mL) (11558232, Fisherbrand)
- Glass coverslips (10 mm of diameter) (CB00100RA020MNZ0, Epredia) and glass slides (11572203, Fisherbrand)
- Methacrylate box (54039454, Cultek)
- Micropipettes (2–20 μL, 20–200 μL, 200–1000 μL) (F167900, Gilson)
- Parafilm (HS234526B, Bemis)
- Pipet-aid (4099, Corning) and serological pipettes (10 and 25 mL) (11839660/11517752, Fisherbrand)
- Polystyrene flask (75 cm^2) for cell culture (156499, Thermo Scientific)

2.2 Reagents

- 2-[4-(2-hydroxyethyl) piperazin-1-yl] ethanesulfonic acid (HEPES) (15630056, Gibco)
- Antibiotic solution: penicillin G (10^4 U/mL) and streptomycin (10^4 μg/mL) (15140122, Gibco)
- Bovine serum albumin (BSA) (A9647, Sigma-Aldrich)
- Dulbecco's Modified Eagle Medium (DMEM) (41966052, Gibco)
- Ethanol 70% (6365, Dismadel)
- Etoposide (0219391880, MP Biomedicals)
- Fetal bovine serum (FBS) (10270106, Gibco)
- Hoechst 33342 Solution (H3570, Invitrogen)
- Formaldehyde solution (F8775, Sigma-Aldrich)
- Phosphate-buffered saline (PBS) (10010015, Gibco)
- ProLong glass antifade mountant (P36984, Invitrogen)
- Staurosporine (0219140090, MP Biomedicals)
- Triton X-100 (T8787, Sigma-Aldrich)
- Trypsin solution (12604039, Gibco)

2.3 Antibodies

- Mouse monoclonal anti-acetylated tubulin primary antibody (T6793, Sigma-Aldrich)
- Rabbit polyclonal anti-γ-tubulin primary antibody (ab11317, Abcam)
- Goat anti-Mouse IgG Alexa Fluor™488 secondary antibody (A11029, Invitrogen)
- Goat anti-Rabbit IgG Alexa Fluor™568 secondary antibody (A11036, Invitrogen)

2.4 Cell line

HepG2 is a cell line isolated from a hepatocellular carcinoma and exhibits epithelial-like morphology. These cells have adherent properties and grow as monolayers in small aggregates (HB-806, ATCC).

2.5 Softwares and equipments

- Adobe illustrator
- GraphPad Prism 8
- Image J
- Leica DMI 3000B microscope
- Leica Application Suite Version 3.4.1

3 Treatments to induce ciliogenesis *in vitro*

Cell culture is a convenient and widely used method in experimental research. For many techniques in different fields of science, large numbers of cells may be required, and such numbers are generally provided only when cultures are grown on established cell lines. Therefore, the development of cell lines is very important. Here, we detail different protocols to induce ciliogenesis in cell culture (Fig. 1).

3.1 Cell line thawing and cell culture

(1) HepG2 cell line aliquot preserved in −80 °C freezer is placed in ice to thaw (see note 1).

(2) Once the aliquot is thawed (see note 2), the entire volume is mixed with 9 mL of fresh culture medium (DMEM supplemented with FBS (10%), HEPES (10 mM), and antibiotics) into 15 mL plastic tube (see note 3).

(3) Cells are centrifuged at 1200 revolutions per minute (r.p.m.) for 5 minutes (min). The resultant pellet is resuspended in complete medium, and cells are seeded in a 25 cm^2 flask and cultured in a 5% CO_2 incubator at 37 °C (see note 4).

(4) After 24 hours (h), once the cells reach approximately 70–80% confluence, the medium is completely discarded and the cells are incubated at room temperature (RT) for 2–3 min with trypsin (see notes 5, 6 and 7).

After the indicated time and once all cells are detached, two volumes of pre-warmed complete growth media are added to inactivate trypsin and cells are centrifuged at 1200 r.p.m. for 5 min at RT.

(5) Supernatant medium is discarded and the whole cell pellet is resuspended in complete fresh medium and seeded in a 75 cm^2 flask (see note 8).

(6) After this initial passage, cell culture passages are performed each time the cells reach approximately 80% confluence in 75 cm^2 flask (see notes 9 and 10)

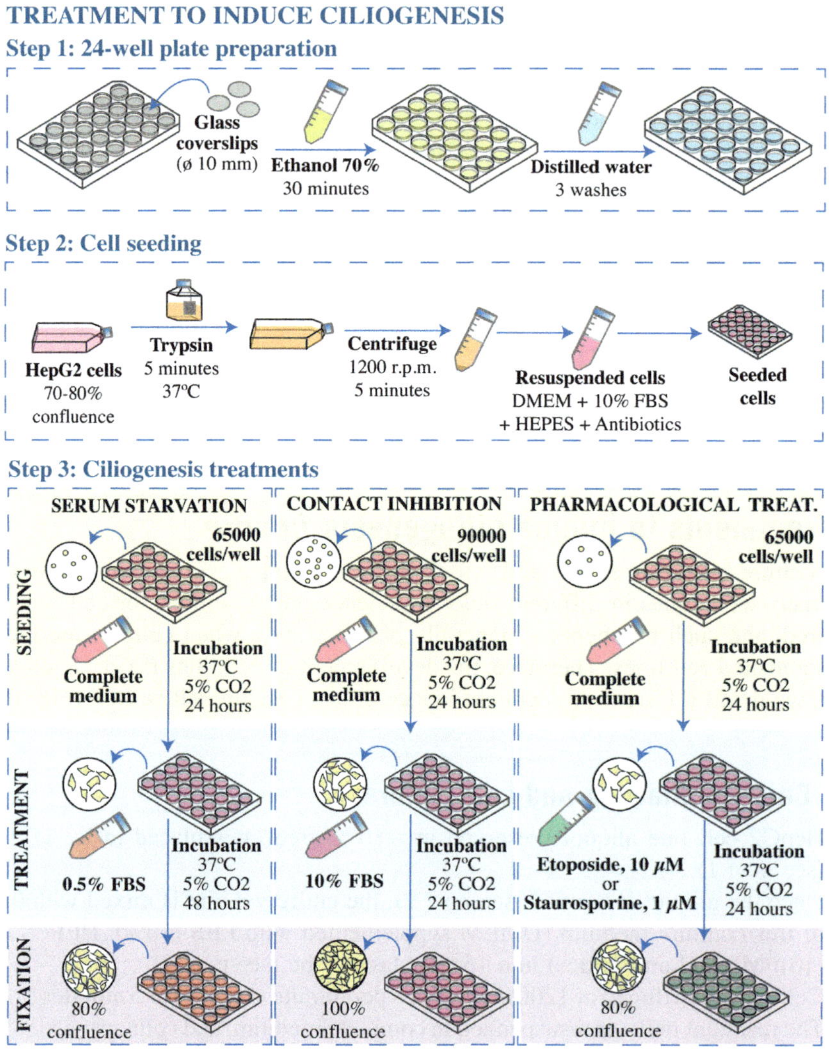

FIG. 1

Detailed protocol for performing treatments to induce ciliogenesis *in vitro*. FBS (Fetal bovine serum), HEPES (4-(2-hydroxyethyl)-1-piperazineethanesulfonic acid), r.p.m. (Revolutions per minute), RT (Room temperature).

3.2 Ciliogenesis induction treatments

Step 1. Before starting the experiment, the 24-well plates are prepared to introduce glass coverslips where cells will be seeded:

(1) One 10mm diameter coverslip is introduced in each well.

(2) Add 300 μL of 70% ethanol per well (see note 11) and wait for at least 30 min to sterilize the coverslips.
(3) After 30 min, ethanol is removed, and coverslips are washed three times with autoclaved distilled water to remove any remaining of ethanol.

Step 2. Cells are seeded:

(1) A 75 cm^2 flask with HepG2 cells with about 70–80% confluence is used.
(2) Culture medium is removed; cells are incubated with trypsin for 5 min (neutralized with fresh medium) and centrifuged at 1200 r.p.m.
(3) The supernatant is discarded, and cell pellet is resuspended in complete medium to obtain the cells at the appropriate concentration for the required treatments (see note 12).
(4) The different treatments for the induction of ciliogenesis are detailed in step 3. Thereby, we can carry out a treatment by serum starvation, a treatment by contact inhibition or a treatment by pharmacological cell cycle inhibitors.

Step 3. Ciliogenesis induction:

1. Ciliogenesis induced by serum starvation:
   (a) 24 h after seeding, the culture medium from each well is replaced with a serum-reduced media (DMEM supplemented with FBS (0.5%), HEPES (10 mM) and antibiotics).
   (b) After 48 h of treatment with FBS serum-reduced media, cells are fixed (see note 13).
2. Ciliogenesis induced by contact inhibition:
   (a) 24 h after seeding, the culture medium from each well is changed to a fresh complete medium (DMEM medium supplemented with FBS (10%), HEPES (10 mM) and antibiotics).
   (b) After 24 h of treatment, cells are fixed when they reach 100% confluence.
3. Ciliogenesis induced by pharmacological cell cycle inhibitors:
   (a) 24 h after seeding, the culture medium from each well is changed to DMEM medium supplemented with FBS (10%), HEPES (10 mM), antibiotics and staurosporine (1 μM) or etoposide (10 μM).
   (b) Cells are fixed after 24 h of pharmacological treatment.

4 Cilia visualization

Cilia are visualized through the use of immunofluorescence techniques, labeling different parts of the cilia (Fig. 2). Both the ciliary axoneme and the basal body are stained following the protocol outlined below:

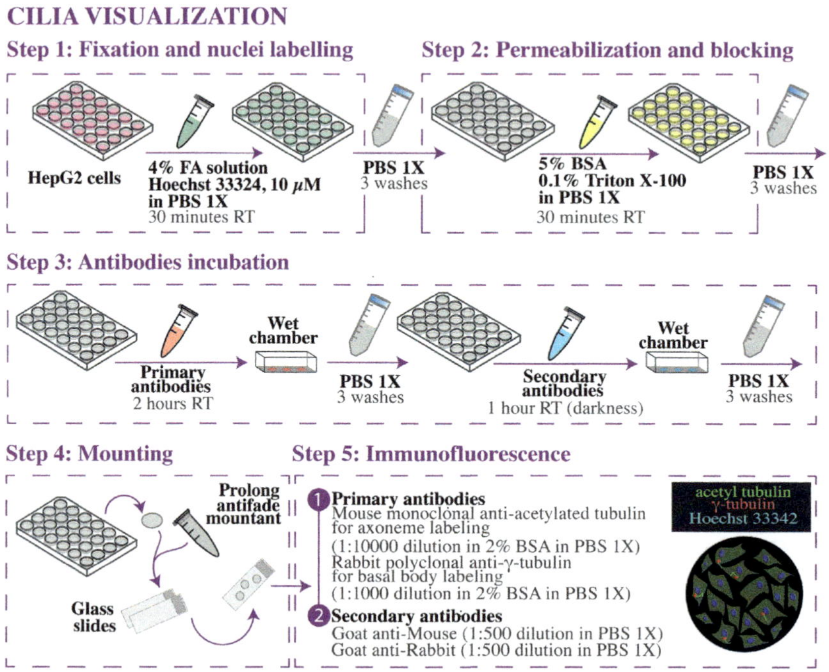

FIG. 2

Detailed protocol to visualize cilia by immunofluorescence. BSA, Bovine serum albumin; FA, Formaldehyde; PBS, Phosphate-buffered saline; RT, Room temperature.

4.1 Fixation and nuclei staining

(1) After the treatments, medium is removed from each well and coverslips with cells are washed twice with 1× PBS (500 μL per well) (see note 14).

(2) Cells are fixed and nuclei are stained by adding (for 30 min at RT) a mixed solution of 4% of formaldehyde (4%) and Hoechst 33342 (10 μM) in 1× PBS (see notes 15 and 16).

(3) Coverslips with cells are washed with 1× PBS three times (adding 500 μL per well) (see notes 16 and 17).

4.2 Permeabilization and blocking

(1) Cells are permeabilized and blocked with a solution composed of 1× PBS, Triton X-100 (0.1%) and BSA (5%) (300 μL per well) for 30 min at RT (see notes 18, 19 and 20).

(2) Cells are washed with 1× PBS three times (500 μL per well).

4.3 Incubation with antibodies

1. Primary antibodies mix is prepared in 1× PBS solution with BSA (2%) (see note 21):
   - Mouse monoclonal anti-acetylated tubulin primary antibody for axoneme visualization (1:10000 dilution).
   - Rabbit polyclonal anti-γ-tubulin primary antibody for basal body visualization (1:1000 dilution).
2. A wet chamber is prepared, it consists of a closed methacrylate box in which the filter paper is inserted and moistened with distilled water, and a piece of parafilm is placed on top (see note 22).
3. Coverslips are transferred back, with the help of clamps, in order to wet the chamber and 35 μL of primary antibodies solution are added and incubated for 2 h at RT.
4. Transfer coverslips back to 24-well plate using clamps; coverslips with cells are washed with 1× PBS three times and incubated minimum 2 min between washes.
5. Secondary antibodies mix are prepared in 1× PBS (see note 23):
   - Secondary antibody anti-mouse IgG Alexa Fluor™488 (1:500 dilution) (see note 24).
   - Secondary antibody anti-rabbit IgG Alexa Fluor™568 (1:500 dilution) (see note 25).
6. Coverslips are transferred back, with the help of clamps, to wet the chamber and 35 μL of secondary antibodies solution are added per coverslip and incubated for 1 h at RT in the dark.
7. Transfer coverslips back to 24-well plate using clamps and coverslips washed with cells with 1× PBS three times and incubated minimum 2 min between washes.

4.4 Mounting of coverslips on slides

After incubated cells with antibodies, coverslips are mounted on properly labeled slides:

(1) 5 μL of ProLong glass antifade mountant is added to the slides per coverslip.
(2) Using clamps, place coverslips with surface where the cells are toward each ProLong drop.
(3) Coverslips are pressed to be well bonded and to allow cilia visualization.
(4) Wait until ProLong is completely dry to visualize cells and their cilia by fluorescence microscopy.

4.5 Visualization by fluorescence microscope

Cilia are visualized and photographed with a fluorescence microscope 40× objective, results are analyzed by GraphPad Prism 8, and figures are assembled by Adobe Illustrator 2021 (Fig. 3).

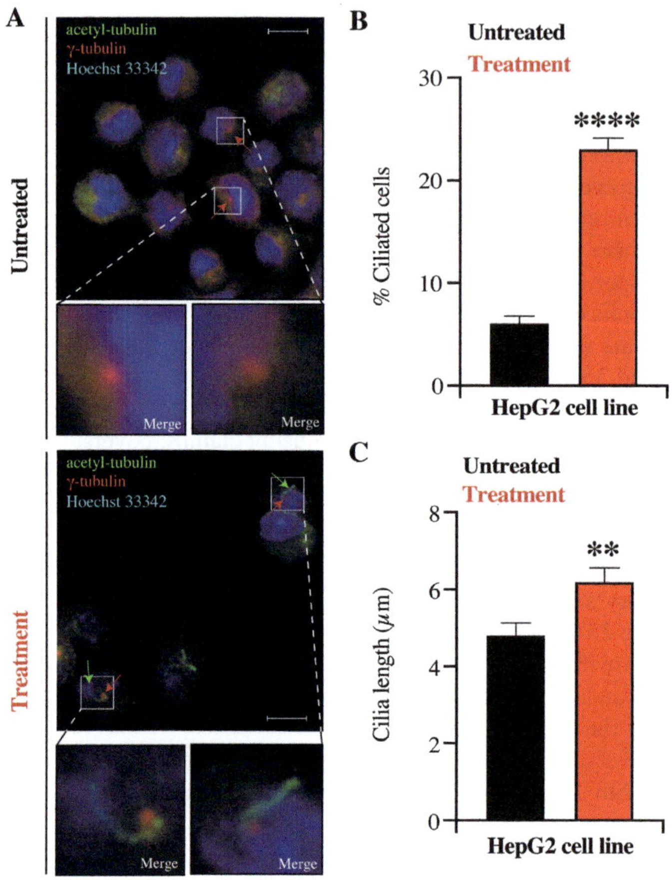

FIG. 3

Ciliogenesis in HepG2 cell line induced by serum starvation. Representative immunofluorescence images (A) in which cilia are labeled with acetylated tubulin antibody in green (cilia axoneme) and γ-tubulin antibody in red (cilia basal body). Analysis of ciliogenesis by percentage of ciliated cells (B) or cilia length (C) in HepG2 cells incubated 48h with complete medium (DMEM+FBS 10%; untreated) or serum reduced medium (DMEM+FBS 0.5%; treatment). Student's *t*-test: $**P<0.001$, $****P<0.00001$. FBS (Fetal bovine serum). Scale bar 10μm.

5 Quantification and statistical analysis

(1) Cells and their cilia were visualized and photographed at 40× objective with a Leica DMI 3000B microscope operated with Leica Application Suite to obtain at least 10 images/coverslip using A4 (BP 360/40nm), GFP (BP 470/40nm) and N2.1 (BP 515–560nm) filters (Fig. 3A) (see note 26).

(2) Ciliogenesis analysis is performed in the following way:
   (a) Number of cells is quantified by counting labeled nuclei.
   (b) Number of ciliated cells is quantified by counting the number of cilia, which are labeled both in axoneme and basal body.
   (c) Percentage of ciliated cells/total cells is calculated by using the data obtained in the previous points. Fig. 3B shows a significant increase in the % of ciliated cells after ciliogenesis treatment compared to the untreated cells.
   (d) Cilia length is measured by using ImageJ software. Fig. 3C shows a significant increase in the cilia length in treated cells compared to the untreated cells.
   (e) All these data are analyzed using GraphPad Prism 8 software (see note 27).
(3) Ciliogenesis is evaluated by percentage of ciliated cells/total cells and cilia length in 10 different regions of each coverslip with cells.
(4) As expected, the number of ciliated cells/total cells and the cilia length increase significantly for each of these methods applied to induce ciliogenesis.

6 Conclusion remarks

Although different, these protocols enable us to stimulate ciliogenesis and accurately measure the number and length of cilia in the HepG2 cell line. Each of these methods can be applied to other cell lines. However, ciliogenesis is cell type-dependent and it will be crucial to adjust these protocols (stimulation time and cell density) for each cell line. Apart from tubulin proteins, cilia can also be observed by studying other proteins, such as Arl13b protein and ALMS1 protein, respectively, located in the ciliary membrane and cilia basal body.

It is important to note that the proteins of the basal body and the axoneme must be close and correctly located in the PC, in order to avoid considering other cellular components outside the PC. Adequate detection of PC in different cell lines will help to study ciliopathies and understand PC function in cells.

7 Notes

1. Cells are cryopreserved in DMEM+FBS (10%)+DMSO (10%), which acts as cryoprotectant.
2. Cell thawing should occur as quickly as possible to avoid possible adverse effects.
3. MEM culture medium (Minimum Essential Media, Gibco) can be used as well.
4. The optimate cell density to avoid cellular stress is between 50% and 80%.
5. It is necessary pre-warmed both trypsin and the culture medium at 37 °C before the cell passage.

6. Flask with cells and trypsin can be placed in the incubator to facilitate cells detachment.
7. Recommended amount of trypsin used to detach cells is: 25 cm^2 flask: 1.5 mL; 75 cm^2 flask: 3 mL; 175 cm^2 flask: 5 mL.
8. Recommended amount of culture medium used to seed cells is: 25 cm^2 flask: 7–10 mL; 75 cm^2 flask: 15 mL; 175 cm^2 flask: 30 mL.
9. To obtain the optimal confluence, at 80%, HepG2 cells are split at 1:10–1:15 every 2–3 days approximately and the cell confluence is analyzed using light microscopy.
10. After thawing the cells, it is convenient to wait at least 1 week before treating the cells so that the cells recover from the freezing period, reactivate their metabolism, and grow normally.
11. Maximum volume capacity by well of 24-well plates is 1.5 mL and the optimal volume is 500 μL.
12. Seeding conditions for 24-well plates to get optimal confluence after treatment is 6.5×10^4 cells/well for 48 h serum starvation treatment and for 24 h pharmacological treatments or 9×10^4 cells/well for 24 h treatment by contact.
13. After 48 h of treatment, cells have a confluence between 60% and 80%.
14. PBS is a balanced saline solution that will be used to wash cells to maintain osmotic balance between internal and external environment of cell.
15. Formaldehyde is a fast cell penetration fixator whose function is to preserve cells as closely as possible to their living state.
16. The fixation with formaldehyde solution and subsequent washing with 1× PBS should be performed under a laminar flow hood due to the toxicity of formaldehyde. In addition, all formaldehyde residues must be placed in a specific container.
17. After fixation and nuclei staining, cells can be stored in 1× PBS solution at 4 °C. Always covering the wells to avoid long-term contamination.
18. Triton X-100 is a non-ionic detergent used to permeabilize living cell membranes, thus allowing antibodies to be introduced into the cell for immunofluorescence labeling.
19. BSA is used as a blocking buffer to prevent nonspecific binding of antibodies.
20. In this step, sodium azide can be added in permeabilization and blocking solution to prevent microbial contamination.
21. Primary antibodies can be modified, but it is essential that they are produced by different host organisms in order to visualize them separately with different secondary antibodies.
22. Wet chamber prevents, on the one hand, evaporation of antibody solution with wet filter paper and on the other hand, expansion outside the covers with cells of antibody solution with parafilm.
23. The secondary antibodies can be modified, but it is mandatory that anti-mouse and anti-rabbit antibodies have different fluorophores to emit a different wavelength.

24. Secondary antibody for mouse is conjugated with AlexaFluor™488 and it is going to bind primary antibody for acetylated tubulin, in such a way, we will be able to observe ciliary axoneme in green color with fluorescence microscopy.
25. Secondary antibody for rabbit is conjugated with AlexaFluor™568 and it is going to bind primary antibody for γ-tubulin, in such a way, we will be able to observe cilia basal body in red color with fluorescence microscopy.
26. Nuclei are stained with Hoechst 33342 (excitation/emission peaks at 350–461 nm), so they are visualized using A4 filter. Cilia axonemes are labeled with AlexaFluor™488 (peaks at 490–525 nm) so they are visualized using GFP filter. Cilia basal bodies are labeled with AlexaFluor™568 (peaks at 578–603 nm) so they are visualized using N2.1 filter.
27. For statistical analyses, P values are calculated by two-tailed unpaired Student's t and symbols indicate statistical comparisons with the untreated condition.

Acknowledgments

J.M.B.-S.P. is funded by "Ramon y Cajal Program" (RYC-2018-025099-I) and supported by Spain's Ministerio de Ciencia e Innovacion (PID2019-108827RA-I00). P.M.-C. is funded by the MINECO Spanish Ministry (FPI grant, PRE2020-092668). Y.C.N. and L.M.G. are funded by Community of Madrid (CT5/21/PEJ-2020-TL/BMD-17685 and CT36/22-41-UCM-INV, respectively).

References

Albrecht-Buehler, G. (1977). Phagokinetic tracks of 3T3 cells: Parallels between the orientation of track segments and of cellular structures which contain actin or tubulin. *Cell*, *12*(2), 333–339. https://doi.org/10.1016/0092-8674(77)90109-x.

Alieva, I. B., Gorgidze, L. A., Komarova, Y. A., Chernobelskaya, O. A., & Vorobjev, I. A. (1999). Experimental model for studying the primary cilia in tissue culture cells. *Membrane & Cell Biology*, *12*(6), 895–905. Retrieved from https://www.ncbi.nlm.nih.gov/pubmed/10512057.

Antonsson, A., & Persson, J. L. (2009). Induction of apoptosis by staurosporine involves the inhibition of expression of the major cell cycle proteins at the G(2)/m checkpoint accompanied by alterations in Erk and Akt kinase activities. *Anticancer Research*, *29*(8), 2893–2898. Retrieved from https://www.ncbi.nlm.nih.gov/pubmed/19661292.

Corbit, K. C., Aanstad, P., Singla, V., Norman, A. R., Stainier, D. Y., & Reiter, J. F. (2005). Vertebrate smoothened functions at the primary cilium. *Nature*, *437*(7061), 1018–1021. https://doi.org/10.1038/nature04117.

Fan, S., Hurd, T. W., Liu, C. J., Straight, S. W., Weimbs, T., Hurd, E. A., et al. (2004). Polarity proteins control ciliogenesis via kinesin motor interactions. *Current Biology*, *14*(16), 1451–1461. https://doi.org/10.1016/j.cub.2004.08.025.

Fliegauf, M., Benzing, T., & Omran, H. (2007). When cilia go bad: Cilia defects and ciliopathies. *Nature Reviews. Molecular Cell Biology*, *8*(11), 880–893. https://doi.org/10.1038/nrm2278.

Francis, S. S., Sfakianos, J., Lo, B., & Mellman, I. (2011). A hierarchy of signals regulates entry of membrane proteins into the ciliary membrane domain in epithelial cells. *The Journal of Cell Biology*, *193*(1), 219–233. https://doi.org/10.1083/jcb.201009001.

Goncalves, J., & Pelletier, L. (2017). The ciliary transition zone: Finding the pieces and assembling the gate. *Molecules and Cells*, *40*(4), 243–253. https://doi.org/10.14348/molcells.2017.0054.

Ishikawa, H., & Marshall, W. F. (2017). Intraflagellar transport and ciliary dynamics. *Cold Spring Harbor Perspectives in Biology*, *9*(3). https://doi.org/10.1101/cshperspect.a021998.

Kim, S., & Dynlacht, B. D. (2013). Assembling a primary cilium. *Current Opinion in Cell Biology*, *25*(4), 506–511. https://doi.org/10.1016/j.ceb.2013.04.011.

Kukic, I., Rivera-Molina, F., & Toomre, D. (2016). The IN/OUT assay: A new tool to study ciliogenesis. *Cilia*, *5*, 23. https://doi.org/10.1186/s13630-016-0044-2.

Liang, A., Ruiz, F., Heckmann, K., Klotz, C., Tollon, Y., Beisson, J., et al. (1996). Gamma-tubulin is permanently associated with basal bodies in ciliates. *European Journal of Cell Biology*, *70*(4), 331–338. Retrieved from https://www.ncbi.nlm.nih.gov/pubmed/8864661.

Liu, L., Sheng, J. Q., Wang, M. R., Gan, Y., Wu, X. L., Liao, J. Z., et al. (2019). Primary cilia blockage promotes the malignant behaviors of hepatocellular carcinoma via induction of autophagy. *BioMed Research International*, *2019*, 5202750. https://doi.org/10.1155/2019/5202750.

Lutz, M. S., & Burk, R. D. (2006). Primary cilium formation requires von hippel-Lindau gene function in renal-derived cells. *Cancer Research*, *66*(14), 6903–6907. https://doi.org/10.1158/0008-5472.CAN-06-0501.

Morleo, M., & Franco, B. (2019). The autophagy-cilia axis: An intricate relationship. *Cell*, *8*(8). https://doi.org/10.3390/cells8080905.

Nechipurenko, I. V. (2020). The enigmatic role of lipids in cilia signaling. *Frontiers in Cell and Development Biology*, *8*, 777. https://doi.org/10.3389/fcell.2020.00777.

Pan, J., Wang, Q., & Snell, W. J. (2005). Cilium-generated signaling and cilia-related disorders. *Laboratory Investigation*, *85*(4), 452–463. https://doi.org/10.1038/labinvest.3700253.

Pitaval, A., Tseng, Q., Bornens, M., & Thery, M. (2010). Cell shape and contractility regulate ciliogenesis in cell cycle-arrested cells. *The Journal of Cell Biology*, *191*(2), 303–312. https://doi.org/10.1083/jcb.201004003.

Rieder, C. L., Jensen, C. G., & Jensen, L. C. (1979). The resorption of primary cilia during mitosis in a vertebrate (PtK1) cell line. *Journal of Ultrastructure Research*, *68*(2), 173–185. https://doi.org/10.1016/s0022-5320(79)90152-7.

Roberson, E. C., Dowdle, W. E., Ozanturk, A., Garcia-Gonzalo, F. R., Li, C., Halbritter, J., et al. (2015). TMEM231, mutated in orofaciodigital and Meckel syndromes, organizes the ciliary transition zone. *The Journal of Cell Biology*, *209*(1), 129–142. https://doi.org/10.1083/jcb.201411087.

Rohatgi, R., & Snell, W. J. (2010). The ciliary membrane. *Current Opinion in Cell Biology*, *22*(4), 541–546. https://doi.org/10.1016/j.ceb.2010.03.010.

Sanchez, I., & Dynlacht, B. D. (2016). Cilium assembly and disassembly. *Nature Cell Biology*, *18*(7), 711–717. https://doi.org/10.1038/ncb3370.

Santos, N., & Reiter, J. F. (2008). Building it up and taking it down: The regulation of vertebrate ciliogenesis. *Developmental Dynamics*, *237*(8), 1972–1981. https://doi.org/10.1002/dvdy.21540.

Satir, P., & Christensen, S. T. (2007). Overview of structure and function of mammalian cilia. *Annual Review of Physiology*, *69*, 377–400. https://doi.org/10.1146/annurev.physiol.69.040705.141236.

Schneider, L., Stock, C. M., Dieterich, P., Jensen, B. H., Pedersen, L. B., Satir, P., et al. (2009). The Na+/H+ exchanger NHE1 is required for directional migration stimulated via PDGFR-alpha in the primary cilium. *The Journal of Cell Biology*, *185*(1), 163–176. https://doi.org/10.1083/jcb.200806019.

Shakya, S., & Westlake, C. J. (2021). Recent advances in understanding assembly of the primary cilium membrane. *Faculty Reviews*, *10*, 16. https://doi.org/10.12703/r/10-16.

Singla, V., & Reiter, J. F. (2006). The primary cilium as the cell's antenna: Signaling at a sensory organelle. *Science*, *313*(5787), 629–633. https://doi.org/10.1126/science.1124534.

Teng, Y. N., Chang, H. C., Chao, Y. Y., Cheng, H. L., Lien, W. C., & Wang, C. Y. (2021). Etoposide triggers cellular senescence by inducing multiple centrosomes and primary cilia in adrenocortical tumor cells. *Cells*, *10*(6). https://doi.org/10.3390/cells10061466.

Tucker, R. W., Pardee, A. B., & Fujiwara, K. (1979). Centriole ciliation is related to quiescence and DNA synthesis in 3T3 cells. *Cell*, *17*(3), 527–535. https://doi.org/10.1016/0092-8674(79)90261-7.

Valente, E. M., Logan, C. V., Mougou-Zerelli, S., Lee, J. H., Silhavy, J. L., Brancati, F., et al. (2010). Mutations in TMEM216 perturb ciliogenesis and cause Joubert, Meckel and related syndromes. *Nature Genetics*, *42*(7), 619–625. https://doi.org/10.1038/ng.594.

Wheatley, D. N., Feilen, E. M., Yin, Z., & Wheatley, S. P. (1994). Primary cilia in cultured mammalian cells: Detection with an antibody against detyrosinated alpha-tubulin (ID5) and by electron microscopy. *Journal of Submicroscopic Cytology and Pathology*, *26*(1), 91–102. Retrieved from https://www.ncbi.nlm.nih.gov/pubmed/8149337.

Wloga, D., Joachimiak, E., Louka, P., & Gaertig, J. (2017). Posttranslational modifications of tubulin and cilia. *Cold Spring Harbor Perspectives in Biology*, *9*(6). https://doi.org/10.1101/cshperspect.a028159.

CHAPTER

A flow cytometry-based approach for the study of primary cilia

2

Katja Baur, Gabriele Hölzl-Wenig, and Francesca Ciccolini*

Department of Neurobiology, Universität Heidelberg, Heidelberg, Germany

**Corresponding author: e-mail address: ciccolini@nbio.uni-heidelberg.de*

Chapter outline

Abstract

Primary cilia provide a specialized subcellular environment favoring ordered and timely interaction and modification of signaling molecules, necessary for the sensing and transduction of extracellular signals and environmental conditions. Crucial to the understanding of ciliary function is the knowledge of the signaling molecules composing the ciliary compartment. While proteomes of primary cilia have been published recently, the selective isolation of primary cilia from specific cell types and whole tissue still proves difficult, and many laboratories

Methods in Cell Biology, Volume 175, ISSN 0091-679X, https://doi.org/10.1016/bs.mcb.2022.07.018
Copyright © 2023 Elsevier Inc. All rights reserved.

instead resort to the analysis of cultured cells, which may introduce experimental artifacts. Here we present a flow cytometry-based method to isolate and characterize primary cilia from the murine ventricular-subventricular zone. After deciliation, primary cilia are immunolabeled with antibodies against ciliary markers. As an example, we here use a double-staining with acetylated tubulin, which stains the ciliary axoneme, and ciliary membrane protein ADP-ribosylation-like factor 13b (Arl13b); additionally, we triple-labeled primary cilia using the ciliary marker adenylate cyclase 3 (AC3). Besides analysis at the single particle level, fluorescence activated cell sorting (FACS) allows collection of pure preparations of primary cilia suited for subsequent proteomic analyses like mass spectrometry or western blot. As an example of analytical application, we performed triple immunostaining and FACS analysis to reveal cilia heterogeneity. Thus, our cilia isolation method, which can readily be applied to other tissues or cell culture, will facilitate the study of this key cellular organelle and shed light on its role in normal conditions and disease.

Abbreviations

AC3	adenylate cyclase 3
acT	acetylated tubulin
Arl13b	ADP-ribosylation-like factor 13b
FACS	fluorescence-activated cell sorting
NSC	neural stem cell
SVZ	ventricular-subventricular zone

1 Introduction

Primary cilia are tiny, rod-shaped cellular organelles present on most human cells. Once thought to be vestigial, primary cilia were established as key signaling hubs of the cell in the last decades (Berbari, O'Connor, Haycraft, & Yoder, 2009; Davenport & Yoder, 2005; Eggenschwiler & Anderson, 2007), critically involved in signal transduction, regulation of cell proliferation and sensing of physical parameters. Defects in cilia-related proteins cause several diseases collectively termed ciliopathies (Afzelius, 2004; Escudier, Duquesnoy, Papon, & Amselem, 2009; Reiter & Leroux, 2017).

Depending on function and location, mammalian primary cilia can be between 1 and 30 μm long and have an average diameter of around 0.4 μm (McGlashan et al., 2010; Wheatley, 2005; Yoder, 2007), thicker at the base and thinning towards the tip. The base of a cilium is made up of the basal body, made up by the mother centriole of the cell, which is anchored to the plasma membrane with so-called transition fibers (Davenport & Yoder, 2005; Garcia-Gonzalo & Reiter, 2017). From the basal body, nine highly acetylated microtubule doublets extend outward, forming the ciliary axoneme, which is surrounded by the plasma membrane (Davenport & Yoder, 2005). Underscoring the need for a better understanding of primary cilia structure, recent findings have revealed that the organized axoneme structure may not extend for the whole length of the cilia. Indeed, it was found that in some cases towards the tip of the cilium, the rigid organized structure of microtubule doublets is discontinued embedded in a mesh of actin filaments, leading to a thinning the diameter of the cilium (Kiesel et al., 2020). Contrary to motile cilia, primary cilia lack a central pair of microtubules and dynein arms which would allow them to move, rendering them immotile (Kiesel et al., 2020; Perkins, Hedgecock, Thomson, & Culotti, 1986; Porter & Sale, 2000). Although the plasma membrane covering the cilium is part of the cellular membrane, it is compartmentalized from the overall cell membrane and harbors several proteins targeted specifically to the cilium (Pazour & Bloodgood, 2008). Such proteins are readily used as cilia markers, such as adenylate cyclase 3 (AC3) and ADP-ribosylation-like factor 13b (Arl13b) (McGlashan et al., 2010; Ou et al., 2009).

The primary cilium is vital for several signaling pathways, especially the sonic hedgehog (Shh) pathway, which controls proliferation and migration of stem and progenitor cells (Breunig et al., 2008; Ihrie et al., 2011; Ruat, Roudaut, Ferent, & Traiffort, 2012). Hence, proteins for this pathway are also present, such as the Gli proteins in activator or repressor form, as well as Patched-1 (Ptc) and, depending on activation, smoothened (Smo) (Corbit et al., 2005; Haycraft et al., 2005; Rohatgi, Milenkovic, & Scott, 2007). The presence of Smo at the base of the cilium or at the cilium tip indicates whether Shh has bound to Ptc, a process which triggers translocation of Smo into the cilium.

Cilia are studied mostly using immunohistochemistry or electron microscopy, with most ciliary proteins being found by visual analysis confirming localization to the cilium. Furthermore, mass spectrometric analyses of primary cilia are usually performed using cilia derived from cultured cells and by purifying them by ultracentrifugation in sucrose density gradients (Ishikawa, Thompson, Yates 3rd, & Marshall, 2012; Mitchell, Szabo, & Otero Ade, 2009; Narita et al., 2012), a lengthy and complicated procedure. Therefore, analysis of the protein content of the cilium has been slow, especially with cilia from specific animal tissues or subtypes of cilia, e.g., separating motile from primary cilia in the same sample (Monaco et al., 2019).

Here we describe a method to isolate cilia from tissue, specifically from neural stem cells in the ventricular-subventricular zone (SVZ) from murine brain (Monaco et al., 2019). This method uses immunofluorescently labeled antibodies to identify ciliary particles shed from the tissue, which are then sorted by a flow cytometer. The ciliary particles are then not only sorted based on marker expression but may also be separated by size, upon calibration with fluorescent beads. Deciliation of the tissue is achieved by calcium shock, which triggers cilium shedding, likely catalyzed by katanin (Ishikawa et al., 2012; Mayer et al., 2009; Mirvis, Siemers, Nelson, & Stearns, 2019; Quarmby & Hartzell, 1994).

We optimized this method to provide pure cilia samples for proteomic analyses such as mass spectrometry or western blot, as well as for fast analyses of labeled proteins and size, which might help characterize ciliary mutants.

Isolating cilia with flow cytometry will enable researchers to quickly and reliably isolate vast amounts of cilia for proteomic analysis, leading to better insight into primary cilia function.

2 Tissue dissection and preparation

2.1 Materials and equipment

- Cell culture medium of choice containing no chelators, approx. 2mL/animal or well is needed for the entire procedure; we here use Euromed-N (Biozol, #ECM0883L) containing 1 × GlutaMAX (Gibco, ##35,050,061) and 10U/mL Penicillin/Streptomycin (Gibco, #15140) supplemented with 2% B27 (Invitrogen, #17504044)
- Dissection medium of choice; we here use a sucrose solution containing 150mM Sucrose, 125mM NaCl, 3.5mM KCl, 1.2mM $NaH_2PO_4 \times H_2O$, 2.4mM $CaCl_2 \times 2H_2O$, 1.3mM $MgCl_2 \times 6H_2O$, 2mM HEPES buffer, 6.65mM D-(+)-glucose
- NP-40 (IGEPAL CA-630, Sigma Aldrich, #18896)
- Fetal calf serum (FCS; Gibco, #10720)
- Ice
- Dissection tools (forceps, scissors)
- Dissection dish
- 24-well-plate
- 1.5mL Eppendorf tubes
- 15mL Falcon tubes
- Stereomicroscope
- 4°C refrigerator

2.2 Method

a. Preparation of tissue or cell culture:
   i. Tissue: The SVZ or ciliated tissue of interest is dissected whole in dissection sucrose, in a way that the cilia are freely accessible from the outside, and placed into a well of a standard 24-well plate containing an appropriate amount of ice-cold cell culture medium (e.g., 500µL neurobasal medium for brain tissue) on ice.
   ii. Adherent cultured cells: Cells should be grown in a 24-well-plate; cells grown in a flask should be plated into an appropriately coated 24-well-plate the day before to guarantee proper attachment of cells. The deciliation protocol is not applicable to nonadherent cells.

The procedure of immunostaining will depend on the localization of the epitope, i.e., intracellular or extracellular:

b. *Intracellular*: If an intracellularly binding antibody is used, the tissue/cells are incubated in an appropriate amount (e.g., 500µL) of ice-cold medium containing 0.1% NP-40 for 10min on ice.

c. *Extracellular*: If extracellularly binding antibodies are used, wells are incubated with medium containing 2% FCS and the appropriate dilution of antibody for 30min at 4°C in the dark. In this case, keep one well without antibodies and one single stained well each for antibody used **as controls** (E.g., if labeling with antibody A and B, you need at least 4 wells: no antibody, A, B, and A+B; this applies also to secondary antibodies). The plate may not be agitated to prevent premature shedding of cilia. The wells are then rinsed one time with fresh medium for 5 min at 4 °C.

3 Deciliation

3.1 Materials and equipment

- Cell culture medium of choice (500 μL per well) containing 2% FCS
- 2.5 M $CaCl_2$ solution
- 1.5 mL Eppendorf tubes
- 5 mL round-bottom polystyrene tubes for flow cytometry (e.g., StemCell, #100-0088)
- Rotary shaker (e.g., Thermo Scientific MaxQ 2000) in a cold room (4 °C)
- Tabletop centrifuge in a cold room (4 °C)

3.2 Method

a. Prepare deciliation medium (500 μL per well of a standard 24-well-plate): add 2 μL 2.5 M $CaCl_2$ solution to 500 μL cell culture medium of choice containing 2% FCS (final concentration = 10 mM $CaCl_2$).
b. Take off the medium from the wells and add 500 μL per well deciliation medium.
c. Deciliate for 5 min at 4 °C on a rotary shaker at 200 rpm.
d. Transfer supernatant to 1.5 mL Eppendorf tubes and spin down at 2000 rpm in a tabletop centrifuge for 5 min at 4 °C.
e. Transfer supernatant containing ciliary particles to a fresh 5 mL round-bottom polystyrene tube for flow cytometry.

4 Staining

4.1 Materials and equipment

- Cell culture medium of choice (500 μL per well) containing 2% FCS
- 25 × EDTA-free Protease Inhibitor (prepared with phosphate-buffered saline (PBS); e.g. cOmplete EDTA-free Protease Inhibitor Cocktail, Sigma Aldrich, #11873580001)
- 5 mL round-bottom polystyrene tubes for flow cytometry
- 1.5 mL Eppendorf tubes
- Antibodies of choice (see also Table 1)
- 4 °C refrigerator

4.2 Method

a. Controls: For each control, use 30–50 μL cilia-containing medium and fill up to 150 μL with medium containing 2% FCS. Then prepare the following controls:

Table 1 List of antibodies validated for cilia isolation and analysis by flow cytometry.

Antigen	Host	Company, order number	Conjugate
Adenylate cyclase 3[a]	Rabbit	ThermoFisher, PA5-35382	Manually conjugated
Arl13b[b]	Mouse	NIH NeuroMab, N295B/66	Manually conjugated
Acetylated α-tubulin[b]	Mouse	Sigma Aldrich, T6793	Manually conjugated
CXCR4/CD184	Rat	Invitrogen, 12-9991-82	PE
PDGFRα/CD140a	Rat	Invitrogen, 12-1401-81	PE
Prominin-1/CD133	Rat	Miltenyi Biotech, 130-102-197	APC
Smoothened	Rabbit	LifeSpan Biosciences, LS-A2668	Manually conjugated
Conjugation kit Dylight488	N/A	Abcam, ab201799	N/A
Conjugation kit PE/ R-phycoerythrin	N/A	Abcam, ab102918	N/A
Conjugation kit Alexa Fluor 647	N/A	Abcam, ab269823	N/A

[a]*May be used as marker for primary cilia.*
[b]*May be used as marker for several types of cilia.*

i. Autofluorescence: no further treatment. If 2.2*c* was performed, use cilia from the well without antibody.
ii. Single staining (one for each antibody used): add intracellularly binding antibody to cilia solution in the appropriate concentration (if necessary, prepare pre-dilutions) and incubate for 30 min at 4 °C before FACS analysis. If secondary antibodies are used, a separate control with only secondary antibody is necessary. If 2.2*c* was performed, use 150 μL of each sample treated with only one antibody, and only the non-treated sample for the remaining intracellularly binding antibody controls.

b. Samples: Prepare one of the following steps, depending on whether the populations should only be analyzed (i) or if a larger amount is needed to sort cilia for downstream applications like western blot or mass spectrometry (ii):

i. Analysis only: if the cilia are prepared to be analyzed but not sorted, prepare 150 μL per sample with all antibodies at the appropriate concentration and incubate for 30 min at 4 °C. If *2.2c* was performed, use the sample treated with all extracellularly binding antibodies, and add any remaining intracellularly binding antibodies.

OR

ii. Sorting: if the cilia are prepared to be sorted, prepare like 4.2*bi*., but use the entire supernatant of the sample retrieved in 3.2*e* (~450 μL). Add 18 μL 25 × EDTA-free protease inhibitor (to 1 × final concentration) to prevent protein degradation.[a]

5 FACS analysis

5.1 Materials and equipment

- Rainbow calibration particles (RRID: AB_2869258)
- Accudrop beads (BD Biosciences, catalogue No. 345249)
- Flow cytometry cell sorter (e.g., BD FACSAria II, Becton Dickinson)

5.2 Method

a. Gate setting

For this step, all controls prepared in step 4.2a are needed. At this point, one should have the following controls:

- Autofluorescence: one cilia sample without antibody
- Secondary control: sample containing only cilia and secondary antibody; one for each secondary antibody used
- Single staining control: sample containing only one primary antibody (with secondary if applicable); one for each primary antibody used when using more than one antibody
- Double staining control: sample containing two primary antibodies (with secondaries if applicable); one for each possible combination when using three antibodies

i. Forward and side scatter: Size

To sort the cilia according to their size and to account for different fluorescence profiles caused by differences in granularity, the cilia are first gated in forward and side scatter (FSC/SSC).

To estimate particle size, the gates are set according to the FSC of fluorescent beads, i.e., Rainbow calibration particles, which have a diameter of 3 μm, and Accudrop beads, 6 μm. The gates are then set according to Fig. 1A.

Next, the Autofluorescence sample is loaded, and 10.000 events recorded. All high-SSC/low-FSC events are excluded (Fig. 1B). Fig. 1B shows the typical FSC/SSC profile of isolated cilia.

ii. Fluorescence

The different size gates, i.e., 3, 6, and 8 μm, are each plotted on a separate graph displaying fluorescence intensities. The gates are set according to the respective negative control (Fig. 1C–F).

[a]Phosphatase inhibitors can also be added at this step if needed.

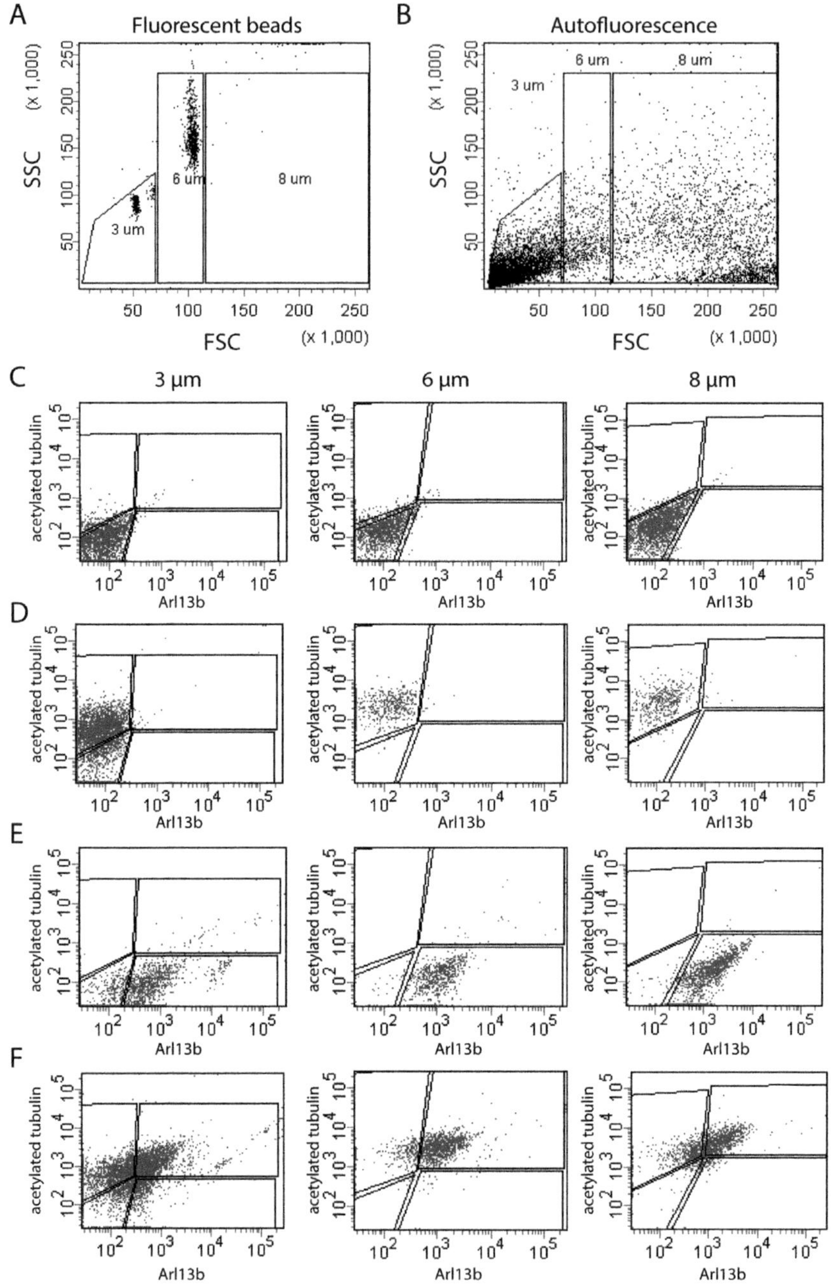

FIG. 1

Gate setting for flow cytometry of double-stained primary cilia. (A) Rainbow and Accudrop beads in forward (FSC) and side scatter (SSC) with gating for 3, 6 and 8 μm particles. (B) Isolated unlabeled cilia in FSC/SSC with size gates. (C) Autofluorescence of unlabeled ciliary particles in 488 and PE channels. (D) Acetylated tubulin-PE-labeled ciliary particles displayed in PE and 488 channels. (E) Arl13b-Dylight488-labeled ciliary particles displayed in PE and 488 channels. (F) Cilia labeled with Arl13b-Dylight488 and acetylated tubulin-PE displayed in the respective channels, divided by size.

For samples stained with three separate antibodies, the gates are set in two separate steps as illustrated in Fig. 2.

b. Analysis

i. Meaningful data can be extracted usually with around 30.000 events. This includes percentage of immunopositive events as well as differences in size distribution. To obtain a statistically relevant result, preparations from multiple animals should be analyzed on the same day and if possible, stained simultaneously.

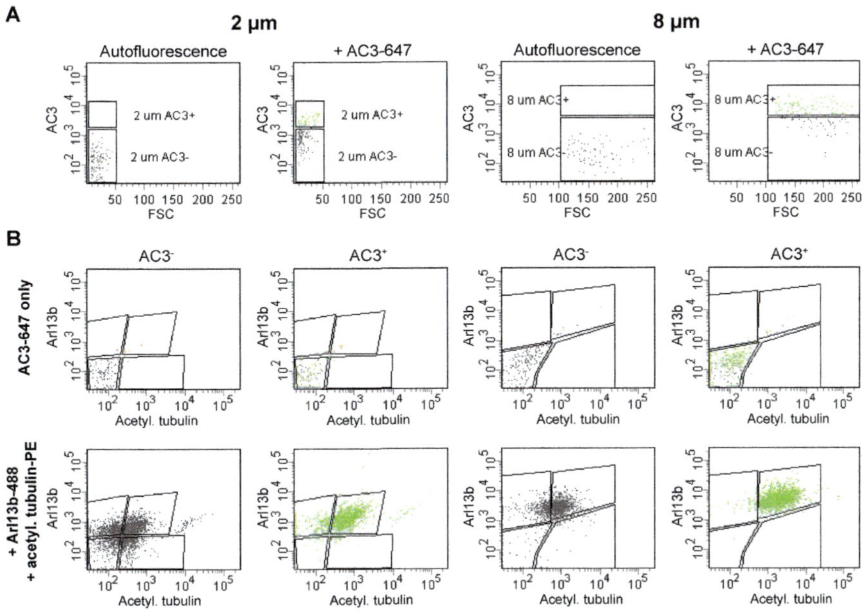

FIG. 2

Gate setting for triple-stained cilia, examples for 2 and 8 μm particles. (A) After gating for the size of the particles as illustrated above in Fig. 1A, particles of the given size positive to a single antigen, in this case AC3-Alexa Fluor 647, are gated as illustrated in left and right panels on the basis of autofluorescence control and an AC3-single-stained control, as indicated. (B) Immunopositive particles from (A) are plotted in a separate FACS plot according to fluorescence levels corresponding to immunoreactivity to the other two antibodies (here: Arl13b-Dylight488 and acetylated tubulin-PE). Gates shown on the plots were set based on single and double immunostaining as described above in Fig. 1D and E. Samples immunostained only with AC3 were used to set compensation for eventual fluorescence spillover (top row). For demonstration purposes, the AC3$^-$ (gray, each left) and AC3$^+$ particles (green, each right) are plotted in separate graphs; they can however be plotted in a single graph. The triple-stained cilia are displayed in the bottom row. Note the higher percentage of Arl13b$^+$acetylated tubulin$^+$ events within the AC3$^+$ particles (green, each right) compared to within the AC3$^-$ population (gray, each left).

6 Cilia sorting and treatment for subsequent proteomic analysis

6.1 Materials and equipment

- Phosphate-buffered saline (PBS)
- 25 × EDTA-free Protease Inhibitor (prepared with PBS; e.g., cOmplete EDTA-free Protease Inhibitor Cocktail, Sigma Aldrich, #11873580001)
- 1.5 mL Eppendorf tubes
- Benchtop ultracentrifuge (e.g. Beckman Coulter Optima TL-100 with TLA-55 rotor)
- −80 °C freezer

6.2 Method

a. Add 40 μL 25 × Protease inhibitor to 60 μL PBS to a final concentration of 10 ×.
b. The cilia of the desired fractions are sorted into Eppendorf tubes containing 100 μL of 10 × protease inhibitor in PBS.[b]
c. *Optional:* After sorting, the cilia are spun down with an ultracentrifuge at 80,000 × *g* for 40 min at 4 °C.[c] The supernatant is mostly removed, leaving around 10 μL.
d. Freeze the samples at −80 °C until further processing.

7 Troubleshooting

a. Unexpected population in SSC/FSC

It may happen that all or most events will show up inside of the red-circled area displayed in Fig. 3. This may be due to clumping of the cilia. This problem is usually fixed by further diluting the sample with medium containing 2% FCS. After appropriate dilution, the red-circled population should become less or disappear completely, while the events disperse further along the FSC-axis.

b. Very few events

If the cytometer detects only very few events (less than ~20/s), there might be several reasons:

[b] For western blot, we found that 200,000–400,000 individual events are suitable. For mass spectrometry, a minimum of 750,000 events is recommended.

[c] If the samples are to be used for mass spectrometry and your facility supports processing of larger volumes (~1 mL), ultracentrifugation can be skipped to prevent loss of soluble proteins.

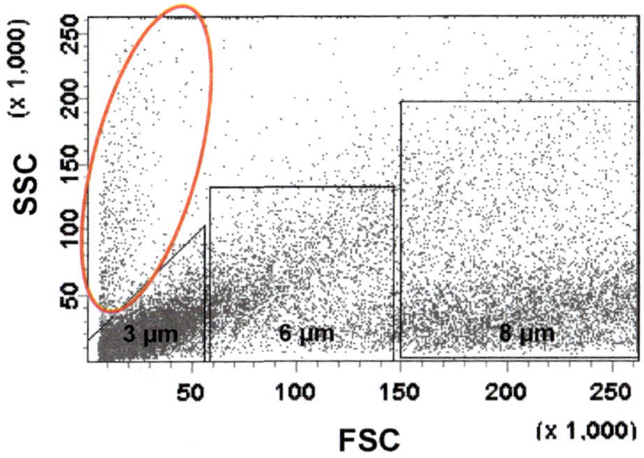

FIG. 3

Unexpected population in FSC/SSC, indicated by a red circle.

i. Deciliation did not work properly

- Not enough $CaCl_2$ in the deciliation medium
  Make sure the final concentration of $CaCl_2$ in the medium is at least 10 mM for deciliation.
- Not enough shaking (not fast enough, not long enough)
  Test multiple speed settings and different deciliation times on your rotary shaker.
- Presence of an agent which binds or chelates $CaCl_2$ (e.g. EDTA)
  Use medium which does not contain chelating agents.

ii. Cilia are clumping together

- Too little medium for amount of cilia
  Dilute with medium containing 2% FCS.
- No/not enough FCS in the sort medium
  Add FCS to a concentration of 2%.

c. Immunopositive population too large/too small

If the immunolabeled population is much bigger or smaller than what is expected, this could have several reasons:

i. Antibody

- Antibody does not bind properly/is binding unspecifically
  Test antibody in immunofluorescence to check specificity and proper binding; use different antibody.
- Concentration too high/too low
  Test multiple antibody concentrations when using a new antibody for this method.
- Missing correction for secondary antibody
  If a secondary antibody is used, the negative gates should always be set using a control stained with secondary but not primary antibody.

ii. Cilia concentration too high—small immunopositive population
If there are too many cilia in the medium, the antibody will not be enough to stain them appropriately; dilute the sample with medium containing 2% FCS and use the appropriate amount of antibody.
iii. Cilia clumping—high immunopositive population
Dilute with medium containing 2% FCS.

8 Concluding remarks

Here we present a method to quickly and reliably label and isolate cilia from tissue or cell culture, and to efficiently separate them based on particle size and marker expression. However, since the particle size is based on round objects and the process of flow cytometry damages the ciliary particles, the size is not an accurate representation of the actual length or thickness of individual cilia; the method can however be used to determine difference in size across populations, e.g. ependymal cilia (>10 μm) have a higher percentage of 8 μm particles than primary cilia (Monaco et al., 2019). Although the here described protocol was designed for cilia isolation and subsequent proteomic analysis such as Western Blot (Monaco et al., 2019), it may also be used to detect the amount of certain proteins present in the cilia, as well as changes in overall number. We have used this method to show an increase in $AC3^{+}Smo^{+}$ particles in the amount of $AC3^{+}$ particles as a result of treatment with smoothened agonist, which triggers the translocation of smoothened into the cilium (Monaco et al., 2019). Analysis by flow cytometry allows the detection of such changes within a few hours instead of with several days of immunostaining and microscopy. Using markers for primary and motile cilia, it is also possible to separate the different organelles for individual analysis (Monaco et al., 2019); this is especially useful for tissues where both cilia types play a prominent role, like in the brain ventricles.

Since this method relies on calcium for deciliation, it may however not be feasible to analyze calcium-dependent proteins or pathways that rely on calcium signaling. Therefore, when attempting analysis of Ca^{2+}-sensitive proteins, another mechanism of deciliation, e.g., mechanical, should be considered. Additionally, while this deciliation method is highly effective, it is not specific for cilia only and may remove also cellular debris, therefore we recommend using multiple highly specific cilia markers to ensure the ciliary nature of the particles. Although the membrane closure after deciliation should prohibit separation of axoneme and membrane, one should consider using antibodies against both membrane and axoneme proteins to ensure isolation of whole cilia instead of separated sub-components. However, the cilium is usually shed without the basal body, so basal body proteins may not be analyzed by this method, and since the cilia are damaged during flow cytometry, this method may not be used to isolate cilia for subsequent morphological analysis. Furthermore, we rely on labeling of cilia markers with fluorescent antibodies. If no sufficiently specific antibody is available, a viable alternative might be expression of a tagged protein under the endogenous promoter. Finally, we validated some cilia-specific antibodies that can be used for isolation, such as adenylate cyclase 3 (AC3), acetylated tubulin, and Arl13b (see Table 1). When using these markers, one should be aware of their differences in expression depending on tissue and age, e.g., Arl13b is not expressed in neuronal cilia in the hippocampus (Sterpka & Chen, 2018) while on the other hand it is present in neuronal growth cones at early developmental stages (Ferent et al., 2019). All in all, using flow cytometry to analyze and isolate cilia gives researchers a new tool to quickly analyze large amounts of cilia, thereby detecting protein expression or localization differences; furthermore, this method yields pure cilia samples for western blot or mass spectrometric analysis from primary tissue without the need of long sucrose density gradient ultracentrifugation or protein purification steps.

References

Afzelius, B. A. (2004). Cilia-related diseases. *The Journal of Pathology*, *204*(4), 470–477. https://doi.org/10.1002/path.1652.

Berbari, N. F., O'Connor, A. K., Haycraft, C. J., & Yoder, B. K. (2009). The primary cilium as a complex signaling center. *Current Biology*, *19*(13), R526–R535. https://doi.org/10.1016/j.cub.2009.05.025.

Breunig, J. J., Sarkisian, M. R., Arellano, J. I., Morozov, Y. M., Ayoub, A. E., Sojitra, S., et al. (2008). Primary cilia regulate hippocampal neurogenesis by mediating sonic hedgehog signaling. *Proceedings of the National Academy of Sciences of the United States of America*, *105*(35), 13127–13132. https://doi.org/10.1073/pnas.0804558105.

Corbit, K. C., Aanstad, P., Singla, V., Norman, A. R., Stainier, D. Y., & Reiter, J. F. (2005). Vertebrate smoothened functions at the primary cilium. *Nature*, *437*(7061), 1018–1021. https://doi.org/10.1038/nature04117.

Davenport, J. R., & Yoder, B. K. (2005). An incredible decade for the primary cilium: A look at a once-forgotten organelle. *American Journal of Physiology. Renal Physiology*, *289*(6), F1159–F1169. https://doi.org/10.1152/ajprenal.00118.2005.

Eggenschwiler, J. T., & Anderson, K. V. (2007). Cilia and developmental signaling. *Annual Review of Cell and Developmental Biology*, *23*, 345–373. https://doi.org/10.1146/annurev.cellbio.23.090506.123249.

Escudier, E., Duquesnoy, P., Papon, J. F., & Amselem, S. (2009). Ciliary defects and genetics of primary ciliary dyskinesia. *Paediatric Respiratory Reviews*, *10*(2), 51–54. https://doi.org/10.1016/j.prrv.2009.02.001.

Ferent, J., Constable, S., Gigante, E. D., Yam, P. T., Mariani, L. E., Legue, E., et al. (2019). The ciliary protein Arl13b functions outside of the primary cilium in Shh-mediated axon guidance. *Cell Reports*, *29*(11), 3356–3366. e3353. https://doi.org/10.1016/j.celrep.2019.11.015.

Garcia-Gonzalo, F. R., & Reiter, J. F. (2017). Open sesame: How transition fibers and the transition zone control ciliary composition. *Cold Spring Harbor Perspectives in Biology*, *9*(2). https://doi.org/10.1101/cshperspect.a028134.

Haycraft, C. J., Banizs, B., Aydin-Son, Y., Zhang, Q., Michaud, E. J., & Yoder, B. K. (2005). Gli2 and Gli3 localize to cilia and require the intraflagellar transport protein polaris for processing and function. *PLoS Genetics*, *1*(4), e53. https://doi.org/10.1371/journal.pgen.0010053.

Ihrie, R. A., Shah, J. K., Harwell, C. C., Levine, J. H., Guinto, C. D., Lezameta, M., et al. (2011). Persistent sonic hedgehog signaling in adult brain determines neural stem cell positional identity. *Neuron*, *71*(2), 250–262. https://doi.org/10.1016/j.neuron.2011.05.018.

Ishikawa, H., Thompson, J., Yates, J. R., 3rd, & Marshall, W. F. (2012). Proteomic analysis of mammalian primary cilia. *Current Biology*, *22*(5), 414–419. https://doi.org/10.1016/j.cub.2012.01.031.

Kiesel, P., Alvarez Viar, G., Tsoy, N., Maraspini, R., Gorilak, P., Varga, V., et al. (2020). The molecular structure of mammalian primary cilia revealed by cryo-electron tomography. *Nature Structural & Molecular Biology*, *27*(12), 1115–1124. https://doi.org/10.1038/s41594-020-0507-4.

Mayer, U., Kuller, A., Daiber, P. C., Neudorf, I., Warnken, U., Schnolzer, M., et al. (2009). The proteome of rat olfactory sensory cilia. *Proteomics*, *9*(2), 322–334. https://doi.org/10.1002/pmic.200800149.

McGlashan, S. R., Knight, M. M., Chowdhury, T. T., Joshi, P., Jensen, C. G., Kennedy, S., et al. (2010). Mechanical loading modulates chondrocyte primary cilia incidence and length. *Cell Biology International*, *34*(5), 441–446. https://doi.org/10.1042/CBI20090094.

Mirvis, M., Siemers, K. A., Nelson, W. J., & Stearns, T. P. (2019). Primary cilium loss in mammalian cells occurs predominantly by whole-cilium shedding. *PLoS Biology*, *17*(7), e3000381. https://doi.org/10.1371/journal.pbio.3000381.

Mitchell, K. A., Szabo, G., & Otero Ade, S. (2009). Methods for the isolation of sensory and primary cilia—An overview. *Methods in Cell Biology*, *94*, 87–101. https://doi.org/10.1016/S0091-679X(08)94004-8.

Monaco, S., Baur, K., Hellwig, A., Hölzl-Wenig, G., Mandl, C., & Ciccolini, F. (2019). A flow cytometry-based approach for the isolation and characterization of neural stem cell primary cilia. *Frontiers in Cellular Neuroscience*, *12*(519). https://doi.org/10.3389/fncel.2018.00519.

Narita, K., Kozuka-Hata, H., Nonami, Y., Ao-Kondo, H., Suzuki, T., Nakamura, H., et al. (2012). Proteomic analysis of multiple primary cilia reveals a novel mode of ciliary development in mammals. *Biology Open*, *1*(8), 815–825. https://doi.org/10.1242/bio.20121081.

Ou, Y., Ruan, Y., Cheng, M., Moser, J. J., Rattner, J. B., & van der Hoorn, F. A. (2009). Adenylate cyclase regulates elongation of mammalian primary cilia. *Experimental Cell Research*, *315*(16), 2802–2817. https://doi.org/10.1016/j.yexcr.2009.06.028.

Pazour, G. J., & Bloodgood, R. A. (2008). Targeting proteins to the ciliary membrane. *Current Topics in Developmental Biology*, *85*, 115–149. https://doi.org/10.1016/S0070-2153(08)00805-3.

Perkins, L. A., Hedgecock, E. M., Thomson, J. N., & Culotti, J. G. (1986). Mutant sensory cilia in the nematode Caenorhabditis elegans. *Developmental Biology*, *117*(2), 456–487. Retrieved from https://www.ncbi.nlm.nih.gov/pubmed/2428682.

Porter, M. E., & Sale, W. S. (2000). The 9 + 2 axoneme anchors multiple inner arm dyneins and a network of kinases and phosphatases that control motility. *The Journal of Cell Biology*, *151*(5), F37–F42. Retrieved from https://www.ncbi.nlm.nih.gov/pubmed/11086017. https://www.ncbi.nlm.nih.gov/pmc/articles/PMC2174360/pdf/0007057.pdf.

Quarmby, L. M., & Hartzell, H. C. (1994). Two distinct, calcium-mediated, signal transduction pathways can trigger deflagellation in Chlamydomonas reinhardtii. *The Journal of Cell Biology*, *124*(5), 807–815. Retrieved from https://www.ncbi.nlm.nih.gov/pubmed/8120101. https://www.ncbi.nlm.nih.gov/pmc/articles/PMC2119959/pdf/jc1245807.pdf.

Reiter, J. F., & Leroux, M. R. (2017). Genes and molecular pathways underpinning ciliopathies. *Nature Reviews. Molecular Cell Biology*, *18*(9), 533–547. https://doi.org/10.1038/nrm.2017.60.

Rohatgi, R., Milenkovic, L., & Scott, M. P. (2007). Patched1 regulates hedgehog signaling at the primary cilium. *Science*, *317*(5836), 372–376. https://doi.org/10.1126/science.1139740.

Ruat, M., Roudaut, H., Ferent, J., & Traiffort, E. (2012). Hedgehog trafficking, cilia and brain functions. *Differentiation*, *83*(2), S97–104. https://doi.org/10.1016/j.diff.2011.11.011.

Sterpka, A., & Chen, X. (2018). Neuronal and astrocytic primary cilia in the mature brain. *Pharmacological Research*, *137*, 114–121. https://doi.org/10.1016/j.phrs.2018.10.002.

Wheatley, D. N. (2005). Landmarks in the first hundred years of primary (9+0) cilium research. *Cell Biology International*, *29*(5), 333–339. https://doi.org/10.1016/j.cellbi.2005.03.001.

Yoder, B. K. (2007). Role of primary cilia in the pathogenesis of polycystic kidney disease. *Journal of the American Society of Nephrology*, *18*(5), 1381–1388. https://doi.org/10.1681/ASN.2006111215.

CHAPTER

3

Microscopic observation of human airway ciliary movement using wheat germ agglutinin

Ryosuke Nakamura[a], Seiji Oyagi[b], Tatsuya Katsuno[b], Yo Kishimoto[b,*], and Koichi Omori[b]

[a]*New York University Grossman School of Medicine, New York, NY, United States*

[b]*Kyoto University Graduate School of Medicine, Kyoto, Japan*

**Corresponding author: e-mail address: y_kishimoto@ent.kuhp.kyoto-u.ac.jp*

Chapter outline

Abstract

Ciliated cells in the airway epithelium generate mucus streams to remove extraneous particles and microorganisms by beating the motile cilia. This defense mechanism is crucial for maintaining homeostasis and preventing infection in the airway. Conventional methods to assess ciliary beating have revealed that rapid (>10 times per second) and metachronal beating of cilia enables efficient mucus transport. Cilia are oriented to excrete mucus toward the outside of the body. However, conventional methods to directly observe ciliary movements uses transmitted light, which requires translucent samples. Sliced or fragmented tissues are used to observe ciliary movements in thick human airway tissues. Therefore, conventional methods are unsuitable for assessing *in situ* orientation of ciliary movements. The orientation of ciliary beating can be indirectly analyzed by tracking particles spread onto the epithelium; however, the particles are not efficiently transported by immature cilia. To address this issue, we developed a method for labeling airway motile cilia with fluorescently labeled wheat germ

Methods in Cell Biology, Volume 175, ISSN 0091-679X, https://doi.org/10.1016/bs.mcb.2022.07.019
Copyright © 2023 Elsevier Inc. All rights reserved.

agglutinin (FL-WGA). The new method enables microscopic observation of ciliary movements without slicing or fragmenting the airway tissues. Since the airway epithelium is observed from the apical side, *in situ* orientation of ciliary beating can be analyzed using this method. Additionally, epithelial damage, and the number and maturity of cilia can be assessed during the observation of ciliary beating. The new method, in combination with other methods, can provide more comprehensive data regarding ciliary movements.

1 Introduction

In the airway epithelium, ciliated cells generate mucus streams to remove extraneous particles and microorganisms by beating motile cilia on their apical surfaces (Randell et al., 2006; Whitsett & Alenghat, 2015). This mechanism, known as mucociliary clearance, contributes to the maintenance of homeostatic conditions in the airway. Impairment of mucociliary clearance function increases the risk of airway inflammation, and in turn, inflammatory signals can lead to an undesirable cycle of inflammation and dysfunctional mucociliary clearance. Genetic disorders of cilium motility are associated with chronic pulmonary infections and, in severe cases, lead to airway stenosis due to subepithelial hypertrophy (Afzelius, 1976; Ferkol & Leigh, 2012). Mucus accumulation itself is also problematic because it makes breathing harder and thus, decreases the quality of life (Cook et al., 2019).

To understand the pathology and physiology of airway motile cilia, microscopic observations using a high-speed camera (high-speed video microscopy) have been performed (Kim et al., 2010; Nakamura et al., 2009; Thomas, Rutman, & O'Callaghan, 2009). This technique enables the observation of tiny (0.2–0.3 μm in diameter), and rapidly beating (>10 times per second) cilia (Tilley, Walters, Shaykhiev, & Crystal, 2015). Conventionally, transmitted light has been used to observe ciliary movements. Since translucent samples are required for observation using transmitted light, airway tissues are typically sliced perpendicularly to the epithelial plane prior to the observation (Kim et al., 2010; Nakamura et al., 2009; Thomas et al., 2009); epithelial cells in the cross section are observed at the perpendicular view to the epithelial plane. Alternatively, fragments of epithelium stripped with a brush have been used for observations. During observation, tissues are laid in a buffer containing nutrients, such as a cell culture medium, to support cell survival. Conventional methods have revealed that the ciliary beating frequency (CBF) is regulated by various factors, such as respiratory diseases, inflammatory cytokines, pH, and temperature (Rhee et al., 1999; Salathe, 2007; Veale, Rodgers, Griffiths, Ashcroft, & Gibson, 1993; Yaghi, Zaman, Cox, & Dolovich, 2012). In combination with histological analyses, structural abnormalities of cilia due to gene mutations or inflammation have been shown to cause decelerated or no ciliary beating (Tilley et al., 2015). Additionally, metachronal ciliary movement, which is crucial for forward transport of mucus, can be observed by high-speed video microscopy using sliced tissues (Bustamante-Marin & Ostrowski, 2017). However, despite their great contribution to the current knowledge, these methods are unsuitable for assessing *in situ* orientations of ciliary beating, because epithelia are either observed at a perpendicular view to the plane of the epithelial surface or by using fragmented tissues.

Kunimoto et al. succeeded in observing ciliary movements from the apical side of the mouse tracheal epithelium without slicing the trachea (Kunimoto et al., 2012). Due to a thin tracheal wall in mice, sufficient light is transmitted from the outside of the trachea to the luminal side for microscopic observations. This elegant study clarified that the orientation of ciliary beating is coordinated in distal-to-proximal direction of the trachea. Moreover, this study also revealed that genetic disturbance in coordinated ciliary orientation led to sneezing- and coughing-like symptoms, further confirming the importance of ciliary orientation. This method is promising for further analyses of the ciliary movements. Nevertheless, it remains challenging to assess the orientation of ciliary beating in most mammals and humans, wherein the tracheal walls are thick. A method using fluorescent microspheres can be employed for the analysis of the orientation of ciliary beating in thick samples by tracking the movement of the microspheres (Kunimoto et al., 2012). However, cilia themselves cannot be observed using this method. This method cannot be used to illustrate the properties of ciliary beating in areas with a weak mucus flow.

To address this issue, we developed a method to fluorescently label motile airway cilia using wheat germ agglutinin (WGA). WGA binds to sialic acid and *N*-acetylglucosamine (Baos, Phillips, Wildling, McMaster, & Berry, 2012; Bhavanandan & Katlic, 1979; Grulich-Henn, Weksler, Watanabe, & Jaffe, 1988; Lindahl, Couchman, Kimata, & Esko, 2015), which consist of sugar chains of membrane-tethered mucins (MUC1 and MUC4) and proteoglycans abundantly present on airway cilia (Button et al., 2012; Lillehoj, Kato, Lu, & Kim, 2013). Therefore, WGA preferentially binds to cilia rather than to other parts of the airway epithelium (Okada, Nicholas, Kreda, Lazarowski, & Boucher, 2006). Using fluorescently labeled WGA (FL-WGA), we succeeded in observing ciliary movements in the human trachea (Nakamura et al., 2020). Since light illuminated onto the luminal surface of the trachea generates a reflection of fluorescence, ciliary movements are observable from the apical side regardless of the thickness of the tracheal wall.

Herein, we describe a method for observing ciliary movements in human tracheae using WGA. This method is likely to be applicable across most mammalian species because FL-WGA has enabled the observation of ciliary movement in mouse, rat, rabbit, and canine tracheae (Nakamura et al., 2019, 2020; Tsuji et al., 2018; Ueda et al., 2020). Therefore, in addition to its advantage for observing human cilia, the new method is useful for research regarding surgical interventions, which often employ animals larger than mice. Moreover, ciliated cells in the nasal and bronchial epithelia are expected to be stained with FL-WGA because of the presence of membrane-tethered mucins (Chen et al., 2019; Kesimer et al., 2013; Voynow, Selby, & Rose, 1998).

2 Preparation of tracheal sample

1. Human tracheal pieces are added to conical tubes filled with pre-warmed Dulbecco's modified Eagle's medium (DMEM) containing 25 mM HEPES (30–37 °C) and transferred to the laboratory from a surgical room (see Note 1).

The temperature of the samples is to be kept at 30–37 °C until the observation of cilia is completed. (see Note 2).

2. The conical tube containing the sample is inverted several times to detach the mucus from the tracheal epithelial surface (see Note 3).
3. (Optional) If necessary, the sample should be separated into smaller pieces or slit at several lines using a razor so that the regions to be observed become flat (Fig. 1A, B).
4. The outer surface of the tracheal sample is attached to a 60-mm dish using adhesives, such as Vetbond (3M, Maplewood, MN, USA).
5. Pre-warmed DMEM is then poured into the dish. The dish is swirled to rinse the surface of the sample and the liquid is poured off (see Notes 4 and 5). This procedure is to be repeated twice.
6. The sample is incubated with DMEM containing 8 μg/mL fluorescein isothiocyanate-conjugated WGA (FITC-WGA; Vector Laboratories, Burlingame, CA, USA) for 1 h. The dish containing the sample is swirled every 15 min during the incubation period (see Notes 6 and 7).
7. Unattached FITC-WGA is removed by washing with DMEM as described in Step 5.

3 Observation of ciliary beating

8. The medium is replaced by phenol red-free DMEM/F-12 containing HEPES (see Notes 8 and 9).
9. The sample is kept on a heating plate attached to the stage of an upright fluorescence microscope and left for 5 min (see Note 10).
10. A pre-warmed objective lens is inserted into the dish such that the lens is immersed in DMEM/F-12 (Fig. 1C) (see Note 11). A 60× water-immersion lens is recommended to discern the orientation of ciliary beating (see Note 12). A lower magnification, such as 10×, is adequate for analyzing the CBF.
11. The fluorescence of FITC-WGA is observed using a high-speed camera (FASTCAM mini UX50; Photron, Tokyo, Japan) connected to the microscope (Fig. 1C).
12. Ciliary beating is captured at rates of $\geq$50 and $\geq$100 frames per second to analyze the orientations of cilia and CBF, respectively (Fig. 2A, B) (see Notes 13–16).
13. (Optional) After observing ciliary beating, the FITC-WGA-stained sample can be used for further examinations, such as the observation of microbead transport and histological analyses (see Notes 17 and 18).

4 Analysis of the orientation of ciliary beating

14. The video file of ciliary beating is opened in ImageJ.
15. A line is drawn along the orientation of beating in each ciliated cell, and the angle of the line is measured (Fig. 2C, D) (see Note 19).

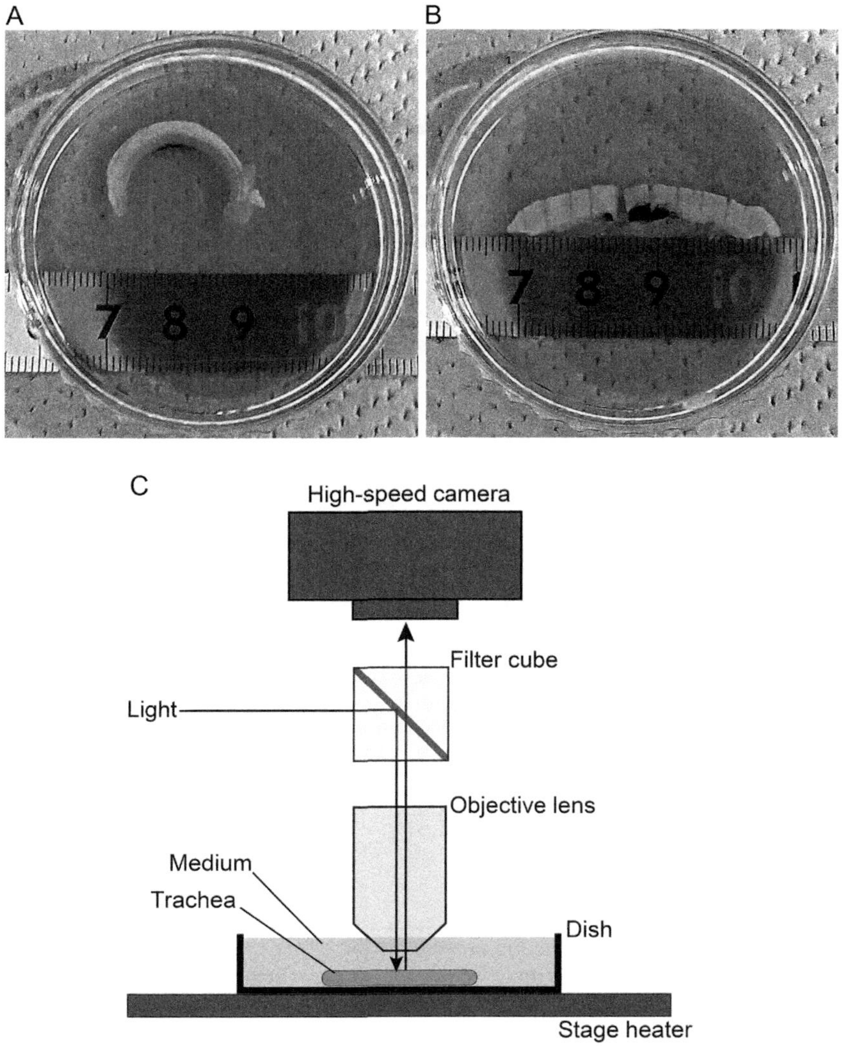

FIG. 1

Preparation of tracheal piece and setting of the microscope for observation of ciliary movements. Since the human trachea is curved (A), making slits helps flatten the trachea (B). Schematic depicting the microscope setting to observe ciliary movements (C). A high-speed camera is mounted onto an epifluorescence microscope. A sample stained with fluorescently labeled wheat germ agglutinin (FL-WGA) is fixed on a dish using an adhesive and laid in cell culture medium. To observe ciliary movements, a water immersion objective lens is inserted into the medium. Excitation light passes through a filter cube, and fluorescence from the sample is detected by the high-speed camera.

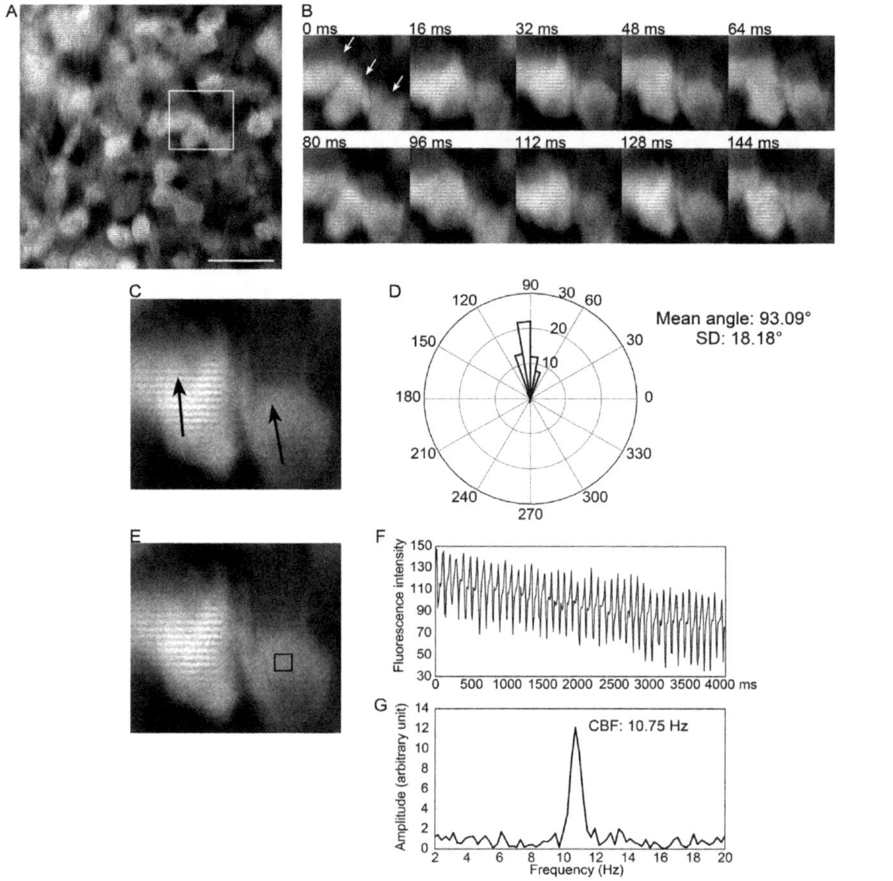

FIG. 2

Movement of human tracheal cilia and analyses of the ciliary movements. Human tracheal cilia stained with fluorescently labeled wheat germ agglutinin (FL-WGA) (A). Scale bar: 20 μm. Time-lapse images of FL-WGA-stained cilia (B). Three ciliated cells are shown in the images (indicated by white arrows). One cycle of power and recovery strokes is apparently completed in 80–96 milliseconds (ms). Orientations of ciliary beating are denoted as black arrows (C). Orientations of ciliary beating were measured in the field shown in A, and their distribution is illustrated as a rose diagram (D). Ciliary beating frequency (CBF) was measured (E–G). Fluorescence intensity in the square was measured (E), and fluorescence intensity shift is illustrated (F). The fast Fourier transform was employed to analyze the wave-form shift, and the frequency with the highest amplitude was determined to be CBF (G).

16. The distribution of orientations of ciliary beating can be illustrated as a rose diagram (Fig. 2) (see Note 20).
17. The angle data are handled in the complex plane for statistical analysis. The mean and standard deviation of the orientation data are determined using the following equations (Fig. 2D):

$$v = \cos\theta + i\sin\theta$$

$$\overline{v} = \frac{1}{n}\sum_{j} v_j \; (j = 1, 2, 3, \ldots, n)$$

$$\overline{\theta} = arg(\overline{v})$$

$$\overline{R} = |\overline{v}|$$

$$S = \sqrt{-2\ln\overline{R}}$$

where the orientations of ciliary beating directions are regarded as unit vectors, v is the unit vector in the complex plane, θ is the orientation, i is the imaginary unit, $\overline{v}$ is the mean of unit vectors, n is the number of measured cells, $\overline{\theta}$ is the mean of orientations, $\overline{R}$ is the length of the mean of unit vectors, and S is the standard deviation of the orientations.

5 Analysis of ciliary beating frequency

18. A video file with 2^n frames is opened in ImageJ (see Notes 21 and 22).
19. A small region is selected for each ciliated cell, and the fluorescence intensity is measured at each frame (Fig. 2E) (see Notes 23 and 24).
20. A wave derived from the shift in fluorescence intensity is analyzed using the fast Fourier transform on other mathematical software or Microsoft Excel (Fig. 2F).
21. The frequency with the highest amplitude is determined to be the CBF (Fig. 2F) (see Notes 25–27).

6 Concluding remarks

The method described here enables observation of ciliary movements from the apical side of the tracheal epithelium. This method is useful for assessing the properties, including *in situ* orientation, of human tracheal cilia. In fact, our preliminary clinical study using this method has suggested that irradiation of the head and neck regions leads to the formation of dysfunctional cilia in terms of length and orientation (Nakamura et al., 2020).

This method is advantageous for enabling direct observation of ciliary movements even in thick samples. The morphology of the epithelium and movement of cilia are simultaneously observable using this method; ciliary movement dependent on epithelial conditions (damaged, with few ciliated cells or immature cilia) can be assessed. However, quantitative analysis is challenging. There is no method to quantitatively analyze the immaturity of cilia observed from the apical side. The orientation of the ciliary movement is measured manually in our method. Tracking of microsphere transportation can provide more reliable quantitative data. Another limitation of our method is the deceleration of ciliary beating due to the binding of WGA to cilia (Nakamura et al., 2020). CBFs of WGA-stained and unstained cilia cannot be compared.

Overall, our method provides an alternative approach to further understanding ciliary movements; more comprehensive data can be obtained in combination with other methods.

7 Notes

1. Pressure put onto the luminal surface with forceps or any instruments can damage ciliated cells.
2. Low temperatures can promote the secretion and aggregation of mucus and prevent WGA from reaching cilia.
3. Too gentle inversion may be insufficient to remove mucus, but too aggressive inversion can damage cilia.
4. Aggregates of excess Vetbond can float in DMEM. The floating aggregate should be removed using a pipette to prevent the aggregate from covering the epithelium.
5. DMEM should not be directly poured onto the samples.
6. We have confirmed Texas Red-conjugated WGA also works for the observation of ciliary movement. Due to the low auto-fluorescence of cilia, any fluorescent dyes are likely feasible.
7. Swirling the dish is recommended to prevent mucus from aggregating and to uniformly stain cilia. A shaker is helpful in continuously swirling the dish.
8. Since phenol red exhibits auto-fluorescence, phenol red-free medium is recommended.
9. Any typically used cell culture medium is likely usable. Cell culture media designed for fluorescent live-cell imaging (e.g., FluoroBrite DMEM Media, Gibco) are commercially available and may improve the visibility of FL-WGA.
10. The temperature of the sample changes after it is set on the heating plate and becomes relatively stabilized in some time.
11. Using a lens heater and/or prewarming the microscope is recommended to avoid a decrease in the temperature of the sample. To warm a microscope, including the microscope in the incubation system would be ideal. We typically warm the microscopy room and turn on the light of the microscope to warm the microscope itself 1 h before collecting a tracheal sample.
12. We use a 60× water immersion objective lens with a 1.0 NA and 2.0 mm WD (Olympus, Tokyo, Japan). The WD should be longer than 1 mm to decrease the risk of the lens touching the sample.

13. Higher speeds improve the temporal resolution but decrease the brightness. We typically capture ciliary beating at 125 fps for analyzing orientation and CBF.
14. In the epithelium with few ciliated cells or immature cilia, FL-WGA staining of the cilia is relatively weak, and ciliary beating tends to be slow. The ciliary movement should be captured at a slow rate (~50 fps) in such epithelium.
15. FL-WGA preferentially stains cilia, but can stain various components in tissues. Non-ciliated epithelial cells are observable at a slow capture speed in the FL-WGA-stained samples. The subepithelial extracellular matrix is also observable in areas lacking epithelium, due to high auto-fluorescence of extracellular fibrous structures and binding of FL-WGA.
16. Given that the temperature of the tracheal epithelium decreases by air exchange *in vivo*, the temperature of the sample does not necessarily need to be 37 °C. However, to reduce variations in CBF data, the temperature should be maintained within a certain range during the observation.
17. Although flesh samples are ideal for the observation of microbead transport and histological analyses, these examinations following the observation of FL-WGA-stained cilia can provide additional information about mucus flow and morphological properties of cilia in areas where FL-WGA observation has been performed.
18. Mucus accumulates during the FL-WGA observations. Observations for microbead transport and histological analyses should be performed after washing the surface of the sample.
19. Since angles are manually measured, a blind test is recommended.
20. We use Octave to draw a rose diagram (https://www.gnu.org/software/octave/index).
21. The fast Fourier transform requires 2^n of fluorescence intensity data.
22. At least 256 frames should be analyzed. Longer duration increases the temporal resolution of CBF data, but leads to fading of the fluorescence. We typically use 512 frames.
23. A region of interest is set at an area where the fluorescence intensity clearly changes upon ciliary beating.
24. The fluorescence intensity in each frame is measured by following procedures in ImageJ: (1) Analyze > Set Measurement > Fill in the box of Mean gray value; (2) Plugins > Stacks > Measure Stack.
25. Waves of nearly zero frequencies may have high amplitudes owing to noise, and should be ignored.
26. The fast Fourier transform generates a vertically symmetrical graph, and the right half of the data (higher frequencies) is meaningless. The lower frequency with the highest amplitude is determined as CBF.
27. CBF measured in FL-WGA-stained samples is not comparable with unstained samples, because ciliary beating is decelerated by binding of WGA to cilia (Nakamura et al., 2020).

References

Afzelius, B. A. (1976). A human syndrome caused by immotile cilia. *Science*, *193*, 317–319.

Baos, S. C., Phillips, D. B., Wildling, L., McMaster, T. J., & Berry, M. (2012). Distribution of sialic acids on mucins and gels: A defense mechanism. *Biophysical Journal*, *102*, 176–184.

Bhavanandan, V. P., & Katlic, A. W. (1979). The interaction of wheat germ agglutinin with sialoglycoproteins. The role of sialic acid. *The Journal of Biological Chemistry*, *254*, 4000–4008.

Bustamante-Marin, X. M., & Ostrowski, L. E. (2017). *Cilia and mucociliary clearance* (p. 9). Cold Spring Harbor Perspectives in Biology.

Button, B., Cai, L. H., Ehre, C., Kesimer, M., Hill, D. B., Sheehan, J. K., et al. (2012). A periciliary brush promotes the lung health by separating the mucus layer from airway epithelia. *Science*, *337*, 937–941.

Chen, Z. G., Wang, Z. N., Yan, Y., Liu, J., He, T. T., Thong, K. T., et al. (2019). Upregulation of cell-surface mucin MUC15 in human nasal epithelial cells upon influenza A virus infection. *BMC Infectious Diseases*, *19*, 622.

Cook, N., Gey, J., Oezel, B., Mackay, A. J., Kumari, C., Kaur, V. P., et al. (2019). Impact of cough and mucus on COPD patients: Primary insights from an exploratory study with an online patient community. *International Journal of Chronic Obstructive Pulmonary Disease*, *14*, 1365–1376.

Ferkol, T. W., & Leigh, M. W. (2012). Ciliopathies: The central role of cilia in a spectrum of pediatric disorders. *The Journal of Pediatrics*, *160*, 366–371.

Grulich-Henn, J., Weksler, B. B., Watanabe, K., & Jaffe, E. A. (1988). Wheat germ agglutinin inhibits thrombin-induced rises in cytosolic free calcium and prostacyclin synthesis by human umbilical vein endothelial cells. *Journal of Cellular Physiology*, *137*, 553–558.

Kesimer, M., Ehre, C., Burns, K. A., Davis, C. W., Sheehan, J. K., & Pickles, R. J. (2013). Molecular organization of the mucins and glycocalyx underlying mucus transport over mucosal surfaces of the airways. *Mucosal Immunology*, *6*, 379–392.

Kim, D. Y., Pyun, J., Choi, J. W., Kim, J. H., Lee, J. S., Shin, H. A., et al. (2010). Tissue-engineered allograft tracheal cartilage using fibrin/hyaluronan composite gel and its in vivo implantation. *Laryngoscope*, *120*, 30–38.

Kunimoto, K., Yamazaki, Y., Nishida, T., Shinohara, K., Ishikawa, H., Hasegawa, T., et al. (2012). Coordinated ciliary beating requires Odf2-mediated polarization of basal bodies via basal feet. *Cell*, *148*, 189–200.

Lillehoj, E. P., Kato, K., Lu, W., & Kim, K. C. (2013). Cellular and molecular biology of airway mucins. *International Review of Cell and Molecular Biology*, *303*, 139–202.

Lindahl, U., Couchman, J., Kimata, K., & Esko, J. D. (2015). Proteoglycans and sulfated glycosaminoglycans. In A. Varki, E. J. D. Cummings, P. Stanley, G. W. Hart, M. Aebi, A. G. Darvill, … P. H. Seeberger (Eds.), *Essentials of Glycobiology* (pp. 207–221). NY: Cold Spring Harbor.

Nakamura, R., Katsuno, T., Kishimoto, Y., Kaba, S., Yoshimatsu, M., Kitamura, M., et al. (2020). A novel method for live imaging of human airway cilia using wheat germ agglutinin. *Scientific Reports*, *10*, 14417.

Nakamura, R., Katsuno, T., Kitamura, M., Yamashita, M., Tsuji, T., Suzuki, R., et al. (2019). Collagen sponge scaffolds containing growth factors for the functional regeneration of tracheal epithelium. *Journal of Tissue Engineering and Regenerative Medicine*, *13*, 835–845.

Nakamura, T., Sato, T., Araki, M., Ichihara, S., Nakada, A., Yoshitani, M., et al. (2009). In situ tissue engineering for tracheal reconstruction using a luminar remodeling type of artificial trachea. *The Journal of Thoracic and Cardiovascular Surgery*, *138*, 811–819.

Okada, S. F., Nicholas, R. A., Kreda, S. M., Lazarowski, E. R., & Boucher, R. C. (2006). Physiological regulation of ATP release at the apical surface of human airway epithelia. *The Journal of Biological Chemistry*, *281*, 22992–23002.

Randell, S. H., Boucher, R. C., & University of North Carolina Virtual Lung G. (2006). Effective mucus clearance is essential for respiratory health. *American Journal of Respiratory Cell and Molecular Biology*, *35*, 20–28.

Rhee, C. S., Hong, S. K., Min, Y. G., Lee, C. H., Lee, K. S., Ahn, S. H., et al. (1999). Effects of IL-1 beta, TNF-alpha, and TGF-beta on ciliary beat frequency of human nasal ciliated epithelial cells in vitro. *American Journal of Rhinology*, *13*, 27–30.

Salathe, M. (2007). Regulation of mammalian ciliary beating. *Annual Review of Physiology*, *69*, 401–422.

Thomas, B., Rutman, A., & O'Callaghan, C. (2009). Disrupted ciliated epithelium shows slower ciliary beat frequency and increased dyskinesia. *The European Respiratory Journal*, *34*, 401–404.

Tilley, A. E., Walters, M. S., Shaykhiev, R., & Crystal, R. G. (2015). Cilia dysfunction in lung disease. *Annual Review of Physiology*, *77*, 379–406.

Tsuji, T., Nakamura, R., Katsuno, T., Kishimoto, Y., Suehiro, A., Yamashita, M., et al. (2018). Long-term preservation of planar cell polarity in reversed tracheal epithelium. *Respiratory Research*, *19*, 22.

Ueda, Y., Sato, T., Nakamura, R., Tamari, S., Muranishi, Y., Yutaka, Y., et al. (2020). Evaluation of regenerated tracheal cilia function on a collagen-conjugated scaffold in a canine model. *Interactive Cardiovascular and Thoracic Surgery*, *31*, 644–649.

Veale, D., Rodgers, A. D., Griffiths, C. J., Ashcroft, T., & Gibson, G. J. (1993). Variability in ciliary beat frequency in normal subjects and in patients with bronchiectasis. *Thorax*, *48*, 1018–1020.

Voynow, J. A., Selby, D. M., & Rose, M. C. (1998). Mucin gene expression (MUC1, MUC2, and MUC5/5AC) in nasal epithelial cells of cystic fibrosis, allergic rhinitis, and normal individuals. *Lung*, *176*, 345–354.

Whitsett, J. A., & Alenghat, T. (2015). Respiratory epithelial cells orchestrate pulmonary innate immunity. *Nature Immunology*, *16*, 27–35.

Yaghi, A., Zaman, A., Cox, G., & Dolovich, M. B. (2012). Ciliary beating is depressed in nasal cilia from chronic obstructive pulmonary disease subjects. *Respiratory Medicine*, *106*, 1139–1147.

CHAPTER

4

Time-lapse imaging of primary cilium behavior with physiological expression of fluorescent ciliary proteins

Ryota Nakazato†, Hiroshi Otani†, Faryal Ijaz†, and Koji Ikegami*

Department of Anatomy and Developmental Biology, Graduate School of Biomedical and Health Sciences, Hiroshima University, Hiroshima, Japan

**Corresponding author: e-mail address: k-ikegami@hiroshima-u.ac.jp*

Chapter outline

†These authors contributed equally to this work.

Methods in Cell Biology, Volume 175, ISSN 0091-679X, https://doi.org/10.1016/bs.mcb.2022.10.003
Copyright © 2023 Elsevier Inc. All rights reserved

Abstract

Almost all cell types of mammals have a small protrusion named a primary cilium on their surface. Primary cilia are enriched by cilia-specific ion channels and G-protein-coupled receptors. They are known to regulate various cellular functions that contribute to the development and homeostasis of living organisms by receiving extracellular signals and transfusing them to the cell body. All functions are performed when the structure of the primary cilia is maintained properly. Abnormalities in primary cilia or their signaling can lead to a collection of diseases in various organs called ciliopathies. The primary cilium is dynamic, static, or fixed. The length of primary cilia varies as the cell cycle progresses and is also altered by extracellular stimuli. Ligand binding to cilia-specific receptors is also known to alter the length. Thus, there is a need for a method to study the morphological changes of the primary cilium in a time-dependent manner, especially under stimuli or mechanical shocks. Time-lapse imaging of primary cilia is one of the most powerful methods to capture the time-dependent behavior of primary cilia. Overexpression of ciliary proteins fused to fluorescent proteins is commonly used for the time-lapse imaging of primary cilia. However, overexpression has drawbacks in terms of artifacts. In addition, the time-lapse imaging of the tiny primary cilia requires some technical tricks. Here, we present a detailed description of the methods for time-lapse imaging of primary cilium, from the generation of cell lines that stably express fluorescent protein-labeled cilia-localized proteins at the physiological level to image analysis, including quantification through image acquisition.

1 Introduction

The primary cilium protrudes from the cell surface like an antenna into the extracellular environment surrounding the cell. The primary cilium comes in diverse forms that have been modified to respond to different kinds of stimuli. These are found on almost all mammalian cells including cells of the renal tubule, photoreceptor cells of the retina, cells of the olfactory epithelium, fibroblasts, neurons, and cholangiocytes of the biliary tree. The decade's progress in experimental technologies has provided us with accumulating evidence that primary cilia are present on most vertebrate cells and work as signal receivers and transducers (Marshall & Nonaka, 2006; Reiter & Leroux, 2017; Singla & Reiter, 2006), since K.W. Zimmermann described the primary cilium in 1898.

Primary cilia harbor cilia-specific ion channels, signaling molecules along with several G-protein-coupled receptors (GPCRs) that can sense extracellular chemical, osmotic, and mechanical stimuli (Ferreira, Fukui, Chow, Vilfan, & Vermot, 2019; Poole, Flint, & Beaumont, 1985). In addition, primary cilia regulate the cell cycle (Pan, Seeger-Nukpezah, & Golemis, 2013), cell differentiation (Ezratty et al., 2011; Forcioli-Conti, Lacas-Gervais, Dani, & Peraldi, 2015), cell-cell communication (Viau et al., 2018), and autophagy (Orhon et al., 2016; Pampliega et al., 2013). The multiple functions of primary cilia along with their importance make primary cilia crucial for organism development and homeostasis (Badano, Mitsuma, Beales, & Katsanis, 2006).

Ciliopathies are a multitude of human diseases arising from mutations that impair the formation of cilia or its signaling activity and cause changes in ciliary length differences and ciliary trafficking defects (Alazami, Seidahmed, Alzahrani, Mohammed, & Alkuraya, 2014; Bredrup et al., 2011; Duran et al., 2017; Halbritter et al., 2013; Shaheen et al., 2020; Zhang et al., 2016). Ciliopathies can manifest renal, hepatic, and pancreatic cyst formation; situs abnormalities; retinal degeneration; anosmia; cerebellar or other brain anomalies; postaxial polydactyly; bronchiectasis; and infertility (Badano et al., 2006; Hildebrandt, Benzing, & Katsanis, 2011; Reiter & Leroux, 2017). Mutations in the IFT80 gene have been shown to cause bone anomalies, a very narrow rib cage, failure of the lungs, and retinal degeneration (Beales et al., 2007; Cavalcanti et al., 2011; Moran et al., 2018). Similarly, mutations in another IFT gene Tg737/Ift88 have been associated with abnormal craniofacial and photoreceptor development, retinal degeneration, polycystic kidney disease, and human cleft palate (Pazour, San Agustin, Follit, Rosenbaum, & Witman, 2002; Schrick et al., 1995; Tian et al., 2017). Mutations in the lipid-metabolizing enzyme, inositol polyphosphate 5-phosphatase E (INPP5E) are associated with Joubert syndrome and mental retardation, truncal obesity, retinal dystrophy, and micropenis (MORM) syndrome. In Joubert syndrome, point mutations in the catalytic domain of the INPP5E change the phosphatase activity of the enzyme. In MORM syndrome, C-terminus-truncating mutations of the INPP5E affect its ciliary localization. These mutations cause abnormally shortened cilia and ciliary disassembly (Bielas et al., 2009; Jacoby et al., 2009; Travaglini et al., 2013). Bardet-Biedl syndrome (BBS) results from disruption of BBSome components, or disruption of BBSome trafficking or assembly (Novas, Cardenas-Rodriguez, Irigoín, & Badano, 2015). BBSome plays a crucial role in the transport of diverse ciliary cargoes and disruption of these results in the broad phenotypic range of BBS which includes retinal degeneration, cystic kidneys, obesity, polydactyly, and cognitive impairment (Forsythe & Beales, 2013). Furthermore, in diseases not caused by mutations in primary ciliary genes, such as type 2 diabetes, obesity, and renal injury, cilia lengths have also been observed to be lengthened or shortened on affected cells (Han et al., 2014; Ritter et al., 2018; Verghese, Weidenfeld, Bertram, Ricardo, & Deane, 2008; Yu et al., 2019). The lengths of renal primary cilia are increased in human transplants that suffered acute tubular necrosis (Verghese et al., 2008).

Primary cilia are morphologically and structurally dynamic. They elongate, shrink, or sometimes completely disappear upon the cell cycle. Primary cilia are assembled when cells become quiescent and enter the G1 and G0 phases of the cell cycle. Cilia disassembly begins before cells re-enter the cell cycle (Malicki & Johnson, 2017; Nigg & Stearns, 2011; Plotnikova, Golemis, & Pugacheva, 2008; Zhu, Shi, Wang, & Liao, 2009). Since protein synthesis is not known to occur at the cilium, ciliary assembly and maintenance are done by a specialized system of bidirectional transportation called intraflagellar transport (IFT) (Kozminski, Johnson, Forscher, & Rosenbaum, 1993). The structural and signaling molecules are transported from the cytoplasm to the ciliary axoneme through a transition zone followed by the transport along the microtubular axoneme via IFT. The IFT system consists of proteins and motors. IFT proteins form two network complexes named IFT-A, which bind to dynein-2 motors to transport proteins from the cilium tip to the cell body (retrograde) (Pazour, Dickert, & Witman, 1999; Pazour, Wilkerson, & Witman, 1998; Porter, Bower, Knott, Byrd, & Dentler, 1999; Signor et al., 1999); and IFT-B complex, that interacts with kinesin-II motors to transport proteins from the basal body to the cilium tip (anterograde) (Cole et al., 1998; Katoh et al., 2016). IFT transport system also plays key role in the transport of receptors and signaling molecules in and out of the primary cilia (Nachury, 2018). IFT complexes are connected to signaling molecules via another multiprotein complex, the BBSome, which is also required for the ciliary export of activated signaling complex (Nachury, 2018; Ye, Nager, & Nachury, 2018).

The cycle of cilia assembly and disassembly also regulates the length of primary cilia (Marshall, Qin, Brenni, & Rosenbaum, 2005; Marshall & Rosenbaum, 2001). Cilium length regulation depends on the coordinated regulation of IFT particle trafficking, size, frequency, and speed of IFT train ciliary entry (Chien et al., 2017; Fai, Mohapatra, Kar, Kondev, & Amir, 2019; Marshall et al., 2005; Wemmer, Ludington, & Marshall, 2020; Wren et al., 2013). Primary cilium length is also influenced by the actin filaments and microtubule dynamics through soluble tubulin levels available in the cytosol for primary cilia extension (Kim et al., 2010; Sharma, Kosan, Stallworth, Berbari, & Yoder, 2011). Recent works reveal that primary cilia also actively change their lengths upon ligand stimulation and changes in the extracellular environment. For example, in renal epithelial cells, fluid shear-mediated deflection shortens the primary cilia by decreasing the cyclic AMP levels, leading to a subsequent reduction in mechanosensitivity to fluid shear (Besschetnova et al., 2010). In synoviocytes, inhibition of cilia-localized adenylyl cyclase 3 (AC3) induces primary cilia elongation (Ou et al., 2009). A GPCR ligand, melanin-concentrating hormone, induces shortening of cilia length in hRPE1 cells via ciliary melanin-concentrating hormone receptor 1 (Hamamoto et al., 2016). In addition to ciliary length regulation, primary cilia also modify their shape depending on the external signals. IFT and second messengers play key roles in the adjustment of cilia shape in response to extracellular stimuli. For instance, in decreased sensory signaling, specialized cilia of *C. elegans* olfactory neurons enhance their olfactory receptivity by changing their morphology from finger-like to fan-shaped (Mukhopadhyay, Lu, Shaham, & Sengupta, 2008). The tip of primary cilia can become the balloon-like shape and be severed in response to serum stimulation or ligand binding (Nager et al., 2017; Phua et al., 2017).

Time-lapse imaging of primary cilia is a powerful tool to study those dynamic and drastic changes of primary cilia morphology, especially upon stimuli or mechanical stresses. So far, expressing fluorescent protein-tagged cilia-localized proteins, such as Arl13b, serotonin receptor 6 (5HT6), somatostatin receptor 3 (SSTR3) etc., are used to visualize primary cilium in the time-lapse imaging. However, transient overexpression of those cilia-localized proteins causes artifacts of cilia morphology such as abnormally elongated cilia (Fig. 1, A and B). Here, we describe

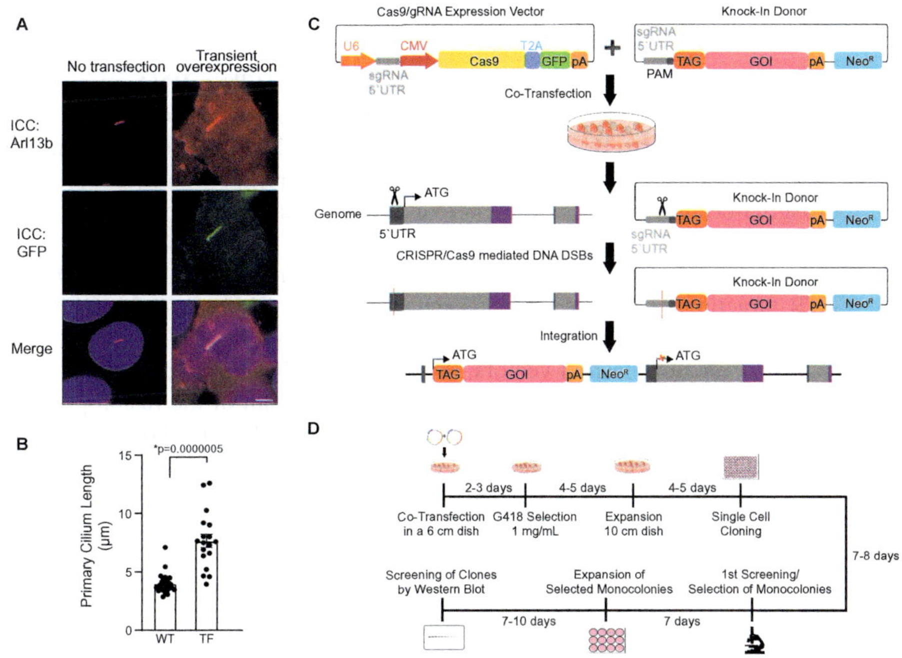

FIG. 1

Generation of stable knock-in cells by 5′UTR-targeting genome editing. (A) Images representative of the primary cilium morphology in control (no knock-in and no transfection of Arl13b-Venus), and cells with transient overexpression of Arl13b-Venus as described in Section 1. Ciliary marker (Arl13b; red), Arl13b-Venus (GFP; green) and the nucleus (DAPI; blue). Scale bar = 10 μm. (B) Quantitative analysis of the primary cilium length from A. Data are shown as the mean ± s.e.m. (WT: wildtype; TF: transfection; *, $p < 0.05$). (C) Schematics of the donor plasmid and targeting strategy for CRISPR/Cas9-mediated homology-independent knock-in of tagged gene of interest (GOI) at the 5′UTR of the endogenous gene locus as described in Section 2. After co-transfection, Cas9 induce a double-strand break (DSBs; red dotted lines) upstream of the protospacer adjacent motif (PAM) sequence (dark gray) in the 5′UTR of genomic DNA and tagged GOI in the donor plasmid. NHEJ-directed repair results in the integration of the cassette into the genome. (NeoR: neomycin resistance gene cassette) (D) Timeline for the preparation of stable cell lines.

Panels (B–D) are reprinted and modified by permission from Ijaz, F., & Ikegami, K. (2021). Knock-in of labeled proteins into 5′UTR enables highly efficient generation of stable cell lines. Cell Structure and Function, 21002.

a method for time-lapse imaging of primary cilium, including generation of stable cell lines expressing a primary ciliary marker protein Arl13b tagged with fluorescent marker Venus, steps of image acquisition, and analysis to measure primary cilium shape. With minor variations, this approach can easily be adapted to several cell lines and further fluorescent proteins to study cilia length, shapes, and behaviors.

2 Generation of knock-in cells

1. The 5′ untranslated regions (5′UTR) of mouse Arl13b mRNA sequences are retrieved using UTRdB (http://utrdb.ba.itb.cnr.it/). Retrieved 5′UTR sequences are put into Broad Institutes GPP sgRNA Designer to acquire a list of sgRNA candidates (https://portals.broadinstitute.org/gpp/public/analysis-tools/sgrna-design). A target sequence of sgRNA is selected from the list (see Notes 1–2).
2. To construct a Cas9-sgRNA-5′UTR-expression plasmid vector, each a 20-bp 5′UTR-targeting sgRNA sequence is sub-cloned into CMV-Cas9-2A-GFP plasmid backbone (ATUM, CA, USA) (see Notes 3–4).
3. To construct donor plasmid vectors, the 20-bp 5′UTR-targeting sgRNA sequence and 3-bp PAM sequence are inserted upstream of the fusion protein Arl13b-Venus sequences (see Notes 5–7). The donor plasmid should contain an antibiotic-resistance gene and lack mammalian expression promoter sequence (see Note 8). In our plasmids, a neomycin-resistance gene is used (see Note 9).
4. NIH/3T3 cells are seeded in a standard 6-cm cell culture dish and cultured for one to 2 days until transfection (see Note 10). Details about cell culture are described in Section 3. Suitable confluency for transfection is checked on a light microscope before transfection (see Note 11).
5. Cas9/gRNA-expression plasmids and Arl13b-Venus sequence-harboring donor plasmid are mixed at 1:1 in PBS (Solution A) (see Note 12). For a 6-cm dish, 2 μg of each plasmid, i.e., total of 4 μg plasmids are diluted in 150 μL of PBS (Solution A). Twelve micrograms of polyethylenimine (PEI) (3 μL of 4 μg/μL stock solution) are added into 150 μL of PBS (Solution B) (see Note 13).
6. Solution A (150 μL) and B (150 μL) are carefully mixed and incubated at room temperature for 20 min to form transfection complexes (see Note 14). The ratio of DNA and PEI is 1:3 in weight in this case (see Note 15). Transfection complexes (300 μL) are then added to the culture medium in the 6-cm dish containing cells and returned to the incubator (37 °C, 5% CO_2) (Fig. 1D).
7. Six to 12 h later, the culture medium is replaced with 3 mL of fresh medium to remove PEI and DNA. The culture medium is detailed in Section 3.
8. Two to 3 days later, the culture medium is replaced with a fresh one that contains 1 mg/mL G418 (Fig. 1D) (see Notes 16–19).
9. After four to 5 days, live cells are transferred to a standard 10-cm cell culture dish to expand the culture (Fig. 1D). Eight milliliter of 1 mg/mL G418-containing culture medium is placed in the 10-cm dish.

10. Further four to 5 days later, G418-resistant cells are detached using 0.05% trypsin/EDTA and resuspended in the culture medium (see Note 20). After counting the number of cells in the cell suspension, the cell suspension is diluted in DMEM High Glucose supplemented with 10% FBS and 0.5 mg/mL G418 to have a cell suspension with the cell concentration of ~750 cells/50 mL (see Notes 21). For single cell cloning, the limiting dilution is performed (Fig. 1D); one hundred microliters of the cell suspension is added per well of a 96-well plate so that one well contains theoretically ~1.5 cells (see Notes 21–22). A total of approximately 500 wells are seeded, i.e., five 96-well plates are used.
11. Cells are allowed to proliferate for 1–2 weeks. For the selection of monocolonies, microscopic observations are done to monitor single cell colony formation and confluency (Fig. 1D) (see Notes 23–27).
12. After 2–3 weeks of seeding, selected monocolonies are expanded into a 12-well plate (Fig. 1D) (see Note 28). Cells are gently washed with pre-warmed PBS and are detached with 20 μL of 0.05% trypsin/EDTA at 37 °C for 3 min. Eighty microliters of DMEM High Glucose supplemented with 10% FBS is added to the well to inactivate trypsin/EDTA, and a total of 100 μL of the solution is mixed vigorously by pipetting in the well to dissociate the monocolonies. A hundred microliters of the cell suspension is taken into a well of a 12-well plate (i.e., "lower dilution") where 1 mL of DMEM High Glucose supplemented with 10% FBS and 0.25 mg/mL G418 is present. The remaining 20 μL of the cell suspension is transferred to a well of a duplicate 12-well plate (i.e., "higher dilution") (see Note 29).
13. One to 2 weeks later, knock-in clones are screened with Western blot analyses (Fig. 1D) (see Notes 30–31). Samples are prepared by adding an appropriate volume of 1× SDS-PAGE lysis buffer to the individual wells of a 12-well plate (lower dilution) containing expanded clones after rinsing the cells well with PBS (see Note 32). The cell lysates are collected into microtubes and boiled for 5 min (see Note 33). After the Western blot analyses, a few numbers of the best possible Arl13b-Venus knock-in clones are selected by observing the cell growth and morphology in the duplicate 12-well plate (higher dilution) (see Notes 34–37) (Fig. 2).
14. Following the final selection of clones, the selected clones from duplicate 12-well plates (higher dilution) are expanded to a 10-cm dish by detaching the cells upon incubation with 0.05% trypsin/EDTA and being resuspended in 8 mL DMEM High Glucose supplemented with 10% FBS and 0.2 mg/mL G418 (see Note 38). Cells are allowed to grow until 80% confluency. Cell stocks are prepared for storage and subsequent experiments (move to Section 3).

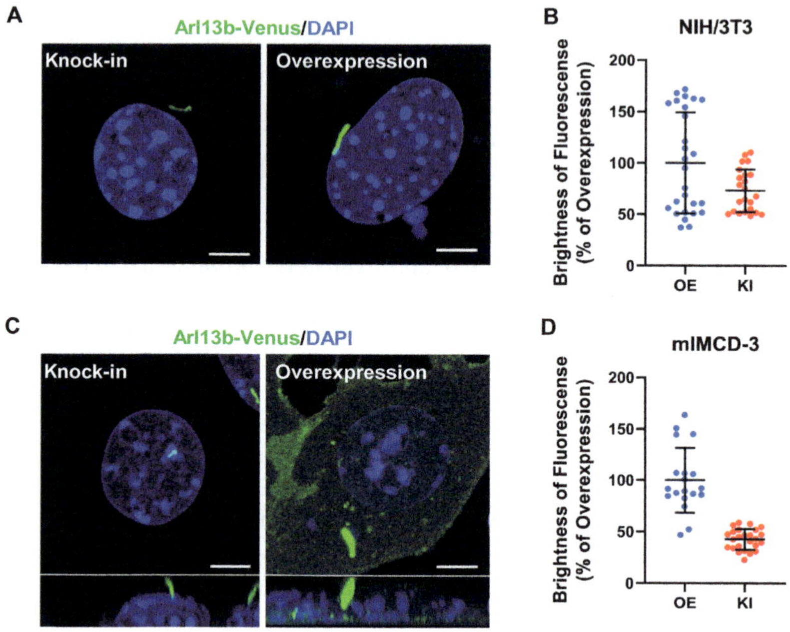

FIG. 2

Comparison of the Arl13b-Venus fluorescence between endogenous promoter- and CMV-driven expression. Arl13b-Venus knock-in cells were fixed with 4% paraformaldehyde (PFA) after serum starvation as described in Section 2. Cells with transient overexpression of Arl13b-Venus were prepared by transfection with the vector expressing Arl13b-Venus upon the CMV promoter as described in Section 1. Cells with transient overexpression of Arl13b-Venus were starved for serum simultaneously with transfection and fixed with 4% PFA 24h later. (A) Z-projection images representative of the primary cilia in knock-in NIH/3T3 cells and NIH/3T3 cells with transient overexpression of Arl13b-Venus. (B) Quantitative analysis of the fluorescence intensity in Arl13b-Venus from A. (C) Z-projection (upper) and X-projection (bottom) images representative of the primary cilia in knock-in mIMCD-3 cells and mIMCD-3 cells with transient overexpression of Arl13b-Venus. (D) Quantitative analysis of the fluorescence intensity in Arl13b-Venus from C. Arl13b-Venus (green) and the nucleus (DAPI; blue). Scale bar, 5 µm. OE: transient overexpression; KI: knock-in.

3 Cell culture

3.1 NIH/3T3 cells

1. Mouse fibroblast-like cell line, NIH/3T3 is cultured in High-Glucose Dulbecco's Modified Eagle Medium (DMEM) (see Note 39) with 10% fetal bovine serum (FBS) (see Note 40) at 37 °C, 5% CO_2.

2. Once maintenance cultures reach 80–90% confluency, exhausted culture medium is discarded, and cells are gently washed with pre-warmed PBS (Note 41).
3. The cell monolayer is treated with 0.05% trypsin/EDTA for 2–3 min (Note 42). Then, 10-time volume of DMEM/10% FBS is added to inactivate trypsin.
4. Cells are immediately dissociated by pipetting 10–20 times (see Note 43). Detached and dissociated cells are collected in centrifuge tubes and precipitated with centrifugation at $200 \times g$ for 3 min.
5. The precipitated cells are resuspended in culture medium. The number of cells in the cell suspension is measured with a hemocytometer. To maintain the culture, cells are reseeded at 1:5 to 1:10 dilution rate in a fresh culture dish (see Notes 24 and 44).
6. For time-lapse imaging, NIH/3T3 cells stably expressing Arl13b-Venus are seeded at 4.5×10^5 per 35-mm glass-bottom dish and cultured in DMEM with 10% FBS at 37 °C, 5% CO_2 (see Notes 45–46).
7. 24 h later, culture cells are serum-starved; cells are washed with serum-free DMEM and the culture medium is replaced with DMEM containing 1% FBS. Cells are further cultured in the low-serum medium for 24 h (see Note 47). Cells are ready for time-lapse imaging (move to Section 4, step 1).

3.2 mIMCD-3 Cells

8. Mouse Inner Medullary Collecting Duct-3 (mIMCD-3) cells are cultured in DMEM/Ham's F-12 (see Note 48) with 10% FBS at 37 °C, 5% CO_2.
9. Once maintenance cultures reach 80–90% confluency, the exhausted culture medium is discarded, and cells are vigorously washed with pre-warmed PBS (Note 49).
10. The cell monolayer is treated with 0.25% trypsin/EDTA for 3–5 min (Note 50). Then, a 10-time volume of DMEM/Ham's F-12 containing 10% FBS is added to inactivate trypsin.
11. Cells are dissociated, collected, and precipitated as described in Sections 3, step 3 and 3, step 4 (see also Note 43). Then, the precipitated cells are resuspended with a culture medium, and the cell number is measured as described in Section 3, step 5. To maintain the culture, cells are reseeded at a 1:10 to 1:20 dilution rate in a fresh culture dish (see Notes 44 and 51).
12. For time-lapse imaging, mIMCD-3 cells stably expressing Arl13b-Venus are seeded at 5.5×10^5 per 35-mm glass bottom dish and cultured in DMEM/F-12 with 10% FBS at 37 °C, 5% CO_2 (see Notes 46 and 52).
13. 48 h later, culture cells are serum-starved; cells are washed with pre-warmed PBS and the culture medium is replaced with serum-free colorless DMEM/F-12 (ThermoFisher, 11039-021, USA) (see Note 53). Next, the cells are further cultured in the serum-free medium for 24 h (see Note 47). Cells are ready for time-lapse imaging (move to Section 4, step 3).

4 Image acquisition

4.1 2-D Time-lapse imaging (NIH/3T3 cells) (See Note 54)

1. The culture medium is replaced with serum-free colorless DMEM (Wako, 040-30095, Japan) and the 35-mm glass-bottom dish is placed on the stage top incubator (STXG, TOKAIHIT, Japan) attached to an inverted confocal microscope (STELLARIS 5; Leica, Germany) (see Notes 53 and 55–56).
2. Time-lapse images are captured with a 63× oil immersion objective lens (N.A. 1.40) every 3 min for 1 h in channels of Venus (λEx: 515 nm, λEm: 528 nm). The parameters are as follows: image size, 61.45 × 15.32 × 4.48 μm; Z step, 16 steps; Z-step size, 0.30 μm; pixel size, 1024 × 256 pixels; scan speed, 700 Hz; smart gain, 69.2%; smart intensity (laser power), 5.00%; smart wavelength, 507.0 nm: pinhole size, 1.00 AU; zoom, 3.00 (see Notes 57–58). Images are taken using Adaptive Focus Control (AFC) built in STELLARIS 5 to maintain the focal plane. The data are ready for image processing (move to Section 5, step 1) or data quantification (move to Section 6, step 1).

4.2 3-D Time-lapse imaging (mIMCD-3 cells) (See Note 59)

3. The 35-mm glass bottom dish is placed on the stage top incubator and attached to a confocal microscope (STELLARIS 5) (see Note 55–56).
4. Time-lapse images are captured with a 63× oil immersion objective lens (N.A. 1.40) every 3 min for 1.5 h in channels of Venus (λEx: 515 nm, λEm: 528 nm) (see Note 60). The parameters are as follows: image size, 36.9 × 18.64 × 10.74 μm; format, 256 × 150 pixels; scan seed, 800 Hz; scan direction, unidirectional; zoom, 5.00; pinhole, 1.00 AU; Z step, 30 steps; Z-step size, 0.371 μm; laser source, 507 nm; laser power, 85% of White Light Laser, 3.00% of the smart intensity of 507 nm laser line; detection range, HyD S SiPM 512–650 nm; detector sensitivity, smart gain 100.1, offset 0 (see Note 61). Images are taken using the AFC of STELLARIS 5 to maintain the focal plane. The data are ready for image processing (move to Section 5, step 4) or data quantification (move to Section 6, step 5 or 6, step 11).

5 Image processing

5.1 2-D Time-lapse imaging (NIH/3T3 cells)

1. Time-lapse images acquired by STELLARIS 5 are processed using LAS X (Leica). Images are deconvoluted by LIGHTNING (Leica) to decrease noise (see Note 62).

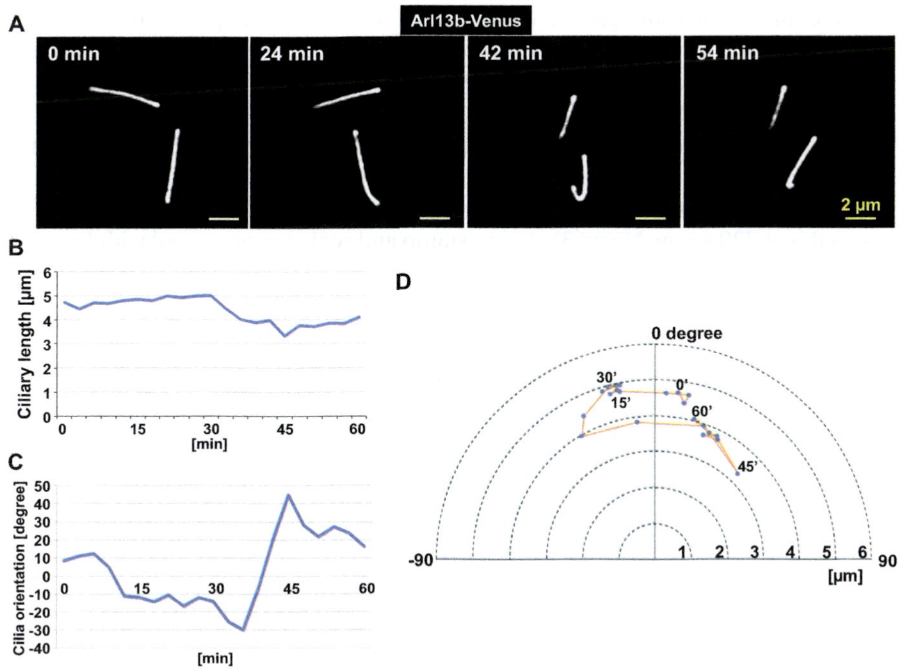

FIG. 3

Two-dimensional time-lapse imaging of primary cilia protruding horizontally on the cell surface. Arl13b-Venus expressing NIH/3T3 cells as described in Section 2 were seeded in a 35-mm glass bottom dish and were cultured for 24h in DMEM with 10% FBS. Then, the culture medium was changed to DMEM with 1% FBS and the cells were incubated for further 24h, as described in Sections 3 and 4. Finally, Z-stack images were taken every 3min at STELLARIS 5, for a total of 1h as described in Sections 4 and 5. The length and base-to-tip direction (angle) of the primary cilia were quantified in time-lapse images, as described in Section 6. Representative images of Arl13b-Venus in NIH/3T3 cells (A), quantitative data on the length of the primary cilia (B), the orientation of the primary cilia (C), and quantitative data on combined length and orientation of the primary cilia (D) are presented. Scale bar, 2µm.

2. All Z-stack images in each time point are compressed into an x-y plane with maximum projection to acquire two-dimensional images (Fig. 3A).
3. Time-lapse motion picture is created at a speed of 4 frames per second. A total of 21 frames results in a video of 5 s (Supplementary Movie 1 in the online version at https://doi.org/10.1016/bs.mcb.2022.10.003). Thus, the playback speed is at 720× (see Note 63). A timestamp and scale bar are added with ImageJ.

5.2 3-D Time-lapse imaging (mIMCD-3 cells)

4. Time-lapse images acquired by STELLARIS 5 are processed and deconvoluted using LAS X (Leica) as described in Section 5, step 1 (see Note 62).

5. Three-dimensional images are created from all Z-stack images with a surface rendering of LAS X 3D Viewer (Leica). Frames, axis scaling, and scale bar are added in LAS X 3D Viewer. The viewpoint is obliquely looking down on the primary cilia from above (Fig. 4A).
6. Time-lapse motion picture is created at a speed of four frames per second. A total of 31 frames results in a video of 7.75 s (Supplementary Movie 2 in the online version at https://doi.org/10.1016/bs.mcb.2022.10.003). Thus, the playback speed is at 720× (see Note 63). A timestamp and scale bar are added with ImageJ.

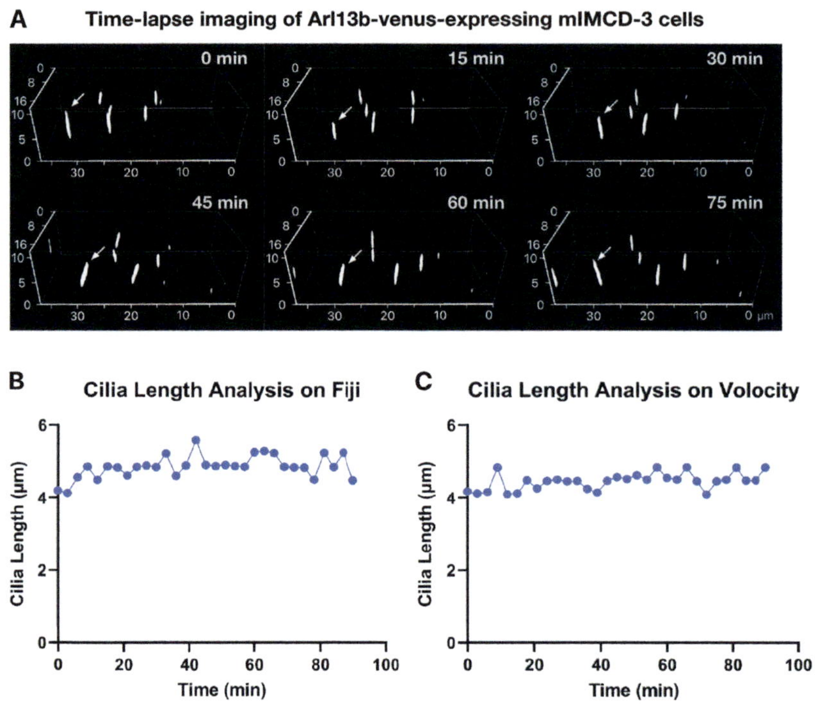

FIG. 4

Three-dimensional time-lapse imaging of primary cilia protruding perpendicularly on the apical surface of cells. Arl13b-Venus-expressing mIMCD-3 cells as described in Section 2 were seeded in a 35-mm glass bottom dish and were cultured for 48 h in DMEM/F12 with 10% FBS. Then, the culture medium was changed to serum-free colorless DMEM/F12 and the cells were incubated for further 24 h, as described in Sections 3 and 4. Finally, Z-stack images were taken every 3 min at STELLARIS 5, for total of 1 h 30 min, as described in Sections 4 and 5. The lengths of the primary cilia were quantified in time-lapse images, as described in Section 6. Representative images of Arl13b-Venus in IMCD-3 cells (A), quantitative data on the length of the primary cilia using Fiji (ImageJ) (B), and quantitative data on the length of the primary cilia using Volocity (C) are presented.

6 Data quantification

6.1 2-D Time-lapse imaging (NIH/3T3 cells)

1. Data obtained with STELLARIS 5 (the extension is ".lif") are imported into Fiji (ImageJ) (see Notes 64–65).
2. Lengths of primary cilia on NIH/3T3 are measured by tracing the primary cilia with the "Segmented Line" in ImageJ. Calculation of the length of primary cilia traced is performed by calculating the number of pixels per μm, given that the size of the image in the y-axis direction is 15.32 μm. Graphs are created in an Excel file (Fig. 3B).
3. The direction of the primary cilia is measured by drawing a line from the base of the primary cilia to the tip by "Straight Line" in ImageJ and using the "Angle" at that point. The angle from the intersection of the XY-axis to the Y-axis is 0 degrees. Graphs are created in an Excel file (Fig. 3C).
4. The combined length and direction of the primary cilia graph shows the intersection of the XY-axis as the base of the primary cilia and the point at each coordinate as the tip of the primary cilia. The coordinate (x, y) of the tip of each primary cilium is calculated using the following formula: x-coordinate = Primary cilia length × cos θ; y-coordinate = Primary cilia length × sin θ, θ is the angle made by the X-axis with the straight line from the base to tip of the primary cilia. Graphs are created using scatter plots in an Excel file (Fig. 3D).

6.2 3-D Time-lapse imaging (mIMCD-3 cells)

6.2.1 ImageJ (see Note 66)

5. Data obtained with STELLARIS 5 (the extension is ".lif") are imported into ImageJ (see Notes 64–65). Bio-Formats Import Options dialogue appears. View stack with Hyperstack is selected. The color is changed with Pseudo color change, "Image > Color > Channels Tool". Further, the image is adjusted with look up table with "Image > Adjust > Brightness/Contrast."
6. Z-projection of time-lapse images is created by using "Image > Stacks > Z Project."
7. Primary cilia are enclosed in a rectangle using the "Rectangle Tool" and primary cilia are labeled using the ROI tool. ROI tool can be used by selecting "Analyze > Tools > ROI Manager."
8. Primary cilia in a rectangle ROI are traced by the "Straight Line" tool (see Note 67).
9. The top and bottom Z-slice images in which the primary cilia are present are took down, and how many Z-stack images the primary cilia signal is present in are calculated (see Note 68).

10. The length of primary cilia is calculated using the following formula according to the Pythagorean theorem (Dummer, Poelma, DeRuiter, Goumans, & Hierck, 2016): (Z Projected length in step8)^2 + ((Step size of Z-axis) × (Z-slice number in step9–1))^2 = (the length of primary cilia)^2. Time-dependent changes in primary cilia length are depicted in a graph with GraphPad Prism 9 version 9.3.1 (471) for Windows 64-bit, (GraphPad.inc, USA, www.graphpad.com) (Fig. 4B).

6.2.2 *Volocity* (see note 69)

11. Data obtained with STELLARIS 5 (the extension is ".lif") are imported into Volocity (Quorum Technologies, UK).
12. Primary cilia are traced by the "Line" tool after selecting "Measurements" (see Note 70).
13. The length of the primary cilia is displayed when "Measurements > Make Measurements Item" is selected. Time-dependent changes of primary cilia length are depicted in a graph with GraphPad Prism 9 (Fig. 4C).

7 Concluding remarks

The methods we presented here enable us to observe and analyze "natural" shapes and lengths of the primary cilium and their changes under chemical or mechanical stimuli in real time. A combination of highly stable and bright fluorescent proteins, such as recently developed StayGold (Hirano et al., 2022), and optimum acquisition settings on the cutting-edge confocal microscope would minimize the fluorescent attenuation time lapse and result in a quite longer acquisition of data (e.g., for 1 day) with more frequent shootings (e.g., every several seconds). A couple of modes of quantifications enable the best possible efficient analysis depending on the types of the primary cilium, available time, and tool. As the morphology of the primary cilium is dynamic and quickly changed under certain conditions, our methods are suitable to study its dynamics under chemical or mechanical stimuli.

8 Notes

1. When deciding on the gRNA-targeting sequence, it is better to choose a sequence that will cause the minimum deletion of the 5′UTR region, as 5′UTR is important for the post-transcriptional regulation of translation (Pickering & Willis, 2005; Van Der Velden & Thomas, 1999). In other words, the sgRNA-guided Cas9-cleaving site is better to be just upstream of the start codon.

2. The gRNA-targeting sequence is to be carefully selected to minimize off-target effects. NHEJ is susceptible to off-target integrations (Maruyama et al., 2015). The frequency of such undesirable events depends on the spontaneous integration frequency of the donor, the cleavage efficiency at the on-target site, and the specificity of the gRNA that is used (Lackner et al., 2015).
3. Other plasmid vectors that have similar structures can also be used. It would be better to use plasmid vectors that can harbor both the Cas9-expression unit and the gRNA-expression unit (Fig. 1C). If separate plasmids are used for the expression of Cas9 and gRNA, transfection efficiencies of three plasmids (Cas9 expression, gRNA expression, and donor vector) would decrease.
4. If the Cas9 expression vector contains a fluorescent protein sequence as a marker of transfection (Fig. 1C), transfection efficiency can be investigated (see also Note 17), though the fluorescence protein expression is optional.
5. In our example, the Arl13b-Venus expression vector is used, which has been constructed previously from pcDNA derivatives for the transient overexpression of Arl13b-Venus, as the materials to construct donor vectors. If common mammalian expression vectors such as pcDNA derivatives are used as the material of donor vectors, the 20-bp 5′UTR-targeting sgRNA sequence and 3-bp PAM sequence can be inserted just by replacing the mammalian expression promoter, such as CMV, with the sgRNA-PAM sequences.
6. 5′UTR sequence of target genes can be added between the sgRNA-PAM sequences and the fluorescent protein-tagged sequence in the donor vector to minimize the changes of the 5′UTR to a few indels resulting from NHEJ repair, though this makes donor vector construction more complex and time-consuming.
7. The donor vector p5′UTRsgRNA-Arl13b-Venus constructed from the Arl13b-Venus expression vector is ready for use as it contained polyadenylation signal sequences to stabilize transcribed RNA and the expression cassette of antibiotic-resistance genes. It is thus better to use common mammalian expression vectors, such as pcDNA derivatives, for the initial source of donor vectors (see also Note 5).
8. The tagged gene of interest is expressed only when the donor sequence is correctly inserted with the correct orientation at the endogenous GOI loci as it is promoter-less.
9. Any antibiotic-resistance genes can be chosen. The choice of antibiotic-resistance genes is important in some cases. Some cell lines are resistant to some antibiotics.
10. Cells with the least number of passages are best for making stable cell lines.
11. In our observation, PEI-based transfection achieves maximal efficiency when NIH/3T3 cells are 50–60% confluency. For other cell types, these may need to be optimized.
12. Intactness of both the Cas9-sgRNA-5′UTR-expression vector and the donor vector are important. Both plasmids should be circular. If plasmid vectors are accidentally linearized during preparation, those linearized plasmids can be integrated into the Cas9-cleaved site, which results in a reduction of knock-in efficiencies.

13. In our experiment, PEI is purchased from Polysciences, Inc., USA (#24765-2). Various other transfection reagents are commercially available and can be used in place of PEI. Though, applying alternative liposomal formations may need some modifications to the transfection protocol. More cautiously, some transfection reagents are not good at co-transfection of multiple plasmids.
14. The incubation time should be appropriately decided. Longer, as well as shorter incubation time, may cause a decrease in transfection efficiencies.
15. The best ratio of DNA and PEI can vary depending on cell types. This need to be optimized for other cell types. In our observation, an increase in the amount of DNA results in the reduction of transfection efficiency in many cell types.
16. Following co-transfection of the Cas9-sgRNA-5′UTR-expression vector and the donor vector p5′UTRsgRNA-Arl13b-Venus, Cas9 endonuclease cleave and linearize the donor vector and the linearized donor vector is integrated at the target site, i.e., 5′UTR of endogenous Arl13b in genomic DNA during the repair of Cas9-mediated DSB through the NHEJ pathway (Fig. 1C).
17. It is possible to check transfection efficiency visually under a fluorescence microscope if fluorescence protein is expressed as a marker of transfection (Fig. 1C) (see also Note 4). 70–80% cells emitting fluorescence after 1–2 days of transfection is a good indicator of optimal transfection and makes getting knock-in clones speedier. Transfection efficiencies vary in cell types.
18. The concentration of antibiotics for cell selection needs to be optimized based on the cell line in use. In our experiment, G418 is purchased from Wako, Japan (#074-05181).
19. To obtain biallelic knock-in clones, the concentration of antibiotic for selection can be increased. However, the increased concentration of antibiotics would slow down the growth of cultured cells and need a longer culture period. Also, the biallelic substitution of endogenous proteins with fluorescent protein-tagged forms may impede the normal function of proteins and causes abnormal cellular behaviors.
20. To have single cells, the cell layer should be certainly dissociated with careful and appropriate trypsinization and trituration. Meanwhile, vigorous and rough handling will result in a drastic reduction of clones. It is thus important to decide the best condition of trypsinization and trituration that can achieve almost complete dissociation and maintenance of cellular healthiness.
21. The number of cells to seed per well for single cell cloning depends on the cell line in use. In the case of NIH/3T3 cells, ~1.5 cells/well seems to work well as NIH/3T3 cells are vulnerable to limiting dilution. When using other cell lines, the number of cells per well needs to be optimized. Some "tough" cells can be seeded at ~0.2 cells/well. Using a conditioned medium is also optional for vulnerable cell types; in this case, the number of cells per well will differ.
22. To avoid cell damage, wide bore tips are used to seed cells into a 96-well plate (see also Note 20).

23. About 1 week later after cloning, the 96-well plates are investigated under a microscope, and wells with single cell colonies are marked. Also, the area covered by a colony is marked in each well, which is a good estimate of cell growth. Ideally, a single colony covering 50% or more of a well is a sign of a healthy clone.
24. The growth of monoclonies takes about 2–3 weeks depending on cell types. NIH/3T3 cells proliferate with a doubling time of 36–45 h.
25. Small single cell colonies that cover a lesser area are checked again after another week or two. If the colony still has not grown further, it usually indicates unhealthy growth. It is better to exclude these from further expansion.
26. Edges of wells need to be carefully observed to examine whether or not the wells contain additional colonies. If two or more colonies grow in the same well, they must be excluded from further selection to avoid mixed clones.
27. During longer incubations in 96-well plates, culture medium tends to evaporate from wells, especially at the corner and the side of the plate. Fresh culture medium needs to be added to these wells to compensate for evaporated medium.
28. In our hands, about three to five dozen monocolonies are acquired for further expansion out of the initial seeding of 500 wells.
29. Monocolonies are expanded into the duplicate of 12-well plates from which one culture is used to make frozen stocks (higher dilution) for subsequent use and the other culture for screening (lower dilution).
30. Suitable confluency for screening is checked every 2 to 3 days on a light microscope. In our observation, it takes about 7–10 days to reach 100% confluency for NIH/3T3 clones with normal cell growth in a 12-well plate with a lower dilution of cells. This factor also depends on the cell type. Some clones will grow faster or slower than others during expansion. The ideal situation for screening is when most of the clones are at 100% confluency or at least at more than 50% confluency.
31. PCR can be also used after extracting genomic DNA for screening.
32. It is important to remove the culture medium completely and wash cells with PBS well. Residual serum may generate non-specific bands in Western blot analyses that hamper evaluation.
33. The volume of 1× SDS-PAGE lysis buffer needs to be adjusted according to the confluency of the cells in a well; e.g., 100 μL of lysis buffer for 100% confluency, 50 μL for 50% confluency, and so on. In our experiment, 1× SDS-PAGE lysis buffer is prepared by diluting Sample Buffer Solution (Wako, Japan, #196-11022) twice with ultrapure water.
34. At least cell shape, growth, and physiology are investigated with reference to untransfected wild-type cells. Knock-in cells closest to wild-type cells' growth and morphology are best.
35. In our experiments, 32 clones among 37 clones selected in the first screening (86%) in the case of NIH/3T3 and 49 among 60 clones selected in the first screening (82%) in the case of mIMCD-3 were positive for Arl13b-Venus knock-in clones.

36. The difference in primary ciliary length between Arl13b-Venus knock-in cells and cells with overexpression of Arl13b-Venus is shown in Ijaz and Ikegami (2021).
37. The brightness of the knock-in Arl13b-Venus expressed under the endogenous Arl13b promoter is approximately 40 to 70% compared to CMV-overexpressed Arl13b-Venus (Fig. 2).
38. G418 is added throughout the single cell cloning and screening steps until the stocks are made from positive clones. The G418 concentration is decreased at each step after selection to 0.2–0.5 mg/mL. G418 can be omitted when the stable cell lines are maintained for experiments.
39. In our experiments, NIH/3T3 cells were purchased from ATCC (CRL-1658). High-glucose DMEM is purchased from Wako, Japan (#044-29765). Any manufacturer can be chosen if the medium with the same composition as ours is available.
40. In our experiments, FBS is purchased from Gibco, USA (#10270-106). It is important to find out a good lot of reliable manufacturers. FBS is incubated at 56 °C for 20 min to inactivate complement components.
41. It is important not to make cells reach over confluency. It results in cell death and a change in cell characteristics.
42. In our experiments, 0.05% trypsin/EDTA solution is purchased from Wako (#202-16931). Any manufacturer can be chosen.
43. Making air bubbles should be avoided. Air bubbles damage cells. Meanwhile, enough dissociation is needed to accurately measure cell numbers.
44. The dilution rate varies depending on the culture period to the next passage.
45. Seeding at this cell count results in a cell density of around 80% after 24 h. The cell density seeded can be optimized depending on experimental schedules and cell types used.
46. 35-mm glass bottom dish is from IWAKI, Japan (#3911-035). The thickness of the glass is 0.13–0.17 mm. The glass of 0.16–0.19 mm in thickness is better. The glass of 0.165–0.175 mm in thickness is more ideal for 3-D imaging.
47. This is important because serum inhibits the assembly of primary cilia. Primary cilia are formed on cells that are in the G_0 phase, the resting phase of the cell cycle.
48. In our experiments, mIMCD-3 cells were purchased from ATCC (CRL-2123). DMEM/Ham's F-12 is purchased from Wako (#048-29785). Any manufacturer can be chosen if the medium with the same composition is available.
49. It is quite important not to make cells reach over confluency. Overconfluent cells are hard to be detached and dissociated even with 5-min trypsinization.
50. In our experiments, 0.25% trypsin/EDTA solution is purchased from Wako (#209-16941). Any manufacturer can be chosen.
51. mIMCD-3 cells proliferate with doubling time of ~24 h.
52. Seeding at this cell count results in a cell density of around 100% after 48 h. The cell density seeded can be optimized depending on experimental schedules and cell types used.

53. Colorless medium may decrease background signals of live cell imaging. Phenol red may increase the background of live cell imaging.
54. 2-D time-lapse imaging is chosen for the NIH/3T3 cells as primary cilia of NIH/3T3 cells grow horizontally to the cell surface. This enables us to capture primary cilia in a narrow focal plane. Other cell types that have similar characteristics can be used. RPE-1 is a popular cell line.
55. When transferring to the confocal microscope, the glass bottom dish is placed on an insulator warmed at 45 °C and covered with a Styrofoam box to avoid stress due to temperature changes in cells.
56. In the stage top incubator, the condition is maintained at 37 °C, 5% CO_2. The warm-up process of the incubator takes 10–20 min. It is important to wait until the temperature becomes stable to avoid fluctuation of temperature of samples and lens that causes focus drifting during the shot.
57. It is better to choose YFP-originated photo-stable and bright fluorescent proteins, such as Venus with 515-nm, not 488-nm, selected as excitation light to avoid photobleaching. Highly sensitive camera or detector enables to decrease the power of excitation light, which also prevents photobleaching. The time required to take a single frame should be as short as possible for further prevention of photobleaching. Increasing the interval of time-lapse imaging is also effective.
58. Parameters need to be optimized depending on the expression level of fluorescent ciliary markers, stability of fluorescent proteins used, the vulnerability of cells to the excitation light, and so on.
59. The 3-D time-lapse imaging is chosen for the mIMCD-3 cells because the primary cilia of mIMCD-3 cells grow perpendicularly on the cell surface into the medium.
60. Silicon immersion objective lens is ideal for 3-D imaging to capture images at a deeper plane from the bottom of the dish.
61. The time taken to acquire a volume, i.e., a Z-stack series, is set to less than 5 s. Less than 3 s are ideal to avoid fluctuation of primary cilia caused by Brownian movement of water during image shot.
62. LIGHTNING can perform deconvolution in parallel with data acquisition. Other algorithms may ask to wait several minutes for deconvolution.
63. The playback speed can be chosen arbitrarily.
64. The version of Fiji (ImageJ) used in this chapter is Version: 2.1.0/1.53q (Build: 5f23140693).
65. Data in proprietary file formats from confocal microscopes (STELLARIS 5) are imported to Fiji (ImageJ) via a plug-in called Bio-Formats. (https://docs.openmicroscopy.org/bio-formats/6.5.1/supported-formats.html#term-Ratings-legend-and-definitions).
66. Fiji (ImageJ) has some limitations though it is free and easy for all to use. For example, the length of curved primary cilia cannot be accurately measured. Volocity, albeit expensive, enables us to easily measure the length of curved primary cilia. Other high-end expensive software may be used for quantification.
67. This process measures the length of the primary cilia on the X-Y plane.

68. This process measures the length of the primary cilia on the Z axis.
69. The version of Volocity used in this chapter is 6.5.1.
70. If the primary cilium is curved, multiple clicks are made to trace the primary cilium in line with the curve.

Acknowledgments

This work was supported in part by the Japan Society for Promotion of Science (JSPS) Grants-in-Aid for Early-Career Scientists 21K15088 (to F.I.) and 20K15792 (to R.N.), and JSPS Grant-in-Aid for Scientific Research (C) 21K06172 and Japan Science and Technology Agency, Precursory Research for Embryonic Science and Technology JPMJPR17H1 (to K.I.). A part of this work was carried out at the Natural Science Center for Basic Research and Development, Hiroshima University, and at Advanced Research Facilities. We thank Madoka Hamada, Mimoko Katoh, and Yoko Hayashi for their technical assistance.

References

Alazami, A. M., Seidahmed, M. Z., Alzahrani, F., Mohammed, A. O., & Alkuraya, F. S. (2014). Novel IFT122 mutation associated with impaired ciliogenesis and cranioectodermal dysplasia. *Molecular Genetics & Genomic Medicine*, *2*, 103–106.

Badano, J. L., Mitsuma, N., Beales, P. L., & Katsanis, N. (2006). The ciliopathies: An emerging class of human genetic disorders. *The Annual Review of Genomics and Human Genetics*, *7*, 125–148.

Beales, P. L., Bland, E., Tobin, J. L., Bacchelli, C., Tuysuz, B., Hill, J., et al. (2007). IFT80, which encodes a conserved intraflagellar transport protein, is mutated in Jeune asphyxiating thoracic dystrophy. *Nature Genetics*, *39*, 727–729.

Besschetnova, T. Y., Kolpakova-Hart, E., Guan, Y., Zhou, J., Olsen, B. R., & Shah, J. V. (2010). Identification of signaling pathways regulating primary cilium length and flow-mediated adaptation. *Current Biology*, *20*, 182–187.

Bielas, S. L., Silhavy, J. L., Brancati, F., Kisseleva, M. V., Al-Gazali, L., Sztriha, L., et al. (2009). Mutations in INPP5E, encoding inositol polyphosphate-5-phosphatase E, link phosphatidyl inositol signaling to the ciliopathies. *Nature Genetics*, *41*, 1032–1036.

Bredrup, C., Saunier, S., Oud, M. M., Fiskerstrand, T., Hoischen, A., Brackman, D., et al. (2011). Ciliopathies with skeletal anomalies and renal insufficiency due to mutations in the IFT-A gene WDR19. *The American Journal of Human Genetics*, *89*, 634–643.

Cavalcanti, D. P., Huber, C., Sang, K. H. L. Q., Baujat, G., Collins, F., Delezoide, A. L., et al. (2011). Mutation in IFT80 in a fetus with the phenotype of Verma-Naumoff provides molecular evidence for Jeune-Verma-Naumoff dysplasia spectrum. *The Journal of Medical Genetics*, *48*, 88–92.

Chien, A., Shih, S. M., Bower, R., Tritschler, D., Porter, M. E., & Yildiz, A. (2017). Dynamics of the IFT machinery at the ciliary tip. *eLife*, *6*, e28606.

Cole, D. G., Diener, D. R., Himelblau, A. L., Beech, P. L., Fuster, J. C., & Rosenbaum, J. L. (1998). Chlamydomonas kinesin-II–dependent intraflagellar transport (IFT): IFT particles contain proteins required for ciliary assembly in Caenorhabditis elegans sensory neurons. *The Journal of Cell Biology*, *141*, 993–1008.

Dummer, A., Poelma, C., DeRuiter, M. C., Goumans, M. J., & Hierck, B. P. (2016). Measuring the primary cilium length: Improved method for unbiased high-throughput analysis. *Cilia*, *5*, 1–9.

Duran, I., Taylor, S. P., Zhang, W., Martin, J., Qureshi, F., Jacques, S. M., et al. (2017). Mutations in IFT-A satellite core component genes IFT43 and IFT121 produce short rib polydactyly syndrome with distinctive campomelia. *Cilia*, *6*, 1–13.

Ezratty, E. J., Stokes, N., Chai, S., Shah, A. S., Williams, S. E., & Fuchs, E. (2011). A role for the primary cilium in notch signaling and epidermal differentiation during skin development. *Cell*, *145*, 1129–1141.

Fai, T. G., Mohapatra, L., Kar, P., Kondev, J., & Amir, A. (2019). Length regulation of multiple flagella that self-assemble from a shared pool of components. *eLife*, *8*, e42599.

Ferreira, R. R., Fukui, H., Chow, R., Vilfan, A., & Vermot, J. (2019). The cilium as a force sensor-myth versus reality. *The Journal of Cell Science*, *132*, jcs213496.

Forcioli-Conti, N., Lacas-Gervais, S., Dani, C., & Peraldi, P. (2015). The primary cilium undergoes dynamic size modifications during adipocyte differentiation of human adipose stem cells. *Biochemical and Biophysical Research Communications*, *458*, 117–122.

Forsythe, E., & Beales, P. L. (2013). Bardet–Biedl syndrome. *European Journal of Human Genetics*, *21*, 8–13.

Halbritter, J., Bizet, A. A., Schmidts, M., Porath, J. D., Braun, D. A., Gee, H. Y., et al. (2013). Defects in the IFT-B component IFT172 cause Jeune and Mainzer-Saldino syndromes in humans. *The American Journal of Human Genetics*, *93*, 915–925.

Hamamoto, A., Yamato, S., Katoh, Y., Nakayama, K., Yoshimura, K., Takeda, S., et al. (2016). Modulation of primary cilia length by melanin-concentrating hormone receptor 1. *Cellular Signalling*, *28*, 572–584.

Han, Y. M., Kang, G. M., Byun, K., Ko, H. W., Kim, J., Shin, M. S., et al. (2014). Leptin-promoted cilia assembly is critical for normal energy balance. *The Journal of Clinical Investigation*, *124*, 2193–2197.

Hildebrandt, F., Benzing, T., & Katsanis, N. (2011). Ciliopathies. *New England Journal of Medicine*, *364*, 1533–1543.

Hirano, M., Ando, R., Shimozono, S., Sugiyama, M., Takeda, N., Kurokawa, H., et al. (2022). A highly photostable and bright green fluorescent protein. *Nature Biotechnology*, *40*, 1–11.

Ijaz, F., & Ikegami, K. (2021). Knock-in of labeled proteins into 5′UTR enables highly efficient generation of stable cell lines. *Cell Structure and Function*, *46*, 21–35.

Jacoby, M., Cox, J. J., Gayral, S., Hampshire, D. J., Ayub, M., Blockmans, M., et al. (2009). INPP5E mutations cause primary cilium signaling defects, ciliary instability and ciliopathies in human and mouse. *Nature Genetics*, *41*, 1027–1031.

Katoh, Y., Terada, M., Nishijima, Y., Takei, R., Nozaki, S., Hamada, H., et al. (2016). Overall architecture of the intraflagellar transport (IFT)-B complex containing Cluap1/IFT38 as an essential component of the IFT-B peripheral subcomplex. *Journal of Biological Chemistry*, *291*, 10962–10975.

Kim, J., Lee, J. E., Heynen-Genel, S., Suyama, E., Ono, K., Lee, K., et al. (2010). Functional genomic screen for modulators of ciliogenesis and cilium length. *Nature*, *464*, 1048–1051.

Kozminski, K. G., Johnson, K. A., Forscher, P., & Rosenbaum, J. L. (1993). A motility in the eukaryotic flagellum unrelated to flagellar beating. *Proceedings of the National Academy of Sciences of the United States of America*, *90*, 5519–5523.

Lackner, D. H., Carré, A., Guzzardo, P. M., Banning, C., Mangena, R., Henley, T., et al. (2015). A generic strategy for CRISPR-Cas9-mediated gene tagging. *Nature Communications*, *6*, 1–7.

Malicki, J. J., & Johnson, C. A. (2017). The cilium: Cellular antenna and central processing unit. *Trends in Cell Biology*, *27*, 126–140.

Marshall, W. F., & Nonaka, S. (2006). Cilia: Tuning in to the cell's antenna. *Current Biology*, *16*, R604–R614.

Marshall, W. F., Qin, H., Brenni, M. R., & Rosenbaum, J. L. (2005). Flagellar length control system: Testing a simple model based on intraflagellar transport and turnover. *Molecular Biology of the Cell*, *16*, 270–278.

Marshall, W. F., & Rosenbaum, J. L. (2001). Intraflagellar transport balances continuous turnover of outer doublet microtubules: Implications for flagellar length control. *The Journal of Cell Biology*, *155*, 405–414.

Maruyama, T., Dougan, S. K., Truttmann, M. C., Bilate, A. M., Ingram, J. R., & Ploegh, H. L. (2015). Increasing the efficiency of precise genome editing with CRISPR-Cas9 by inhibition of nonhomologous end joining. *Nature Biotechnology*, *33*, 538–542.

Moran, J., Sanderson, G., Maynes, J. K., Vig, A., Batmanabane, V., Kannu, P., et al. (2018). IFT80 mutations cause a novel complex ciliopathy phenotype with retinal degeneration. *Clinical Genetics*, *94*, 368–372.

Mukhopadhyay, S., Lu, Y., Shaham, S., & Sengupta, P. (2008). Sensory signaling-dependent remodeling of olfactory cilia architecture in C. elegans. *Developmental Cell*, *14*, 762–774.

Nachury, M. V. (2018). The molecular machines that traffic signaling receptors into and out of cilia. *Current Opinion in Cell Biology*, *51*, 124–131.

Nager, A. R., Goldstein, J. S., Herranz-Pérez, V., Portran, D., Ye, F., Garcia-Verdugo, J. M., et al. (2017). An actin network dispatches ciliary GPCRs into extracellular vesicles to modulate signaling. *Cell*, *168*, 252–263.

Nigg, E. A., & Stearns, T. (2011). The centrosome cycle: Centriole biogenesis, duplication and inherent asymmetries. *Nature Cell Biology*, *13*, 1154–1160.

Novas, R., Cardenas-Rodriguez, M., Irigoín, F., & Badano, J. L. (2015). Bardet–Biedl syndrome: is it only cilia dysfunction? *FEBS Letters*, *589*, 3479–3491.

Orhon, I., Dupont, N., Zaidan, M., Boitez, V., Burtin, M., Schmitt, A., et al. (2016). Primary-cilium-dependent autophagy controls epithelial cell volume in response to fluid flow. *Nature Cell Biology*, *18*, 657–667.

Ou, Y., Ruan, Y., Cheng, M., Moser, J. J., Rattner, J. B., & van der Hoorn, F. A. (2009). Adenylate cyclase regulates elongation of mammalian primary cilia. *Experimental Cell Research*, *315*, 2802–2817.

Pampliega, O., Orhon, I., Patel, B., Sridhar, S., Diaz-Carretero, A., Beau, I., et al. (2013). Functional interaction between autophagy and ciliogenesis. *Nature*, *502*, 194–200.

Pan, J., Seeger-Nukpezah, T., & Golemis, E. A. (2013). The role of the cilium in normal and abnormal cell cycles: Emphasis on renal cystic pathologies. *Cellular and Molecular Life Sciences*, *70*, 1849–1874.

Pazour, G. J., Dickert, B. L., & Witman, G. B. (1999). The DHC1b (DHC2) isoform of cytoplasmic dynein is required for flagellar assembly. *Journal of Cell Biology*, *144*, 473–481.

Pazour, G. J., San Agustin, J. T., Follit, J. A., Rosenbaum, J. L., & Witman, G. B. (2002). Polycystin-2 localizes to kidney cilia and the ciliary level is elevated in orpk mice with polycystic kidney disease. *Current Biology*, *12*, R378–R380.

Pazour, G. J., Wilkerson, C. G., & Witman, G. B. (1998). A dynein light chain is essential for the retrograde particle movement of intraflagellar transport (IFT). *Journal of Cell Biology*, *141*, 979–992.

Phua, S. C., Chiba, S., Suzuki, M., Su, E., Roberson, E. C., Pusapati, G. V., et al. (2017). Dynamic remodeling of membrane composition drives cell cycle through primary cilia excision. *Cell*, *168*, 264–279.

Pickering, B. M., & Willis, A. E. (2005). The implications of structured 5′ untranslated regions on translation and disease. *Seminars in Cell and Developmental Biology*, *16*, 9–47.

Plotnikova, O. V., Golemis, E. A., & Pugacheva, E. N. (2008). Cell cycle–dependent ciliogenesis and cancer. *Cancer Research*, *68*, 2058–2061.

Poole, C. A., Flint, M. H., & Beaumont, B. W. (1985). Analysis of the morphology and function of primary cilia in connective tissues: A cellular cybernetic probe? *Cell Motility*, *5*, 175–193.

Porter, M. E., Bower, R., Knott, J. A., Byrd, P., & Dentler, W. (1999). Cytoplasmic dynein heavy chain 1b is required for flagellar assembly in Chlamydomonas. *Molecular Biology of the Cell*, *10*, 693–712.

Reiter, J. F., & Leroux, M. R. (2017). Genes and molecular pathways underpinning ciliopathies. *Nature Reviews Molecular Cell Biology*, *18*, 533–547.

Ritter, A., Friemel, A., Kreis, N. N., Hoock, S. C., Roth, S., Kielland-Kaisen, U., et al. (2018). Primary cilia are dysfunctional in obese adipose-derived mesenchymal stem cells. *Stem Cell Reports*, *10*, 583–599.

Schrick, J. J., Onuchic, L. F., Reeders, S. T., Korenberg, J., Chen, X. N., Moyer, J. H., et al. (1995). Characterization of the human homologue of the mouse Tg737 candidate polycystic kidney disease gene. *Human Molecular Genetics*, *4*, 559–567.

Shaheen, R., Alsahli, S., Ewida, N., Alzahrani, F., Shamseldin, H. E., Patel, N., et al. (2020). Biallelic mutations in tetratricopeptide repeat domain 26 (Intraflagellar transport 56) cause severe biliary ciliopathy in humans. *Hepatology*, *71*, 2067–2079.

Sharma, N., Kosan, Z. A., Stallworth, J. E., Berbari, N. F., & Yoder, B. K. (2011). Soluble levels of cytosolic tubulin regulate ciliary length control. *Molecular Biology of the Cell*, *22*, 806–816.

Signor, D., Wedaman, K. P., Orozco, J. T., Dwyer, N. D., Bargmann, C. I., Rose, L. S., et al. (1999). Role of a class DHC1b dynein in retrograde transport of IFT motors and IFT raft particles along cilia, but not dendrites, in chemosensory neurons of living Caenorhabditis elegans. *Journal of Cell Biology*, *147*, 519–530.

Singla, V., & Reiter, J. F. (2006). The primary cilium as the cell's antenna: Signaling at a sensory organelle. *Science*, *313*, 629–633.

Tian, H., Feng, J., Li, J., Ho, T. V., Yuan, Y., Liu, Y., et al. (2017). Intraflagellar transport 88 (IFT88) is crucial for craniofacial development in mice and is a candidate gene for human cleft lip and palate. *Human Molecular Genetics*, *26*, 860–872.

Travaglini, L., Brancati, F., Silhavy, J., Iannicelli, M., Nickerson, E., Elkhartoufi, N., et al. (2013). Phenotypic spectrum and prevalence of INPP5E mutations in Joubert syndrome and related disorders. *European Journal of Human Genetics*, *21*, 1074–1078.

Van Der Velden, A. W., & Thomas, A. A. (1999). The role of the 5′ untranslated region of an mRNA in translation regulation during development. *The International Journal of Biochemistry & Cell Biology*, *31*, 87–106.

Verghese, E., Weidenfeld, R., Bertram, J. F., Ricardo, S. D., & Deane, J. A. (2008). Renal cilia display length alterations following tubular injury and are present early in epithelial repair. *The Nephrology Dialysis Transplantation*, *23*, 834–841.

Viau, A., Bienaime, F., Lukas, K., Todkar, A. P., Knoll, M., Yakulov, T. A., et al. (2018). Cilia-localized LKB1 regulates chemokine signaling, macrophage recruitment, and tissue homeostasis in the kidney. *The EMBO Journal*, *37*, e98615.

Wemmer, K., Ludington, W., & Marshall, W. F. (2020). Testing the role of intraflagellar transport in flagellar length control using length-altering mutants of Chlamydomonas. *Philosophical Transactions of the Royal Society B*, *375*, 0190159.

Wren, K. N., Craft, J. M., Tritschler, D., Schauer, A., Patel, D. K., Smith, E. F., et al. (2013). A differential cargo-loading model of ciliary length regulation by IFT. *Current Biology*, *23*, 2463–2471.

Ye, F., Nager, A. R., & Nachury, M. V. (2018). BBSome trains remove activated GPCRs from cilia by enabling passage through the transition zone. *Journal of Cell Biology*, *217*, 1847–1868.

Yu, F., Guo, S., Li, T., Ran, J., Zhao, W., Li, D., et al. (2019). Ciliary defects caused by dysregulation of O-GlcNAc modification are associated with diabetic complications. *Cell Research*, *29*, 171–173.

Zhang, W., Taylor, S. P., Nevarez, L., Lachman, R. S., Nickerson, D. A., Bamshad, M., et al. (2016). IFT52 mutations destabilize anterograde complex assembly, disrupt ciliogenesis and result in short rib polydactyly syndrome. *Human Molecular Genetics*, *25*, 4012–4020.

Zhu, D., Shi, S., Wang, H., & Liao, K. (2009). Growth arrest induces primary-cilium formation and sensitizes IGF-1-receptor signaling during differentiation induction of 3T3-L1 preadipocytes. *The Journal of Cell Science*, *122*, 2760–2768.

CHAPTER

5

Evaluation of ciliary-GPCR dynamics using a validated organotypic brain slice culture method

Yuki Kobayashi and Yumiko Saito*

Graduate School of Integrated Sciences for Life, Hiroshima University, Hiroshima, Japan

**Corresponding author: e-mail address: yumist@hiroshima-u.ac.jp*

Chapter outline

Methods in Cell Biology, Volume 175, ISSN 0091-679X, https://doi.org/10.1016/bs.mcb.2022.09.007
Copyright © 2023 Elsevier Inc. All rights reserved.

Abstract

The primary cilium is a structural organelle present in most mammalian cells. Primary cilia are enriched with a unique protein repertoire distinct from that of the cytosol and the plasma membrane. Such a highly organized microenvironment creates effective machinery for translating extracellular cues into intracellular signals. G protein-coupled receptors (GPCRs) are key receptors in sensing environmental stimuli transmitted *via* a second messenger into a cellular response. Recent data has demonstrated that a limited number of non-olfactory GPCRs, including melanin-concentrating hormone receptor 1 (MCHR1), are preferentially localized to ciliary membranes of several mammalian cell types, including neuronal cells. Evidence was accumulated to support the functional importance of ciliary-GPCR signaling accompanying ciliary structural changes using cilia-specific cell and molecular biology techniques. Thus, cilia are now considered to function as a unique sensory platform for the integration of GPCR signaling and various cytoplasmic domains. Dissociated neurons expressing ciliary-GPCRs can be a useful tool for examining ciliary dynamics. However, losing preexisting neuronal connectivity may alter neuronal ciliary morphology, such as abnormal elongation. Brain slices prepared under *ex vitro* conditions are a powerful approach that maintains the cytoarchitecture, enabling researchers to have accurate control over experimental conditions and to study individual cells from subregions of the brain. Here, we present a detailed description of our novel modified method for organotypic culture of rat brain slice and a validated immunostaining protocol to characterize ciliary-GPCR dynamics in coupling with neuropeptides or aminergic activation.

1 Introduction

Primary cilia are solitary, non-motile, microtubule-based organelles found in multiple mammalian cell types. They are found projecting from the apical cell surface into the surrounding environment. These organelles are important regulators of the cell cycle, both structurally—through the anchorage of the mother centriole and cilium disassembly at cell cycle re-entry—and molecularly through the mediation of multiple developmental signaling pathways (Ishikawa & Marshall, 2011; Louvi & Grove, 2011). In addition to their role in cell proliferation, primary cilia are dynamic signaling organelles that sense the diverse cellular microenvironment, propagating signals into juxtaposed cytoplasmic structures (Nachury & Mick, 2019; Wheway, Nazlamova, & Hancock, 2018). Perturbations that impact any of the multiple signaling processes of cilia can cause clinical disorders, collectively known as ciliopathies that can affect the development of multiple organs and include polycystic kidney disease, Alström syndrome, and Bardet-Biedl syndrome (BBS) (Baker & Beales, 2009).

As the cell's antenna, the primary cilium is enriched in molecules typically involved in signal transduction, including selected receptors such as a limited set of G protein-coupled receptors (GPCRs) (Wheway et al., 2018). GPCRs are some of the most common drug targets and represent the largest and most diverse family of cell-surface signaling molecules, with more than 800 known GPCRs identified in the human genome. Following activation by a diverse array of possible ligands, GPCRs interact with a myriad of proteins that modulate intracellular signaling and alter cellular properties. Therefore, GPCRs regulate many key physiological processes such as photoreception, taste, smell, neurotransmission, metabolism, and endocrine and exocrine secretions. In 2008, somatostatin receptor 3 (SSTR3) and melanin-concentrating receptor 1 (MCHR1) were found to be preferentially localized on neuronal primary cilia, whereas they were not found on the plasma membrane (Berbari, Lewis, Bishop, Askwith, & Mykytyn, 2008). Furthermore, a series of experiments using heterologous expression of epitope-tagged GPCRs was performed by using the telomerase-immortalized human retinal pigmented epithelial (hRPE1), rat renal inner medullary collecting duct (IMCD), and mouse NIH3T3 cell lines (Loktev & Jackson, 2013; Omori et al., 2015). These cell lines are often used for *in vitro* cilia studies because primary cilia are significantly decreased or lost in various common cell lines, such as COS, CHO, and HEK293, used for more general GPCR studies. Odorant receptors form a large family of ciliary-GPCRs in the vertebrate genome. To date, among the approximately 210 non-odorant GPCRs, more than 20 are reported to be preferentially localized to primary cilia (Engle, Bansal, Antonellis, & Berbari, 2021; Hilgendorf, Johnson, & Jackson, 2016; Schou, Pedersen, & Christensen, 2015; Tereshko, Turrigiano, & Sengupta, 2022). Notably, the genes encoding ciliary-GPCRs have been shown to be dysregulated in major psychiatric disorders (Alhassen et al., 2021). Among reported ciliary-GPCRs, more than 10 GPCRs such as SSTR3, MCHR1, 5-HT6 serotonin receptor, NPY family receptor 2, and dopamine receptor D1 have been confirmed as ciliary-GPCRs *in vitro* and *in vivo* by independent laboratories (Kobayashi, Hamamoto, & Saito, 2020; Loktev & Jackson, 2013; Omori et al., 2015).

Cilia on each cell type have a proper length range at the right time (Arellano, Guadiana, Breunig, Rakic, & Sarkisian, 2012; Avasthi & Marshall, 2012; Kobayashi et al., 2022; Kobayashi, Okada, et al., 2021; Sipos, Komoly, & Ács, 2018). Recent evidence has highlighted the importance of ciliary dynamics, which affects dendritic and axonal complexity and synaptic functionality (Guadiana et al., 2013; Guo et al., 2017; Kumamoto et al., 2012; Wang, Hu, Sun, & Zhang, 2019). Ciliary length is also linked to the signaling capacity of cells in response to external stimuli. The lengthening of primary cilia enhances cellular mechanosensitivity (Atkinson et al., 2015; Spasic & Jacobs, 2017), whereas reduced cilium length has been associated with decreased sonic hedgehog-dependent signaling (Jurisch-Yaksi et al., 2013; Vorobyeva & Saunders, 2018). Furthermore, atypical neuronal ciliary length in discrete brain regions has been reported for different animal models of neurological disorders (Canterini et al., 2017; Chakravarthy et al., 2012; Hu, Wang, & Zhang, 2017; Kobayashi et al., 2022; Miyoshi et al., 2014). The ciliary length has also been implicated in energy homeostasis; neuronal ciliary shortening is observed in some brain regions of obese and diabetic mouse models (Agassandian et al., 2014; Han et al., 2014) and chronically starved mice (Kobayashi, Okada, et al., 2021). Therefore, studies on cilia length regulation contribute to understanding energy homeostasis in human ciliopathies with comorbid obesity (Baker & Beales, 2009). Of note, modulation of ciliary length is observed through leptin or GPCR ligand binding, some of which are known to be obesity-related metabolic factors (Brodsky et al., 2017; Hamamoto et al., 2016; Hu et al., 2021; Jin et al., 2014; Kang et al., 2015).

The orexigenic neuropeptide, melanin-concentrating hormone (MCH), is mainly synthesized in the lateral hypothalamus. Axons from the neurons of this region project throughout the brain (Bittencourt et al., 1992). Released MCH binds to MCHR1, a GPCR that acts as the sole receptor of MCH in rodents (Chambers et al., 1999; Saito et al., 1999). The MCH-MCHR1 system functions as a key regulator of food intake, energy expenditure, mood, and the sleep-wake cycle (Diniz & Bittencourt, 2017). Rodent MCHR1 is preferentially located in the primary cilia of neurons in the olfactory tubercle, the accumbens, and both the CA1 and CA3 regions of the hippocampal formation (Diniz et al., 2020; Kobayashi, Okada, et al., 2021). Binding of MCH to MCHR1 mediates primary cilia length shortening in a simple cell culture model; hRPE1 cells heterologously expressed epitope- and fluorescent protein-tagged MCHR1 (Hamamoto et al., 2016). Similarly, MCHR1-mediated primary cilia shortening was observed in an endogenous ciliary MCHR1-expressing system in dissociated hippocampal neurons and human-induced pluripotent stem cell-derived cortical neurons (Kobayashi, Okada, et al., 2021; Miki et al., 2019). Thus, the MCH-MCHR1 system is the first to demonstrate the principal aspect of primary cilia shortening through neurotransmitter-initiated ciliary-GPCR activation. In addition, ciliary MCHR1-based transcriptome profiles have been used to define a novel regulatory cascade underlying cilia length control; MCH can actively shorten cilia *via* a GPCR- Gi/o dependent phosphorylation of Akt/JNK and subsequent signaling pathway involving PDZ and LIM domain-containing protein 5 (Kobayashi et al., 2021).

As mentioned before, a particular cell type has the right length at the right time. Investigations using frozen mouse brain sections demonstrated that the mean length of MCHR1-positive neuronal cilia in the hippocampal CA1 region was significantly longer than that in the CA3 region; furthermore, shortening of cilia was exclusively detected in the CA1 region in fasted mice when compared with fed mice (Kobayashi, Okada, et al., 2021). These data indicated that another methodology is needed to understand such a brain subregion-restricted ciliary response comprehensively. Therefore, to bridge the gap between simplified *in vitro* models and overly-complex *in vivo* animal physiology, we employed an *ex vivo* approach to incorporate the impact of three-dimensional structures into our ciliary-GPCR study (Alhassen et al., 2022; Kobayashi, Okada, et al., 2021). Organotypic brain slice cultures have been widely used to preserve region-specific cytoarchitecture and neuronal networks under *ex vivo* conditions, making them a highly attractive tool for investigating complex cellular processes such as synaptic transmission and downstream responses to natural ligands. Here, we describe a detailed protocol for the organotypic culture of slices from several brain regions, and a method to carry out immunohistochemistry staining for ciliary-GPCRs and analysis of primary cilium length. A similar experimental approach can be adapted to study other types of ciliary dynamics, for example, activation by pharmaceutical agents such as lithium (Miyoshi, Kasahara, Miyazaki, & Asanuma, 2009; Spasic & Jacobs, 2017).

2 Slice cultures

2.1 Making and culturing brain slices (rat hippocampus)

2.1.1 Equipment

- Brain chopper (e.g., McIlwain tissue chopper; Mickle Laboratory Engineering, Surrey, UK) (Fig. 1A)
- Water bath (70 °C)
- Stereo microscope

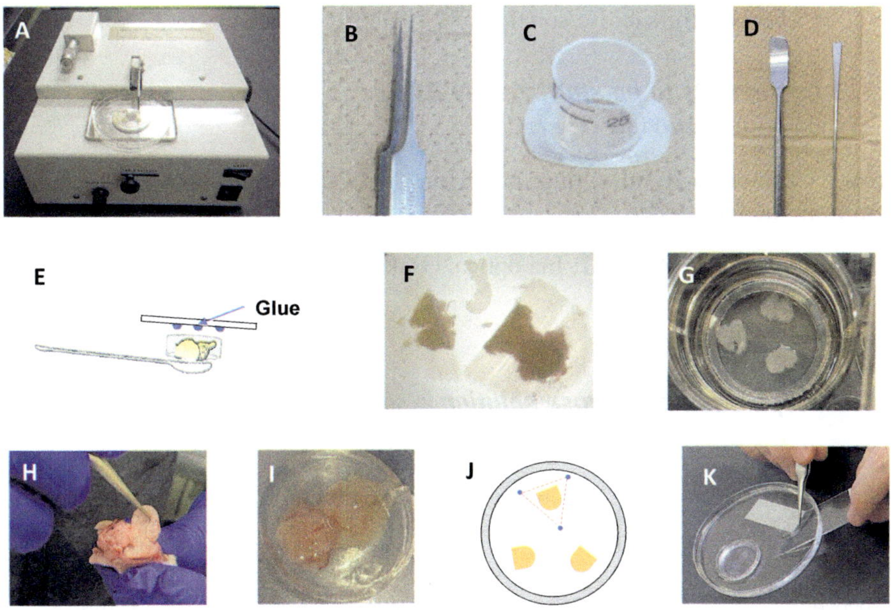

FIG. 1

(A) McIlwain tissue chopper. (B) The tip of tweezers used for brain slice separation. (C) Handmade cylinder for embedding. (D) Spatula. (E) Diagrammatic representation of the method used for adhering the excised brain, after embedding in agarose, to the brain chopper pedestal. (F) Separation of brain slices with a stereo microscope. (G) Whole brain slices cut in the coronal direction placed on the Millicell culture inserts. (H) Photograph of the cerebral cortex after the left and right incisions are removed. (I) Agarose embedding for hypothalamic slices. (J) Coverslip placement for brain slices. Step 1: Cut out the Millicell membrane surrounding the brain slice; it is most efficient to cut the area denoted by the red dotted line with a scalpel, followed by cutting the areas marked with a blue dot. (K) Coverslip placement for brain slices. Step 2: Lifting the edge of the Millicell film with tweezers and sliding it onto the glass slide.

- CO_2 incubator
- 0.4 μm, 30 mm diameter semiporous Millicell culture inserts (e.g., PICM0RG50; Merck Millipore, Germany)

2.1.2 Instruments to be sterilized by dry heat

- Dissecting instruments
- Fine-tip tweezers (e.g., No.5-TITAN; Fontax, Lausanne, Switzerland) (Fig.1B)

This tweezer is used for separating the cut brain slices. Since the tip is weak, it cannot be used for dissection.

- 50-mL beaker (for sucrose solution and ethanol)

2.1.3 Instruments to be sterilized by UV

- Cylinder for embedding (Fig. 1C)

Created by cutting a 20-mL syringe barrel (e.g., ss-20ESz; TERUMO, Tokyo, Japan) 2 cm from the flange side.

- Metal plate (e.g., AB-TC1; Ina optica, Osaka, Japan)
- Spatula (Fig. 1D)
- Razor blade

2.1.4 Preparation of sucrose solution

Reagent	Volume	Manufacturer	Product number
Sucrose	32.03 g	Wako	196-00015
HEPES	1.907 g	Nacalai	17514-86
Glucose	0.793 g	Wako	049-31165
NaCl	0.421 g	Nacalai	31320-05
NaH_2PO_4. $2H_2O$	0.102 g	Wako	190-02811
1 M NaOH	3.2 mL	Wako	192-15985
3 M KCl	333.2 μL	Wako	165-21825
1 mM $CaCl_2$	0.2 mL	Wako	038-24985
Penicillin-Streptomycin	4.0 mL	Nacalai	26253-84
Milli-Q water	Up to 400 mL	–	–

Sucrose solution is easily contaminated with mold, so it is freshly prepared every time it is used.

Perform filter sterilization of the solution using Millex-GS Syringe Filter Unit (0.22 μm).

2.1.5 Preparation of starter medium

Reagent	Volume	Manufacturer	Product number
MEM	26.7 mL	Gibco	12360-038
HBSS	13.3 mL	Gibco	24020-117
Glucose	0.347 g	Wako	049-31165
Penicillin-Streptomycin	2.67 mL	Nacalai	26253-84
Filter sterilization			
Horse serum	Final concentration of 25%	Gibco	16050122

Perform filter sterilization using Millex-GS Syringe Filter Unit, (0.22 μm; Merck, Millipore, Germany). Immediately before use, add horse serum at a final concentration of 25%.

2.1.6 Preparation of culture medium

Reagent	Volume	Manufacturer	Product number
Neuro Basal	39.0 mL	Gibco	21103-049
B-27	800 μL	Wako	17504-044
Gluta Max	100 μL	Gibco	35050-061
Penicillin-Streptomycin	480 μL	Nacalai	26253-84

2.1.7 Preparation of agarose solution for embedding

Reagent	Volume	Manufacturer	Product number
Agarose	0.50 g	Wako	312-01193
NaCl	0.43 g	Nacalai	31320-05
Milli-Q water	Up to 50 mL	–	–

Add the materials to a 50 mL medium bottle and melt them completely in the microwave. Transfer 10 mL to a 15-mL centrifuge tube and heat in a 70 °C water bath until use.

1. Place the Millicell culture inserts in a 6-well plate, add 1 mL of starter medium, and heat in a 5% CO_2 incubator at 37 °C.
2. Anesthetize 8- to 12-day-old Wistar rats (outside this range, the detection rate of primary cilia decreases) with isoflurane and perform decapitation. Disinfect the head by immersing it in ice-cold 70% ethanol.

3. Excise the brain using scissors and tweezers after the head has been submerged in an ice-cold sucrose solution for 3 min.
4. Place the brain (dorsal side up) on a metal plate over ice, and cover it with an embedding cylinder. Pour in an agarose solution to cover about half of the brain and allow it to harden.
5. Remove the solidified agarose and brain from the cylinder and glue the ventral side of brain preparation to the pedestal of the brain chopper with super glue (a dot about 3–4 spots) (Fig. 1E).
6. Set the pedestal of the brain chopper into position and cut slices to a thickness of *200* μm in the *coronal direction*.
7. Remove the agarose and unused brain from the pedestal using a razor and place it in a 10 cm petri dish containing fresh sucrose solution.
8. Under a stereomicroscope, separate brain slices using tweezers and a spatula (Fig. 1F). If necessary, remove unwanted brain regions.
9. Transfer brain sections to the Millicell membrane in the 6-well plate using a spatula (Fig. 1G). Remove any carried-over sucrose solution on the Millicell membrane with micropipette.
10. Place the 6-well plate into the 5% CO_2 incubator and culture at 37 °C.
11. The next day, replace the entire amount of starter medium with culture medium (1 mL). If you gently lift the Millicell culture inserts with tweezers, it will be easier to change the solution.
12. Change one-third (remove 333 μL and add 400 μL for evaporation) of the culture medium every 72 h.
13. On the *7th day* of culture, the drug or ligand (see Section 3.1) is added as needed and is then fixed (see Section 3.2).

*The underlined parts have to be performed depending on the target brain region. See Section 2.2 for details.

2.2 Optimal conditions for specific brain regions (hippocampus, hypothalamus, prefrontal cortex, caudate putamen)

In our laboratory, we have succeeded in organotypic culturing all the listed brain regions. For reference, the conditions are summarized below:

Region	Thickness	Cutting direction	Culture days
Hippocampus	200 μm	Coronal	7 days
Hypothalamus[a]	250 μm	Coronal	7 days
Prefrontal cortex	200 μm	Coronal	7 days
Caudate putamen	200 μm	Sagittal	14 days[b]

[a]*When acquiring the hypothalamic area, the cerebral cortex is opened from the left and right and removed during dissection (Fig. 1H). The brain is trimmed so that the hypothalamus (ventral side) faces upward when embedded in agarose (Fig. 1I). The dorsal side is glued to the pedestal with instant glue and is then cut using the brain chopper.*

[b]*It is important to allow the primary cilia to grow to the same length as that observed* in vivo. *The primary cilia of the Caudate putamen are longer than those of the other areas, so the slices are cultured for 14 days to allow full maturation.*

3 Evaluation of ciliary dynamics

3.1 Treatment with GPCR ligands and/or pharmacological compounds

When adding ligand and/or pharmacological compounds, prepare a high-concentration solution and add it to the medium[a]. To achieve an accurate concentration, it is recommended to prepare wells without Millicell culture inserts with the same solution volume and process them in the same manner as the experimental group, then use this solution volume for your calculations. Use step 1a when using pharmacological compounds and 1b when stimulating GPCRS using a ligand.

1. Lift the Millicell culture inserts with tweezers, add the drug to the medium, and then replace the membrane quickly.
   e.g., 200 ng/mL pertussis toxin (Gi/o protein inhibitor; List Biological Laboratories Inc., Campbell, CA) for 18 h or 10 nM Akti1/2 (Akt phosphorylation inhibitor; ab142088, Abcam, Cambridge, UK) for 0.5 h in Neurobasal medium (Kobayashi, Okada, et al., 2021).
2. Lift the Millicell culture inserts with tweezers and add the ligand to the medium, then replace the membrane quickly.
   e.g., 1–30 nM MCH (4369-v, Peptide Research Institute, Osaka, Japan)[b] for 6 h in 0.1% BSA[c]/Neurobasal (Alhassen et al., 2022; Kobayashi, Okada, et al., 2021; Miki et al., 2019).

3.2 Post-culture immunohistochemical staining

Immunohistochemical staining is performed on 6-well plates. Prepare a 6-well plate for each solution.

Except for the washing process, it is efficient to lift the Millicell culture inserts and pour half of the solution into the well of the 6-well plate. One can then replace the Millicell culture inserts and sprinkle the other half on the brain slices.

1. After removing the medium, pour fresh 4% Paraformaldehyde/PBS (1.4 mL/well) and fix at 4 °C for 3 h.
2. Wash 3 times with PBS (4 mL/well) at room temperature for 10 min.
3. Heat in a water bath at 70 °C for 20 min in 10% Histo VT One (Nacalai, Kyoto, Japan) in PBS (2 mL/well) for antigen retrieval.
4. Wash 3 times with PBS (4 mL/well) at room temperature for 10 min.

---

[a]To yield a final concentration of 1 nM, prepare a 1 μM Neurobasal-ligand solution, and add it to the medium to achieve a 1/1000 dilution.

[b]Neuronal cilia shortening in response to MCH is more sensitive than that in non-neuronal cultured cells (e.g., hRPE1cells), so it is recommended to stimulate at low concentrations (hRPE1: 300–1000 nM, Brain slice: 1–30 nM) (Kobayashi, Okada, et al., 2021; Kobayashi, Tomoshige, et al., 2021).

[c]Add bovine serum albumin to prevent MCH from absorbing plastic material.

5. Block tissue at room temperature for 2 h with blocking solution (PBS containing 5.5% horse serum and 0.1% Triton X-100) (2 mL/well).
6. Incubate slices overnight at 4 °C with rabbit anti-rat adenylate cyclase 3 (AC3, RPCA-ACIII; Encor Biotechnology, Alachua, FL, USA; 1:5000) and goat anti-human MCHR1 (C-17, sc-5534; Santa Cruz Biotechnology; 1:300)[d,e] primary antibodies/blocking solution[f] (1.4 mL/well).
7. Wash 3 times with PBS (4 mL/well) at room temperature for 10 min.
8. The bound primary antibodies are detected by incubating with Alexa Fluor 488-labeled donkey anti-rabbit IgG (1: 300) or Alexa Fluor 546-labeled donkey anti-goat IgG (1: 300) secondary antibody/blocking solution[g] (1.4 mL/well) at room temperature for 2 h while being protected from light exposure.
9. Wash 3 times with PBS (4 mL/well) at room temperature for 10 min.
10. Stain the nuclei with DAPI (1:1000)/PBS (1.4 mL/well) at room temperature for 20 min.
11. Wash with PBS (4 mL/well) at room temperature for 10 min.
12. Transfer the Millicell culture inserts to a 10 cm petri dish lid filled with PBS.
13. Cut the Millicell membrane around the brain slice with a scalpel and place the slice together with the Millicell membrane on a glass slide (Fig.1J and K).
14. Drop a 15-μL Vector shield onto the slice and cover with a coverslip.
15. Turn it over so that the coverslip side is on the bottom, and let it sit on a paper towel for 10 min.
16. Seal the edge of the coverslip with transparent nail polish.

3.3 Image analysis

Images are obtained with a microscope (e.g., FV3000 confocal microscope: Olympus, Tokyo, Japan, BZ-9000 fluorescence microscope: Keyence, Osaka, Japan). A ciliary marker (AC3), representing a positive hair-like appendage longer than 1.5 μm, is counted as a primary cilium. For evaluation of changes in cilia length caused by MCH in each experiment, the lengths of AC3 and MCHR1 double-positive cilia are evaluated. In our laboratory, the primary cilia lengths are measured under a BZ-9000 fluorescence microscope using PhotoRuler Ver. 1.1 software (The Genus Inocybe, Hyogo, Japan). Measuring at least 50 cilia lengths per treatment gives stable results.

[d]The specificity of antibodies for the target GPCR is critically important (Michel, Wieland, & Tsujimoto, 2009). Regarding the validity of our immunohistochemistry, we have verified the specificity of the anti-MCHR1 antibody in the dorsal hippocampus and Caudate putamen from MCHR1 knockout mice (Kobayashi, Okada, et al., 2021); that is, we have defined the typical features of MCHR1 immunoreactivity in primary cilia in wild type brains and the lack of immunoreactivity in the knockout brains.
[e]Unfortunately, this Santa Cruz polyclonal antibody has been discontinued. However, MCHR1 in the primary cilia has been sensitively detected with the rabbit anti-human MCHR1 (PA5-24182, Invitrogen) antibody (Diniz et al., 2020).
[f]The first antibody solution can be reused at least 3 times when stored at −20 °C.
[g]The secondary antibody solution can be reused at least 3 times when stored at −20 °C.

3.4 Characteristics of primary cilia after culturing

The following data confirm that the characteristics are similar to that of the frozen sections following perfusion fixation. See Fig. 2 for the actual state of the primary cilia (Alhassen et al., 2022; Kobayashi, Okada, et al., 2021).

Rat brain region		Cilia length (μm) No MCH	Cilia length (μm) MCH treatment	MCHR1-positive cilia/AC3-positive cilia (%)
Hippocampus	CA1	5.97 ± 0.11	4.42 ± 0.89	86.0%
	CA3	3.83 ± 0.19	No change	84.2%
	Dentate gyrus	3.72 ± 0.46	No change	2.2%
Hypothalamus[a]		4.21 ± 0.25	2.97 ± 0.07	66.3%
Prefrontal cortex		4.39 ± 0.12	3.05 ± 0.06	76.0%
Caudate putamen		10.30 ± 0.23	7.90 ± 0.28	86.2%

[a] *The data from the hypothalamus may only be used as a reference because the number of experiments is not sufficient for statistical analysis.*

All data are presented as mean ± standard error mean (SEM).

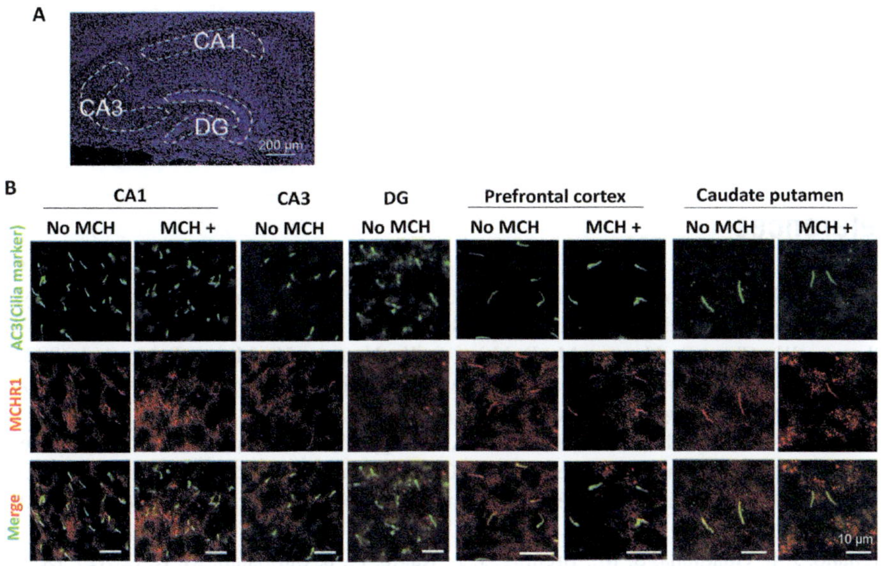

FIG. 2

(A) DAPI-stained image of rat hippocampal slices after culturing. The cell layer is preserved. (B) Representative immunostaining images of MCH responses in each hippocampal region (CA1, CA3, and DG), the prefrontal cortex, and the Caudate putamen.

4 Concluding remarks

This paper presented the detailed methods to conduct organotypic culture of brain slices using a semi-porous membrane insert in a 6-well plate. This modified version of the conventional method is easier to implement and less expensive when used to analyze ligand-initiated ciliary dynamics in specific brain subregions. To achieve dynamic, living brain-slice imaging, a combined method using ciliary-GPCR knock-in mice (Siljee et al., 2018) and the state-of-the-art image-processing tool would enable more efficient analysis of ciliary dynamics and determination of specific processes such as primary cilia excision (Nager et al., 2017; Phua et al., 2017).

Hippocampal slices are relatively fragile when compared to brain slices of other regions. Therefore, the use of a "brain slice-on-a-chip," or microfluidic device, may serve as a superior culture method for hippocampal slices to provide better preservation of the cytoarchitecture of the brain slices (Herreros, Tapia-González, Sánchez-Olivares, Laguna Heras, & Holgado, 2022). Because GPCR-positive ciliary synapses are identified in hippocampal neuron axons (Sheu et al., 2022), the study of ligand-initiated ciliary dynamics in this region, in particular, will lead to new insights into the dynamics of the developing and mature brain functions.

Acknowledgment

This study was supported by research grants from the Ministry of Education, Culture, Sports, Science and Technology of Japan (KAKENHI 18K14742/21K06249 to Y.K. and 15K06775/19K06739/j22K06297 to Y.S.).

References

Agassandian, K., Patel, M., Agassandian, M., Steren, K. E., Rahmouni, K., Sheffield, V. C., et al. (2014). Ciliopathy is differentially distributed in the brain of a Bardet-Biedl syndrome mouse model. *PLoS One*, *9*, e93484.

Alhassen, W., Chen, S., Vawter, M., Robbins, B. K., Nguyen, H., Myint, T. N., et al. (2021). Patterns of cilia gene dysregulations in major psychiatric disorders. *Progress in Neuro-Psychopharmacology & Biological Psychiatry*, *109*, 110255.

Alhassen, W., Kobayashi, Y., Su, J., Robbins, B., Ngyuen, H., Myint, T., et al. (2022). Regulation of brain primary cilia length by MCH signaling: Evidence from pharmacological, genetic, optogenetic and chemogenic manipulations. *Molecular Neurobiology*, *59*, 245–265.

Arellano, J. I., Guadiana, S. M., Breunig, J. J., Rakic, P., & Sarkisian, M. R. (2012). Development and distribution of neuronal cilia in mouse neocortex. *The Journal of Comparative Neurology*, *520*, 848–873.

Atkinson, K. F., Kathem, S. H., Jin, X., Muntean, B. S., Abou-Alaiwi, W. A., Nauli, A. M., et al. (2015). Dopaminergic signaling within the primary cilia in the renovascular system. *Frontiers in Physiology*, *6*, 103.

Avasthi, P., & Marshall, W. F. (2012). Stages of ciliogenesis and regulation of ciliary length. *Differentiation*, *83*, S30–S42.

Baker, K., & Beales, P. L. (2009). Making sense of cilia in disease: The human ciliopathies. *American Journal of Medical Genetics. Part C, Seminars in Medical Genetics*, *151C*, 281–295.

Berbari, N. F., Lewis, J. S., Bishop, G. A., Askwith, C. C., & Mykytyn, K. (2008). Bardet-Biedl syndrome proteins are required for the localization of G protein-coupled receptors to primary cilia. *Proceedings of the National Academy of Sciences of the United States of America*, *105*, 4242–4246.

Bittencourt, J. C., Presse, F., Arias, C., Peto, C., Vaughan, J., Nahon, J. L., et al. (1992). The melanin-concentrating hormone system of the rat brain: An immuno- and hybridization histochemical characterization. *The Journal of Comparative Neurology*, *319*, 218–245.

Brodsky, M., Lesiak, A. J., Croicu, A., Cohenca, N., Sullivan, J. M., & Neumaier, J. F. (2017). 5-HT6 receptor blockade regulates primary cilia morphology in striatal neurons. *Brain Research*, *1660*, 10–19.

Canterini, S., Dragotto, J., Dardis, A., Zampieri, S., De Stefano, M. E., Mangia, F., et al. (2017). Shortened primary cilium length and dysregulated sonic hedgehog signaling in Niemann-pick C1 disease. *Human Molecular Genetics*, *26*, 2277–2289.

Chakravarthy, B., Gaudet, C., Ménard, M., Brown, L., Atkinson, T., Laferla, F. M., et al. (2012). Reduction of the immunostainable length of the hippocampal dentate granule cells' primary cilia in 3xAD-transgenic mice producing human Aβ(1-42) and tau. *Biochemical and Biophysical Research Communications*, *427*, 218–222.

Chambers, J., Ames, R. S., Bergsma, D., Muir, A., Fitzgerald, L. R., Hervieu, G., et al. (1999). Melanin-concentrating hormone is the cognate ligand for the orphan G-protein-coupled receptor SLC-1. *Nature*, *400*, 261–265.

Diniz, G. B., Battagello, D. S., Klein, M. O., Bono, B. S. M., Ferreira, J. G. P., Motta-Teixeira, L. C., et al. (2020). Ciliary melanin-concentrating hormone receptor 1 (MCHR1) is widely distributed in the murine CNS in a sex-independent manner. *Journal of Neuroscience Research*, *98*, 2045–2071.

Diniz, G. B., & Bittencourt, J. C. (2017). The melanin-concentrating hormone as an integrative peptide driving motivated behaviors. *Frontiers in Systems Neuroscience*, *11*, 32.

Engle, S. E., Bansal, R., Antonellis, P. J., & Berbari, N. F. (2021). Cilia signaling and obesity. *Seminars in Cell & Developmental Biology*, *110*, 43–50.

Guadiana, S. M., Semple-Rowland, S., Daroszewski, D., Madorsky, I., Breunig, J. J., Mykytyn, K., et al. (2013). Arborization of dendrites by developing neocortical neurons is dependent on primary cilia and type 3 adenylyl cyclase. *The Journal of Neuroscience*, *33*, 2626–2638.

Guo, J., Otis, J. M., Higginbotham, H., Monckton, C., Cheng, J., Asokan, A., et al. (2017). Primary cilia signaling shapes the development of interneuronal connectivity. *Developmental Cell*, *42*, 286–300.

Hamamoto, A., Yamato, S., Katoh, Y., Nakayama, K., Yoshimura, K., Takeda, S., et al. (2016). Modulation of primary cilia length by melanin-concentrating hormone receptor 1. *Cell Signaling*, *28*, 572–584.

Han, Y. M., Kang, G. M., Byun, K., Ko, H. W., Kim, J., Shin, M. S., et al. (2014). Leptin-promoted cilia assembly is critical for normal energy balance. *The Journal of Clinical Investigation*, *124*, 2193–2197.

Herreros, P., Tapia-González, S., Sánchez-Olivares, L., Laguna Heras, M. F., & Holgado, M. (2022). Alternative brain slice-on-a-chip for organotypic culture and effective fluorescence injection testing. *International Journal of Molecular Sciences*, *23*, 2549.

Hilgendorf, K. I., Johnson, C. T., & Jackson, P. K. (2016). The primary cilium as a cellular receiver: Organizing ciliary GPCR signaling. *Current Opinion in Cell Biology*, *39*, 84–92.

Hu, H. B., Song, Z. Q., Song, G. P., Li, S., Tu, H. Q., Wu, M., et al. (2021). LPA signaling acts as a cell-extrinsic mechanism to initiate cilia disassembly and promote neurogenesis. *Nature Communications*, *12*, 662.

Hu, L., Wang, B., & Zhang, Y. (2017). Serotonin 5-HT6 receptors affect cognition in a mouse model of Alzheimer's disease by regulating cilia function. *Alzheimer's Research & Therapy*, *9*, 76.

Ishikawa, H., & Marshall, W. F. (2011). Ciliogenesis: Building the cell's antenna. *Nature Reviews. Molecular Cell Biology*, *12*, 222–234.

Jin, D., Ni, T. T., Sun, J., Wan, H., Amack, J. D., Yu, G., et al. (2014). Prostaglandin signalling regulates ciliogenesis by modulating intraflagellar transport. *Nature Cell Biology*, *16*, 841–851.

Jurisch-Yaksi, N., Rose, A. J., Lu, H., Raemaekers, T., Munck, S., Baatsen, P., et al. (2013). Rer1p maintains ciliary length and signaling by regulating γ-secretase activity and Foxj1a levels. *The Journal of Cell Biology*, *200*, 709–720.

Kang, G. M., Han, Y. M., Ko, H. W., Kim, J., Oh, B. C., Kwon, I., et al. (2015). Leptin elongates hypothalamic neuronal cilia via transcriptional regulation and actin destabilization. *The Journal of Biological Chemistry*, *290*, 18146–18155.

Kobayashi, Y., Hamamoto, A., & Saito, Y. (2020). Analysis of ciliary status via G-protein-coupled receptors localized on primary cilia. *Microscopy (Oxf)*., *69*, 277–285.

Kobayashi, Y., Kobuchi, S., Koganezawa, N., Sekino, Y., Shirao, T., Saito, T., et al. (2022). Impairment of ciliary dynamics in an APP knock-in mouse model of Alzheimer's disease. *Biochemical and Biophysical Research Communications*, *610*, 85–91.

Kobayashi, Y., Okada, T., Miki, D., Koganezawa, N., Sekino, Y., Shirao, T., et al. (2021). Properties of primary cilia in melanin-concentrating hormone receptor 1-bearing hippocampal neurons in vivo and in vitro. *Neurochemistry International*, *142*, 104902.

Kobayashi, Y., Tomoshige, S., Imakado, K., Sekino, Y., Shirao, T., Koganezawa, N., et al. (2021). Ciliary GPCR-associated transcriptome reveals key regulators as cilia length control. *FASEB BioAdvances*, *3*, 744–767.

Kumamoto, N., Gu, Y., Wang, J., Janoschka, S., Takemaru, K., Levine, J., et al. (2012). A role for primary cilia in glutamatergic synaptic integration of adult-born neurons. *Nature Neuroscience*, *15*, 399–405.

Loktev, A. V., & Jackson, P. K. (2013). Neuropeptide Y family receptors traffic via the Bardet-Biedl syndrome pathway to signal in neuronal primary cilia. *Cell Reports*, *5*, 1316–1329.

Louvi, A., & Grove, E. A. (2011). Cilia in the CNS: The quiet organelle claims center stage. *Neuron*, *69*, 1046–1060.

Michel, M. C., Wieland, T., & Tsujimoto, G. (2009). How reliable are G-protein-coupled receptor antibodies? *Naunyn-Schmiedeberg's Archives of Pharmacology*, *379*, 385–388.

Miki, D., Kobayashi, Y., Okada, T., Miyamoto, T., Takei, N., Sekino, Y., et al. (2019). Characterization of functional primary cilia in human induced pluripotent stem cell-derived neurons. *Neurochemical Research*, *44*, 1736–1744.

Miyoshi, K., Kasahara, K., Miyazaki, I., & Asanuma, M. (2009). Lithium treatment elongates primary cilia in the mouse brain and in cultured cells. *Biochemical and Biophysical Research Communications*, *388*, 757–762.

Miyoshi, K., Kasahara, K., Murakami, S., Takeshima, M., Kumamoto, N., Sato, A., et al. (2014). Lack of dopaminergic inputs elongates the primary cilia of striatal neurons. *PLoS One*, *9*, e97918.

Nachury, M. V., & Mick, D. U. (2019). Establishing and regulating the composition of cilia for signal transduction. *Nature Reviews. Molecular Cell Biology*, *20*, 389–405.

Nager, A. R., Goldstein, J. S., Herranz-Pérez, V., Portran, D., Ye, F., Garcia-Verdugo, J. M., et al. (2017). An actin network dispatches ciliary GPCRs into extracellular vesicles to modulate signaling. *Cell*, *168*, 252–263.

Omori, Y., Chaya, T., Yoshida, S., Irie, S., Tsujii, T., & Furukawa, T. (2015). Identification of G protein-coupled receptors (GPCRs) in primary cilia and their possible involvement in body weight control. *PLoS One*, *10*, e0128422.

Phua, S. C., Chiba, S., Suzuki, M., Su, E., Roberson, E. C., Pusapati, G. V., et al. (2017). Dynamic remodeling of membrane composition drives cell cycle through primary cilia excision. *Cell*, *168*, 264–279.

Saito, Y., Nothacker, H.-P., Wang, Z., Lin, S. H., Leslie, F., & Civelli, O. (1999). Molecular characterization of the melanin-concentrating-hormone receptor. *Nature*, *400*, 265–269.

Schou, K. B., Pedersen, L. B., & Christensen, S. T. (2015). Ins and outs of GPCR signaling in primary cilia. *EMBO Reports*, *16*, 1099–1113.

Sheu, S. H., Upadhyayula, S., Dupuy, V., Pang, S., Deng, F., Wan, J., et al. (2022). A serotonergic axon-cilium synapse drives nuclear signaling to alter chromatin accessibility. *Cell*, *185*, 3390–3407.

Siljee, J. E., Wang, Y., Bernard, A. A., Ersoy, B. A., Zhang, S., Marley, A., et al. (2018). *Nature Genetics*, *50*, 180–185.

Sipos, É., Komoly, S., & Ács, P. (2018). Quantitative comparison of primary cilia marker expression and length in the mouse brain. *Journal of Molecular Neuroscience*, *64*, 397–409.

Spasic, M., & Jacobs, R. C. (2017). Lengthening primary cilia enhances cellular mechanosensitivity. *European Cells & Materials*, *33*, 158–168.

Tereshko, L., Turrigiano, G. G., & Sengupta, P. (2022). Primary cilia in the postnatal brain: Subcellular compartments for organizing neuromodulatory signaling. *Current Opinion in Neurobiology*, *74*, 102533.

Vorobyeva, A. G., & Saunders, A. J. (2018). Amyloid-β interrupts canonical sonic hedgehog signaling by distorting primary cilia structure. *Cilia*, *7*, 1–11.

Wang, B., Hu, L., Sun, Z., & Zhang, Y. (2019). Cilia function is associated with axon initial segment morphology. *Biochemical and Biophysical Research Communications*, *516*, 15–21.

Wheway, G., Nazlamova, L., & Hancock, J. T. (2018). Signaling through the primary cilium. *Frontiers in Cell and Development Biology*, *6*, 8.

CHAPTER

6 High-efficient CRISPR/Cas9-mediated gene targeting to establish cell models of ciliopathies

Kosuke Hosoba[a,b], Tomoka Morita[c], Ying Zhang[c], Hiroko Kishi[c], Takashi Yamamoto[a,b], and Tatsuo Miyamoto[c,*]

[a]*Program of Biomedical Science, Graduate School of Integrated Sciences for Life, Hiroshima University, Hiroshima, Japan*

[b]*Program of Mathematical and Life Science, Graduate School of Integrated Sciences for Life, Hiroshima University, Hiroshima, Japan*

[c]*Department of Molecular and Cellular Physiology, Graduate School of Medicine, Yamaguchi University, Hiroshima, Japan*

[*]*Corresponding author: e-mail address: t-miyamoto@yamaguchi-u.ac.jp*

Chapter outline

Abstract

Primary cilia are antenna-like structures developed on the cell surface of mammalian cells during the quiescent G_0 phase. Primary cilia in mammalian cells receive extracellular signals for early development and cell tissue homeostasis. Ciliopathies characterized with congenital anomalies such as cerebellar hypoplasia, polycystic kidney and polydactyly are caused by germline mutations of ciliary structure- and function-related genes. Gene knock-out

Methods in Cell Biology, Volume 175, ISSN 0091-679X, https://doi.org/10.1016/bs.mcb.2022.07.020
Copyright © 2023 Elsevier Inc. All rights reserved.

techniques in ciliated cultured cells with the uniformed genetic background are useful to evaluate the pathophysiological roles of ciliopathy-related gene products. Genome editing technology has been applied into the gene knock-out in many types of cultured cell lines. However, the frequency of genome editing varies according to cell species and cycle because of dependency on error-free homology-directed repair (HDR) activity. The human telomerase reverse transcriptase-immortalized retinal pigmented epithelial cell line (hTERT-RPE1) is well known for its suitability in cilia research. However, the efficacy of the HDR-mediated knock-out clone isolation was low. Here, we introduce the clustered regularly interspaced short palindromic repeats-obligate ligation-gated recombination (CRISPR-ObLiGaRe) system, which is a nonhomologous end-joining (NHEJ)-mediated gene targeting method, to generate the knock-out clones effectively even in the lower-HDR activity cell lines including hTERT-RPE1 cell. This CRISPR-ObLiGaRe system is a powerful tool for establishing ciliopathy model cell libraries and identifying each gene function in cilia-related phenotypes.

1 Introduction

Almost all mammalian cells have the potential to protrude primary cilia, microtubule-based antenna-like structures on the cell surface. These organelles arise in the quiescent G_0 phase and receive extracellular signals for signal transduction, including sonic hedgehog (Shh), Wnt, and TGF-β signals to regulate embryogenesis and adult tissue homeostasis (Reiter & Leroux, 2017). Ciliopathies are genetic disorders caused by congenital defects of cilia-related gene functions in humans. These disorders are characterized by the cerebellar hypoplasia, polycystic kidney, retinal pigmentosa, polydactyly, mental retardation, obesity, and other unique developmental anomalies (Reiter & Leroux, 2017). At least 35 ciliopathies have been identified, and more than 240 candidate genes related to ciliary structures and functions are reported (Reiter & Leroux, 2017).

Previous studies evaluated cilia-related phenotypes, including ciliogenesis and ciliary functions, using siRNA-mediated gene knock-down and ciliopathy cells obtained from patients (Miyamoto et al., 2011, 2015). However, primary human cells might be affected by various factors, such as gender, age, and genetic background. Furthermore, a siRNA-mediated knock-down is an incomplete tool for loss-of-function analysis due to dependency on cell lines, biological species, target genes, and sequences. Therefore, establishing knock-out cell lines by gene targeting technics in uniform genetic backgrounds is desired to analyze link between genes and phenotypes.

Zinc-finger nucleases, transcription activator-like nuclease (TALENs), and clustered regulatory interspaced short palindromic repeat/CRISPR-associated protein 9 (CRISPR/CAS9)-based RNA-guide DNA endonucleases have been established as genome engineering tools. These systems induce DNA double-strand breaks (DSBs) at target genes and edit the endogenous sequences in a DNA repair activity-dependent manner. However, the availability of conventional gene targeting techniques is restricted due to the dependency on homology-directed repair (HDR) activity in human cultured cell lines. The introduced DSBs by genome editing tools are repaired by error-prone non-homologous end-joining (NHEJ) and HDR. Additionally, NHEJ activity is stable during the cell cycle; however, HDR activity is limited in the late-S and G_2 phases. Therefore, the efficacy of HDR-mediated genome editing is suggested to depend on tissues and cell types.

The human telomerase reverse transcriptase-immortalized retinal pigmented epithelial cell line (hTERT-RPE1) is derived from normal retinal tissue. It serves as a major model cell in the field of primary cilia research (Miyamoto et al., 2015, 2020). Previously, we depleted the *KIF2A* gene encoding a microtubule depolymerizing kinesin molecule in hTERT-RPE1 cells to demonstrate that KIF2A is required for ciliary disassembly following the cell cycle reentry. Although TALEN and targeting vector containing drug cassette and homology arm used in the research, the efficiency of HDR-mediated knock-out clones was low (~1%) in drug-resistant clones (Miyamoto et al., 2015). Therefore, the development of HDR-independent genome editing techniques is necessary for effective gene targeting in the hTERT-RPE1 background.

Importantly, HDR-independent genome editing techniques have been reported in the last few years. TALEN and CRISPR/Cas9-based PITCh (precise integration into target chromosome) system that is independent of HDR activity are a gene knock-in technique using microhomology-mediated end-joining (MMEJ) (Nakade et al., 2014; Sakuma, Nakade, Sakane, Suzuki, & Yamamoto, 2016). These systems can be applied in human cultured cells and animals, including frogs and silkworms (Nakade et al., 2014). Moreover, using TALEN, NHEJ-mediated gene cassette knock-in technique called ObLiGaRe (Obligate Ligation-Gated Recombination) was reported (Maresca, Lin, Guo, & Yang, 2013). TALEN induces DSBs at the endogenous target site and gene cassette, and a linearized targeting vector is inserted into the endogenous target site by NHEJ (Maresca et al., 2013). Previously, we reported CRISPR/ObLiGaRe method for efficient gene knock-out (Miyamoto et al., 2018, 2020; Royba et al., 2017). This method can insert a targeting vector containing the CRISPR/Cas9 target site and drug-resistant cassette into endogenous locus through NHEJ activity. Royba et al. showed that the monoallelic targeting vector-inserted clones were major in drug-resistant clones (more than 70%). Furthermore, the NHEJ-mediated insertion or deletion of bases (in-dels) were induced in 95.4% of inserted clones. In the report, biallelic targeting vector-inserted clones and heterozygous gene knock-out clones were minor fractions (Royba et al., 2017).

Cholesterol is a key molecule for modulating cell membrane fluidity and functions as a precursor of steroid hormone and bile acid. Previously, it was reported that cholesterol regulates ciliary functions, including the Shh signal (Blassberg, Macrae, Briscoe, & Jacob, 2016). Our previous study revealed that peroxisome directly interacted with primary cilia and supplied cholesterol into the cilia membrane to maintain ciliary function (Miyamoto et al., 2020). We also used the CRISPR/ObLiGaRe method to generate gene knock-out cell libraries associated with intracellular cholesterol transport in the hTERT-RPE1 background, thereby identifying peroxisome-mediated cholesterol trafficking molecules including peroxisomal Rabin8-Rab10-KIFC3 complex and ciliary ORP3. Therefore, CRISPR/ObLiGaRe method is a powerful tool for understanding etiological mechanisms of ciliopathies.

Here we describe the detail of the CRISPR/ObLiGaRe method in the hTERT-RPE1 cell line, focusing on *inositol polyphosphate-5-phosphatase E* (*INPP5E*) gene encoding inositol polyphosphate-5-phosphatase E, the causative gene of a typical ciliopathic disorder Joubert syndrome (Qiu, Fujisawa, Nozaki, Katoh, & Nakayama, 2021; Radha Rama Devi, Naushad, & Lingappa, 2020) as an example of a target. INPP5E maintains the ciliary phosphoinositide PI(4)P via the dephosphorylation of PI(4,5)P2, thereby controlling the Hedgehog signal transduction (Garcia-Gonzalo et al., 2015). This study establishes *INPP5E* knock-out clones and validates each clone by immunofluorescence microscopy.

2 Construction of plasmids

1. For constructing sgRNA- and spCas9-expressing plasmid, a pair of oligodeoxynucleotides without PAM sequence (20 bp) that targets the *INPP5E* gene is designed. Here we targeted exon 7 of the *INPP5E* gene (see Notes 1). Annealed oligodeoxynucleotides are inserted into *Bbs*I-restricted px330-U6-Chimeric_BB-CBh-hSpCas9 plasmid (#42230, Addgene, Watertown, MA).
2. The pBluescript SK II+ backbone plasmid, including *hsvTK-2A-Neo* cassette, is used as a targeting vector reported previously (Royba et al., 2017). Annealed oligodeoxynucleotides, including PAM sequence and the upstream 20 nucleotides, are ligated into BbsI-digested plasmid to generate CRISPR/Cas9 recognition site (see Notes 2). Fig. 1A shows a schema of construction for the CRISPR/ObLiGaRe system.
3. Each plasmid is transformed into *E. coli*, and insertion of oligodeoxynucleotides is validated by Sanger sequence analysis.

3 Cell culture

1. Dulbecco's Modified Eagle Medium (DMEM) is supplemented with 10% FBS and 50 mg/mL of gentamicin. hTERT-RPE1 cells purchased from ATCC (ATCC CRL-4000, Manassas, VA) are cultured in the medium at 37 °C (humidified air with 5% CO_2) (see Notes 3). Periodic Mycoplasma check is performed.
2. After confirming about 80% confluence, the culture medium is removed. The cells are washed with PBS and detached with 0.25% trypsin-EDTA. Next, 10% FBS-DMEM is applied to stop the trypsin activity, and cells are collected in centrifugal tubes. After centrifugation to remove the supernatant, the cells are resuspended and seeded to new dishes.
3. For plasmids transfection, 2×10^5 hTERT-RPE1 cells are seeded into one well of six-well plates and cultured for 24 h.

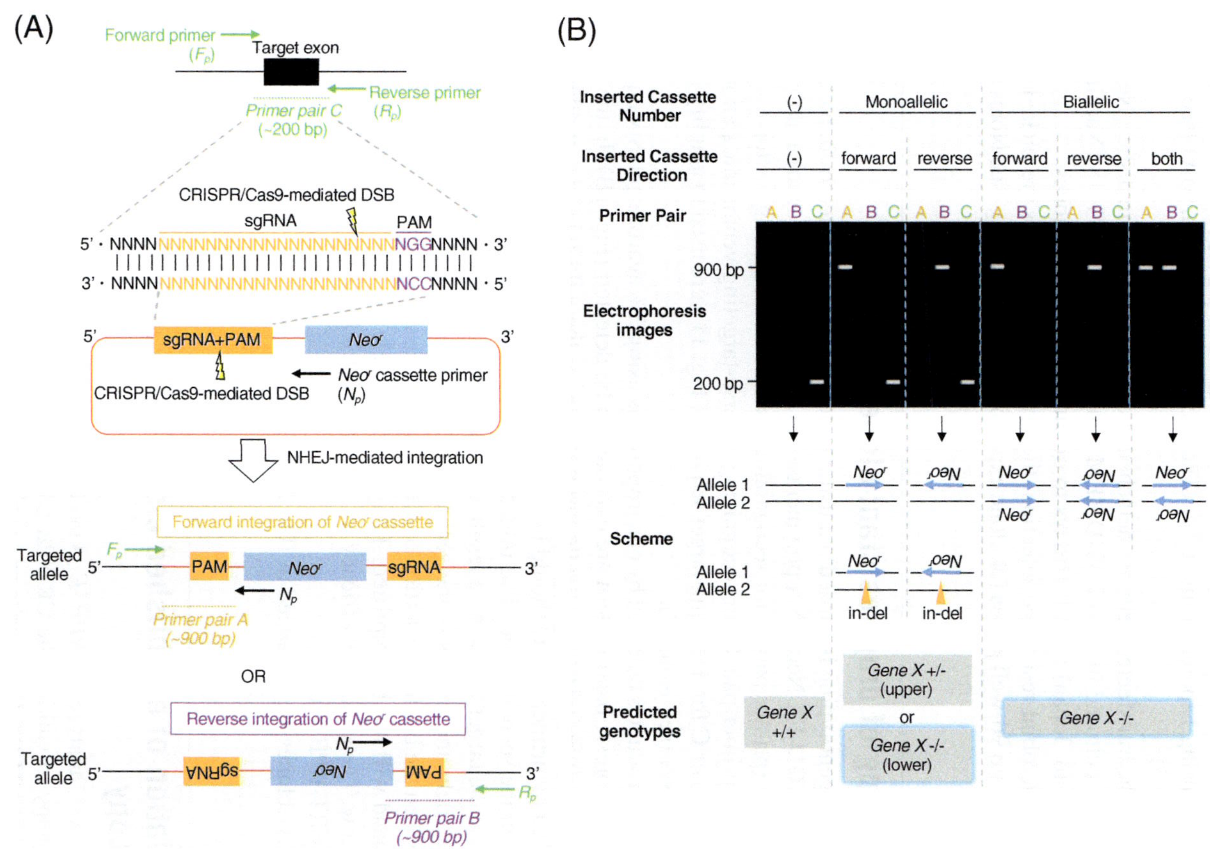

FIG. 1

CRISPR/ObLiGaRe induces a nonhomologous end-joining (NHEJ)-mediated gene knock-out by either or both insertion-deletion mutation (in-del) and insertion of drug-resistant cassette. (A) Overview of the CRISPR/ObLiGaRe system. CRISPR/Cas9-mediated DNA double-strand break (DSB) and linearized targeting vector forward or reverse integration into the target site are induced. (B) Genotyping pattern of gene-targeted cells by CRISPR/ObLiGaRe. Each PCR product using three primer pairs is detected by electrophoresis. Vector integration-derived band can be separated using primer pair A and C. Furthermore, the separation can be done into forward or reverse patterns.

4 Plasmids transfection and drug screening

1. First, 600ng of the px330 plasmid and 20ng of the targeting vector are transfected using Lipofectamine LTX (Thermo Fisher, Waltham, MA) (see Notes 4 and 5).
2. After 48h, the transfected cells are detached and seeded into 15-cm dishes. Next, the cells are cultured in 10% FBS-DMEM containing 2mg/mL G418 (Nacalai Tesque, Kyoto, Japan) for 14–16days to obtain drug-resistant clones.
3. After antimicrobial resistance selection, the drug-resistant clones are picked up and seeded into 96-well plates for clone expansion and genotyping (see Notes 6).

5 Genotyping of drug-resistant clones

1. PCR-based genotyping is carried out using genomic DNA extracted from picked up clones. KOD-FX Neo DNA polymerase (TOYOBO, Osaka, Japan) and three primer pairs (primer pair A for recognizing the forward-targeting vector integration, primer pair B for recognizing reverse-targeting vector integration, and primer pair C for detecting target sequence; Fig. 1) were used to amplify the genomic DNA (see Notes 7).
2. Each PCR product is detected by electrophoresis using 2% agarose gel. Six types of electrophoresis band pattern are expected to be detected (Fig. 1B). In the monoallelic drug- resistant cassette inserted clones, the wild-type size products amplified by primer pair C should be analyzed using direct sequence to confirm in-dels in the second allele. The in-dels induce frameshift mutations and stop codons in the target genes. As a result, clones with targeting vector integration in the one allele and in-dels in the second allele are objective gene knock-out cells. When the biallelic insertion of drug-resistant cassette occurs, the wild-type sized band will not be amplified. To our experiences, the biallelic integration rate is low. The knock-out efficacy was about 45% for *INPP5E* gene targeting in the hTERT-RPE1 cells.
3. Objective clones are expanded in 10% FBS-DMEM without G418 (see Notes 8).

6 Validation of established clones by immunofluorescence microscopy

1. Here, we analyzed *INPP5E* knock-out clones by immunofluorescence microscopy to validate the CRISPR-ObLiGaRe-mediated gene knock-out. Wild-type hTERT-RPE1 cells and each established clone are washed with PBS and detached by 0.25% trypsin-EDTA. After centrifugation, each cell is seeded in a six-well plate, including coverslips (Matsunami Glass, Kishiwada, Japan), and cultured at 37 °C.

2. Cells reach 80–90% confluency, and each well is washed with serum-free DMEM to remove the growth medium (see Notes 9). After washing, cells are cultured in serum-free DMEM at 37 °C for 24 h (see Notes 10).
3. After washing with PBS, cells grown on coverslips are fixed by 4% paraformaldehyde at room temperature for 15 min, permeabilized by 0.2% Triton X-100 for 15 min, and washed with PBS three times. Fixed and permeabilized cells are then blocked with 1% BSA-PBS at room temperature for 30 min.
4. Each cell is incubated with primary antibodies: rabbit anti-INPP5E antibody (Proteintech, Rosemont, IL), mouse anti-ninein antibody (Merck Millipore, Darmstadt, Germany), and mouse anti-acetylated tubulin antibody in 1% BSA-PBS at room temperature for 60 min. After primary antibody treatment, cells are washed with PBS three times (5 min each) and incubated with Alexa Fluor 488, Alexa Fluor 594, and Alexa Fluor 647 conjugated secondary antibodies (Thermo Fisher) in 1% BSA-PBS at room temperature for 30 min (see Notes 11). Next, cells are washed with PBS three times (5 min each) and mounted with ProLong Diamond Antifade medium (Thermo Fisher). Each signal in stained cells is detected using a confocal microscope (LSM800; Carl Zeiss Microimaging Inc., Oberkochen, Germany). In the wild-type hTERT-RPE1 cells, INPP5E signal was localized at the primary cilia, whereas signal was completely disappeared in the established clones (Fig. 2).

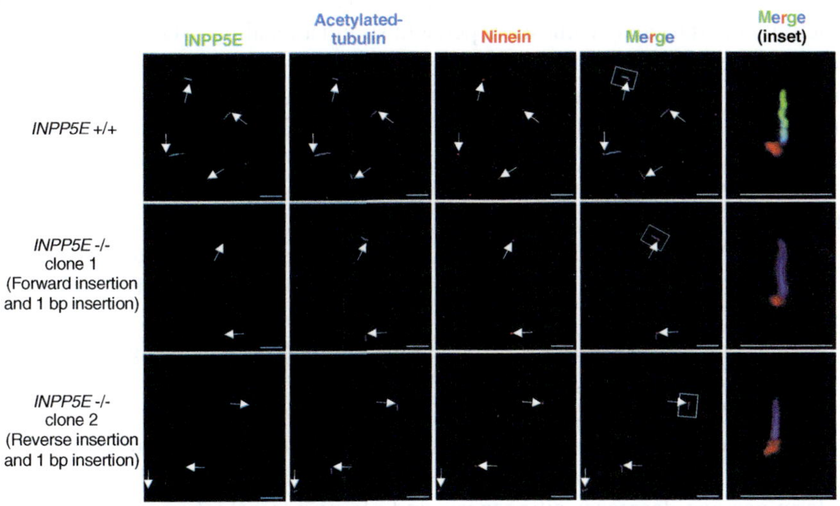

FIG. 2

Validation of *INPP5E* knock-out hTERT-RPE1 cell clones by immunofluorescence microscopy. INPP5E (green), acetylated tubulin (blue), and ninein (red) are stained. Arrows show primary cilia. Magnified images of the boxed areas are shown in inset panels. Scale bar: 5 μm.

7 Concluding remarks

Previous studies had demonstrated low genome editing efficacy in hTERT-RPE1 cells, a general model for primary cilia research (Miyamoto et al., 2015), therefore, we establish a more effective system for generating gene knock-out cell lines in hTERT-RPE1 cells, called CRISPR/ObLiGaRe (Kobayashi et al., 2021; Miyamoto et al., 2020; Royba et al., 2017; Yukimoto et al., 2020). This method effectively induces target-vector integration in one allele by NHEJ and insertion or deletion of a base in the second allele.

More than 340 ciliopathy-related genes have been reported (Reiter & Leroux, 2017). Effective genome editing tools are desired to evaluate gene functions in ciliary structure and their physiological roles. The CRISPR/ObLiGaRe system enables interpret the gene function by effective gene-knock-out, and in addition this method is usually applicable despite of cell species, because the NHEJ activities are conserved in almost all cells. However, low transfection efficacy and the limitation of cell division can reduce the performance of this method (e.g., primary cultured cells).

Considering the diversity of gene mutations, other methodologies than CRISPR/ObLiGaRe system are required. For an instance, to model single nucleotide variants such as missense mutations detected in the patients with ciliopathies, CRISPR/Cas9 system and single-stranded oligodeoxynucleotide (ssODN) mediated knock-in approach is used in mouse fertilized eggs and several human cultured cell lines. However, this method is not practical in the hTERT-RPE1 cells because of the low activity of HDR and synthesis-dependent strand annealing (SDSA). Since recent base-editing technology and prime editing technology are independent on the DNA double-strand breakage repair activities, they will be coming in the ciliary research (Anzalone et al., 2019; Anzalone, Koblan, & Liu, 2020; Gaudelli et al., 2017; Nishida et al., 2016).

Notes

1. For the *INPP5E* targeting, a pair of annealed oligodeoxynucleotides are designed in the exon 7 of *INPP5E* (5′-CGTGGACGCCCTCCTGTGCC-3′) for sgRNA.
2. A targeting vector for *INPP5E* targeting is constructed by designing a pair of oligodeoxynucleotides, including PAM sequence (5′-CGTGGACGCCCTCCTGTGCCAGG-3′). These pairs of oligodeoxynucleotides induce DSBs by CRISPR/Cas9 in the targeting vector.
3. Long-term passage should be avoided. In the case of hTERT-RPE1 cells, long-term passage decreases ciliogenesis and plasmid transfection efficacy, and increases cytotoxicity by plasmid lipofection.
4. Transfection using Lipofectamine LTX (Thermo Fisher) shows low cytotoxicity in hTERT-RPE1 cells. Other transfection reagents should be evaluated from the aspects of transfection efficacy and cytotoxicity.

5. The amount of targeting vector plasmid is critical for the CRISPR/ObLiGaRe method. Transfection of the excess amount of targeting vector causes the random integration to generate drug-resistant clones without the on-target insertion.
6. For cell cloning, custom-order stainless or commercially available cloning cups (RING-04, 05, 07, 09, 12; AGC techno glass, Shizuoka, Japan) are used to avoid mechanical stress on the cells. Before picking up, cell colonies are marked. Briefly, cells are washed with PBS, and cloning cups greased with silicone (335,148, BECKMAN, CA, USA) on the bottom are put on cell colonies. After treatment with 0.25% trypsin-EDTA, cells are collected by pipetting and transferred into 96-well plates.
7. To evaluate genotypes, three types of primer pair are designed: the primer pair C for detecting the target gene site, the primer pair A consisted of the forward primer in the target genes and *Neo*^{r}-reverse primer (5′-GCGGATC TGACGGTTCACTAAACCAGC-3′) for detecting the forward insertion of the targeting vector, and primer pair B consisted of the reverse primer in the target gene site and *Neo*^{r}-reverse primer for detecting the reverse insertion of the targeting vector.
8. In several cases, the targeting donor plasmid-random integrations and DSB inductions in non-targeted locus might cause the off-target effects that affect the cellular phenotypes. Therefore, the analyses using the only one clone are not secured for the biological interpretation. To treat this issue, several clones obtained by the CRISPR/ObLiGaRe method should be used in the phenotypic analyses. The genetic complementary assays that phenotypic recovery is observed in the knock-out cells with the full-length cDNA are better.
9. Before serum starvation, high cell confluency should be confirmed, because cells with low confluency show an incomplete ciliogenesis rate. Therefore, about 80–90% confluency is adequate.
10. Long-term serum starvation is effective (48 h) to increase the ciliogenesis rate.
11. To co-stain mouse anti-ninein antibody (subtype: IgG2a) and mouse anti-acetylated tubulin (subtype: IgG2b), Alexa 647- or 594-labeled anti-mouse IgG2b or IgG2a secondary antibodies are used.

Acknowledgments

This research was supported by Grant-in-Aid for Scientific Research from the Ministry of Education, Culture, Sports, Science and Technology of Japan (to T. Miyamoto: 21H02718 and 20K21845; to K.H.: 19K16521); research grant from the RIKEN-Hiroshima University science and technology hub joint research program (to K.H.); AMED-PRIME from the Japan Agency for Medical Research and Development, AMED (to T. Miyamoto: JP18gm5910011h0004); research grants from the Ono Medical Research Foundation (to T. Miyamoto); Daiichi Sankyo Foundation of Life Science (to T. Miyamoto); and Takeda Science Foundation (to T. Miyamoto).

References

Anzalone, A. V., Koblan, L. W., & Liu, D. R. (2020). Genome editing with CRISPR-Cas nucleases, base editors, transposases and prime editors. *Nature Biotechnology*, *38*, 824–844.

Anzalone, A. V., Randolph, P. B., Davis, J. R., Sousa, A. A., Koblan, L. W., Levy, J. M., et al. (2019). Search-and-replace genome editing without double-strand breaks or donor DNA. *Nature*, *576*, 149–157.

Blassberg, R., Macrae, J. I., Briscoe, J., & Jacob, J. (2016). Reduced cholesterol levels impair smoothened activation in smith-Lemli-Opitz syndrome. *Human Molecular Genetics*, *25*, 693–705.

Garcia-Gonzalo, F. R., Phua, S. C., Roberson, E. C., Garcia, G., 3rd, Abedin, M., Schurmans, S., et al. (2015). Phosphoinositides regulate ciliary protein trafficking to modulate hedgehog signaling. *Developmental Cell*, *34*, 400–409.

Gaudelli, N. M., Komor, A. C., Rees, H. A., Packer, M. S., Badran, A. H., Bryson, D. I., et al. (2017). Programmable base editing of a*T to G*C in genomic DNA without DNA cleavage. *Nature*, *551*, 464–471.

Kobayashi, Y., Tomoshige, S., Imakado, K., Sekino, Y., Koganezawa, N., Shirao, T., et al. (2021). Ciliary GPCR-based transcriptome as a key regulator of cilia length control. *FASEB Bioadv*, *3*, 744–767.

Maresca, M., Lin, V. G., Guo, N., & Yang, Y. (2013). Obligate ligation-gated recombination (ObLiGaRe): Custom-designed nuclease-mediated targeted integration through nonhomologous end joining. *Genome Research*, *23*, 539–546.

Miyamoto, T., Akutsu, S. N., Tauchi, H., Kudo, Y., Tashiro, S., Yamamoto, T., et al. (2018). Exploration of genetic basis underlying individual differences in radiosensitivity within human populations using genome editing technology. *Journal of Radiation Research*, *59*, ii75-ii82.

Miyamoto, T., Hosoba, K., Itabashi, T., Iwane, A. H., Akutsu, S. N., Ochiai, H., et al. (2020). Insufficiency of ciliary cholesterol in hereditary Zellweger syndrome. *The EMBO Journal*, *39*, e103499.

Miyamoto, T., Hosoba, K., Ochiai, H., Royba, E., Izumi, H., Sakuma, T., et al. (2015). The microtubule-depolymerizing activity of a mitotic kinesin protein KIF2A drives primary cilia disassembly coupled with cell proliferation. *Cell Reports*, *10*, 664–673.

Miyamoto, T., Porazinski, S., Wang, H., Borovina, A., Ciruna, B., Shimizu, A., et al. (2011). Insufficiency of BUBR1, a mitotic spindle checkpoint regulator, causes impaired ciliogenesis in vertebrates. *Human Molecular Genetics*, *20*, 2058–2070.

Nakade, S., Tsubota, T., Sakane, Y., Kume, S., Sakamoto, N., Obara, M., et al. (2014). Microhomology-mediated end-joining-dependent integration of donor DNA in cells and animals using TALENs and CRISPR/Cas9. *Nature Communications*, *5*, 5560.

Nishida, K., Arazoe, T., Yachie, N., Banno, S., Kakimoto, M., Tabata, M., et al. (2016). Targeted nucleotide editing using hybrid prokaryotic and vertebrate adaptive immune systems. *Science*, *353*.

Qiu, H., Fujisawa, S., Nozaki, S., Katoh, Y., & Nakayama, K. (2021). Interaction of INPP5E with ARL13B is essential for its ciliary membrane retention but dispensable for its ciliary entry. *Biol Open*, *10*.

Radha Rama Devi, A., Naushad, S. M., & Lingappa, L. (2020). Clinical and molecular diagnosis of Joubert syndrome and related disorders. *Pediatric Neurology*, *106*, 43–49.

Reiter, J. F., & Leroux, M. R. (2017). Genes and molecular pathways underpinning ciliopathies. *Nature Reviews. Molecular Cell Biology*, *18*, 533–547.

Royba, E., Miyamoto, T., Natsuko Akutsu, S., Hosoba, K., Tauchi, H., Kudo, Y., et al. (2017). Evaluation of ATM heterozygous mutations underlying individual differences in radiosensitivity using genome editing in human cultured cells. *Science Report*, *7*, 5996.

Sakuma, T., Nakade, S., Sakane, Y., Suzuki, K. T., & Yamamoto, T. (2016). MMEJ-assisted gene knock-in using TALENs and CRISPR-Cas9 with the PITCh systems. *Nature Protocols*, *11*, 118–133.

Yukimoto, H., Miyamoto, T., Kiyono, T., Wang, S., Matsuura, S., Mizuguchi, A., et al. (2020). A novel CDK-independent function of p27 Kip1 in preciliary vesicle trafficking during ciliogenesis. *Biochemical and Biophysical Research Communications*, *527*, 716–722.

CHAPTER

7

Studying the morphology, composition and function of the photoreceptor primary cilium in zebrafish

Markus Masek[a,b], Jingjing Zang[b], José M. Mateos[c], Marco Garbelli[b], Urs Ziegler[c], Stephan C.F. Neuhauss[b], and Ruxandra Bachmann-Gagescu[a,b,*]

[a]*Institute of Medical Genetics, University of Zurich, Zurich, Switzerland*

[b]*Department of Molecular Life Sciences, University of Zurich, Zurich, Switzerland*

[c]*Center for Microscopy and Image Analysis, University of Zurich, Zurich, Switzerland*

[*]*Corresponding author. e-mail address: ruxandra.bachmann@mls.uzh.ch*

Chapter outline

Methods in Cell Biology, Volume 175, ISSN 0091-679X, https://doi.org/10.1016/bs.mcb.2022.10.004
Copyright © 2023 Elsevier Inc. All rights reserved.

Abstract

Vision is one of our dominant senses and its loss has a profound impact on the life quality of affected individuals. Highly specialized neurons in the retina called photoreceptors convert photons into neuronal responses. This conversion of photons is mediated by light sensitive opsin proteins, which are found in the outer segments of the photoreceptors. These outer segments are highly specialized primary cilia, explaining why retinal dystrophy is a key feature of ciliopathies, a group of diseases resulting from abnormal and dysfunctional cilia. Therefore, research on ciliopathies often includes the analysis of the retina with special focus on the photoreceptor and its outer segment. In the last decade, the zebrafish has emerged as an excellent model organism to study human diseases, in particular with respect to the retina. The cone-rich retina of zebrafish resembles the fovea of the human macula and thus represents an excellent model to study human retinal diseases. Here we give detailed guidance on how to analyze the morphological and ultra-structural integrity of photoreceptors in the zebrafish using various histological and imaging techniques. We further describe how to conduct functional analysis of the retina by electroretinography and how to prepare isolated outer segment fractions for different -omic approaches. These different methods allow a comprehensive analysis of photoreceptors, helping to enhance our understanding of the molecular and structural basis of ciliary function in health and of the consequences of its dysfunction in disease.

1 Introduction

The primary cilium is a microtubule-based organelle protruding from the surface of most cell types and acting as a signaling hub to perceive, regulate and transmit extracellular signals to the cell. The types of signals transmitted depend on the cell type and include developmental signaling pathways such as Hedgehog or Wnt, and/or mechanical or sensory stimuli. Light sensation is mediated by highly specialized neurons called photoreceptors (PRs). The first step of the phototransduction cascade occurs in the outer segments (OSs) of PRs, where photons trigger the conversion of photosensitive opsin proteins, ultimately changing the membrane potential and modifying neurotransmitter release at the photoreceptor synapse signaling to bipolar and horizontal cells. Outer segments are highly modified primary cilia, which are anchored inside the cell through a modified centriole called the basal body (BB) (see Fig. 1). The OS is linked to the inner segment (IS) through the connecting cilium (CC), which is equivalent to the transition zone in other primary cilia and is thought to act as a gatekeeper to the ciliary compartment, controlling the protein and lipid composition of the OS (reviewed by Bachmann-Gagescu and Neuhauss, 2019). The photosensitive opsin proteins and other phototransduction cascade components are embedded within stacks of membrane discs or folds that fill the entire OS. Electrical signals from PRs are integrated and modulated by the bipolar cells and horizontal cells and the resulting signal is transmitted by the ganglion cells to the brain. These different cell types have a defined layered spatial organization described as retinal lamination.

Photoreceptors are divided into two major classes: rods, mediating dim light (scotopic) vision, and cones for bright light (photopic) and color vision. The zebrafish retina possesses four different cone subtypes: UV wavelength-sensitive (UV), short wavelength-sensitive (S) (blue), medium wavelength-sensitive (M) (green) and long wavelength-sensitive (L) (red) cones, expressing opsins activated by the respective light wavelength (Branchek & Bremiller, 1984; Vihtelic, Doro, & Hyde, 1999) (Fig. 1). In addition to their functional differences, rods and cones show distinct morphological characteristics. In mice, the membrane discs of cone outer segments are continuous with the outer plasma membrane and form a cone-shaped structure, whereas in rods the disc membranes are separated from the outer plasma membrane (Carter-Dawson & Lavail, 1979). Similarly, continuity of the lamellar discs with the plasma membrane is proposed for zebrafish cones (Tarboush, Chapman, & Connaughton, 2012). To remove and recycle damaged discs and phototransduction cascade components damaged by radical oxygen species generated through light-induced stress, OS tips are constantly shed and engulfed by the retinal pigment epithelium (RPE). To replenish this apical disc shedding, the photoreceptor continuously produces fresh membrane discs at the base of the OS, resulting in an outer segment turnover of approximately 10 days in mice and 16 days in adult zebrafish (Willoughby & Jensen, 2012; Young, 1967).

The arrangement of cones in the retina of many vertebrate species is not random but follows an exquisite array called the cone mosaic (Ahnelt & Kolb, 2000; Raymond, Barthel, & Curran, 1995). The ratio of cones and rods also differs between

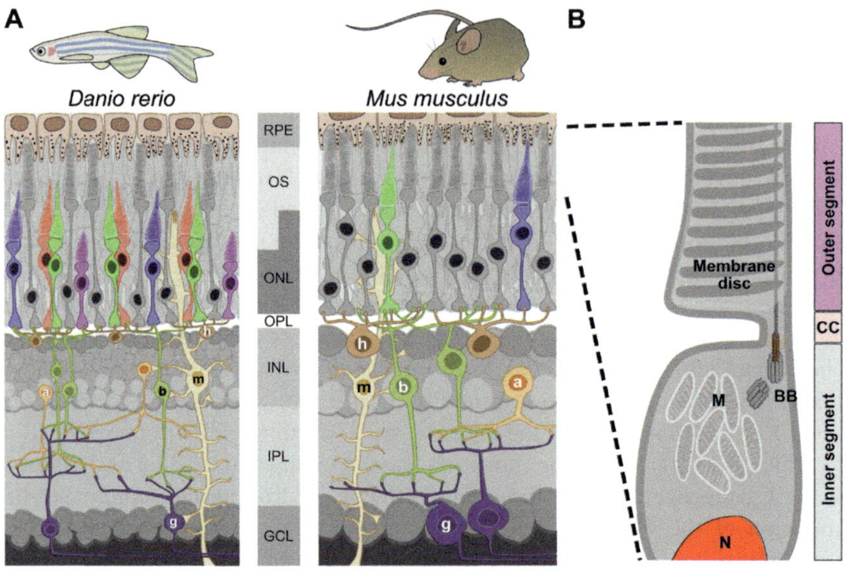

FIG. 1

Schematic overview of the zebrafish retina in comparison to the mouse retina. (A) Schematic showing the comparison of zebrafish and mouse retinal architecture (Note: mouse retina is not to scale. The layers are individually compressed to approximately match the zebrafish layer thickness). Zebrafish and mouse retinae show similar lamination with the retinal pigment epithelium (RPE) engulfing the outer segments (OS) of photoreceptors (PR). These outer segments represent highly specialized primary cilia. The nuclei of PRs form the outer nuclear layer (ONL) while the outer plexiform layer (OPL) is formed by the PR synapses on the dendrites of the ON/OFF bipolar cells and horizontal cells. Horizontal cells (h), amacrine cells (a) and bipolar cells (b) in the inner nuclear layer (INL) integrate and modulate the signals received from PRs and transmit them to dendrites of the ganglion cells (g) in the inner plexiform layer (IPL). Ganglion cells, whose cell bodies lie in the ganglion cell layer (GCL) then transmit the signal to the brain through the optic nerve. Müller glia cells (m) provide homeostatic and metabolic support for retinal neurons. The major difference between the zebrafish and mouse retina lies in the cone:rod ratio as well as in the type of cones present. Zebrafish have a cone-dominant retina with four different subtypes of cones: UV-cones dwell below the other cones while the red/green double cones reside apically of the blue and UV-cones. Mice have only two cone opsins (S-M-Opsin) sensitive for UV/blue and green light and their retina is rod dominated. (B) Schematic of the apical portion of a photoreceptor illustrating the fact that the outer segment is a specialized ciliary compartment of the photoreceptor cell. The basal body (BB), which lies apical to the mitochondrial cluster (M) in the inner segment apical to the nucleus (N), projects the microtubules forming the axoneme into the outer segment. The outer segment is composed of membrane discs containing the proteins required for the phototransduction cascade and is separated from the inner segment through the connecting cilium (CC), which acts as a gatekeeper of the ciliary compartment, similar to the transition zone in other cilia.

species and according to the region within the eye. In contrast to the rod-dominated mouse retina, the zebrafish retina consists of approximately 92% cones at larval stages and ~60% in the adult, thus resembling the cone-rich fovea in humans (Allison et al., 2010; Fadool, 2003; Zimmermann et al., 2018). Therefore, the zebrafish retina provides a useful model system to study the molecular mechanisms involved in the development, function or degeneration of the macula, which is most relevant for human vision. This methods paper will outline the following protocols used to assess the morphology and function of the zebrafish retina:

- Fluorescent immunostainings on cryosections and histology using plastic-embedded sections for determination of protein localization and morphological analyses.
- Ultrastructural analyses by transmission electron microscopy (TEM) and correlative light and electron microscopy (CLEM).
- Functional assays by larval electroretinography (ERG).
- Photoreceptor outer segment isolation for analyses of the proteomic or lipid landscape.

2 Basic buffers and solutions used in multiple protocols

Basic buffers and solutions below can be prepared in advance and stored for several months. Check safety data sheet (SDS) for all chemicals and handle, store and dispose chemicals/solutions accordingly.

- *Phosphate Buffer*:
  - 70 mM $NaH_2PO_4 \cdot 2H_2O$ and 390 mM Na_2HPO_4 solution, adjusted to pH = 7.4.
- *4% Formaldehyde (FA) fixation solution:*
  - Mix 40 g paraformaldehyde (PFA) in 600 mL ddH_2O. (Note: Paraformaldehyde dissolved in water breaks down into its basic building blocks, resulting in a formaldehyde solution which acts as a fixative).
  - Stir and heat up to 60 °C and slowly add 1.5 mL 1 M NaOH to the solution. The solution should clear up. Remove undissolved residues by filtration.
  - Add 250 mL Phosphate Buffer and adjust to pH = 7.4.
  - Fill up to 1000 mL using ddH_2O. Aliquot FA solution and store it at −20 °C for long-term storage. Aliquots stored at 4 °C should be used within 1 week.
- *Trichoroacetic (TCA) fixation solution:*
  - Dissolve 2 g Trichloroacetic acid in 100 mL ddH_2O, aliquot solution and store at −20 °C for long-term storage.
- *Methanol fixation solution*
  - Dilute methanol (MeOH) in DMSO to a final concentration of 80% MeOH.
- *MS-222 (Mesab or Tricaine) euthanizing solution*:
  - 15.3 mM MS222 and 4.8 mM $NaHC0_3$ solution, adjusted to pH = 7.0. Store solution at −20 °C for long-term storage.
- *E3 Medium*
  - 5 mM NaCl, 0.17 mM KCl, 0.33 mM CaCl2 and 0.33 mM $MgSO_4$ solution.

3 Histological analysis of retinal lamination and photoreceptor integrity on plastic-embedded sections and cryosections with immunofluorescence

The characterization of the retina by histology is a first and important step to assess the impact of a mutation/disease on overall retinal integrity. Plastic-embedded sections conserve the structure of the retina well and allow us to study the lamination as well as cell morphology in thin sections with an unspecific staining. Immunofluorescence staining or direct analysis of transgenes allow to specifically study localization of certain proteins within the retina. This labeling technique is of particular utility for targeted approaches. Example images can be found in Fig. 2.

3.1 Cryosectioning and immunofluorescence staining

3.1.1 Key resource table

Name	Company	Catalog #
Solutions		
MS222/Tricaine[a]	Sigma-Aldrich	#886-86-2
Paraformaldehyde[a]	Sigma-Aldrich	#158127
Sucrose	Sigma-Aldrich	#S0389
Tissue Freezing medium	EMS	#72592
20× PBS	VWR life science	#E703-1L
30% H_2O_2	Sigma-Aldrich	#1.07209.0250
Mowiol 4-88	Sigma-Aldrich	#81381
DABCO/1,4-Diazabicyclo[2. 2.2]octan	Sigma-Aldrich	#D27802
Triton X-100	Sigma-Aldrich	#T8787
Dimethyl sulfoxide (DMSO)	Sigma-Aldrich	#472301
Glycerol	Sigma-Aldrich	#G7757
TRIS-HCl	Sigma-Aldrich	#T3253
Material		
27G Needle	Braun	#4665406
Cover slides, Menzel Gläser	VWR	#630-1844
SuperFrost plus slides	Epredia	#J1800AMNZ
Vertical Staining Jar	VWR	#100490-752
Instrument		
Ultra blade MX35	Thermo scientific	#3053835
Cryostar NX50	Thermo scientific	NA

[a]*See basic buffers section for details on preparation.*

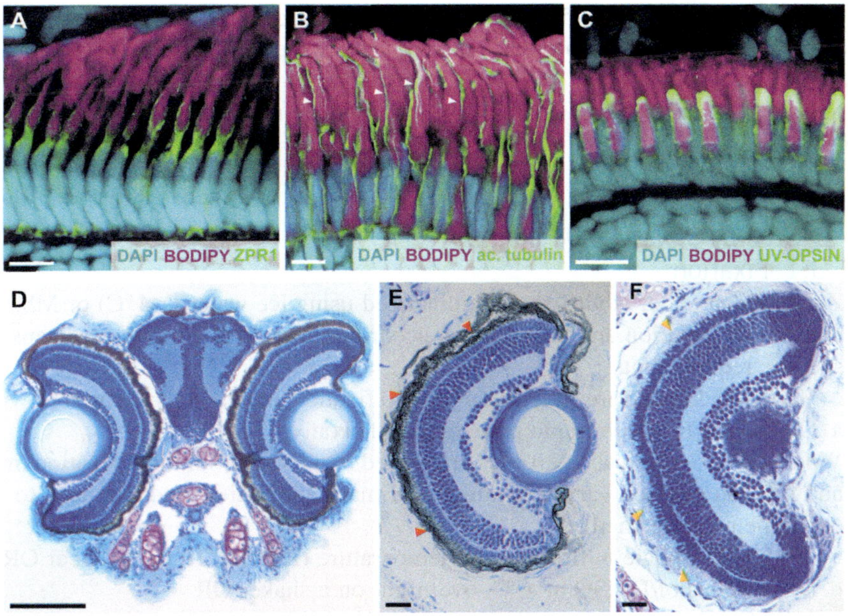

FIG. 2

Examples of Immunofluorescence and histology on plastic sections. (A–C) Immunofluorescence staining of wild-type zebrafish retinae. (A) Retinal cryosection of 8 week old fish stained with ZPR1-AB (green) marking the red/green cone bodies. (B) Retinal cryosection stained with anti-acetylated tubulin antibody (green) marking the axoneme within the outer segment (white arrowheads). (C) Cryosection of 8 dpf retina stained with anti-UV-Opsin antibody (green) highlighting the outer segments of UV-Cones. In (A-C) the indicated specific staining is counter-stained with DAPI (Cyan) to mark the nuclei and with BODIPY (magenta) to highlight the membrane-rich outer segments. Images (D–F) show plastic-embedded sections of 5 dpf larval eyes. (D) Transverse section of the larval head with brain and both eyes visible, stained with Richardson (Romeis) staining solution. (E) Shows one eye at higher magnification. The lamination of the eye can be easily assessed; however, the pigments of the RPE (red arrowheads) cover the PR outer segment layer. In (F) the eye-section was bleached to remove RPE pigments. Now the entire PR layer including the outer segments (orange arrows) can be analyzed. Abbreviations: days post fertilization (dpf); Photoreceptor (PR). Scale bar: A, B, C = 10 μm; D = 100 μm; E, F = 20 μm.

3.1.2 Before you begin

Mowiol 4-88/DABCO solution

- Mix 2.4 g Mowiol with 6 g Glycerol and slowly add 6 mL of ddH_2O while stirring well for 2–3 h at room temperature (RT).
- Add 12 mL 0.2 M TRIS-HCl, pH 8.5 and keep stirring at 50 °C for 20 min. Extend time if Mowiol isn't completely dissolved.
- Transfer solution into a 50 mL falcon tube, centrifuge at 4000 rpm for 20 min and decant supernatant into a clean 50 mL falcon tube.

- Add DABCO to a final 2.5% *w/v* concentration. Once DABCO is dissolved, aliquot solution into light-protected tubes and store the solution at −20 °C for up to 1 year.
- Before use, thaw and heat an aliquot at 37 °C for 10 min and store it at 4 °C for up to 1 month.

3.1.3 Step-by-step method details

3.1.3.1 Fixation

- Zebrafish larvae or adult fish are euthanized using ice water (<4 °C) or MS222 euthanizing solution and decapitation according to animal experiment license specific requirements.

  Eyes of adult fish are enucleated prior to fixation and the sclera is pierced with a 27G needle to ensure rapid penetration of fixative.
- Whole larvae or pierced adult eyes are placed into fixation solution (fixation time and fixation solution vary according to primary antibodies used and need to be experimentally verified):
  - 4% FA/Phosphate buffer at room temperature (RT) for 2 h on a shaker OR
  - 80% Methanol/DMSO at 4 °C, overnight on a shaker OR
  - 2% TCA/Water at RT for 2–4 h on a shaker
- After fixation, wash tissue with 1× PBS three times for 5 min to remove excess fixation solution. For specimens fixed with 80% Methanol (MeOH) rehydrate larvae through decreasing MeOH washes: 50% MeOH/PBS and 25% MeOH/PBS for 5 min each, followed by four times 5 min washes in PBS.

  Note: DNA can be extracted from a tissue biopsy to genotype the larvae. For adult fish it is advised to use a biopsy of the remaining tissue after eye enucleation for genotyping.
- Cryoprotect samples by incubation in 30% sucrose/PBS at 4 °C overnight (until larvae/eyecup sink to the bottom of the tube).
- Tissue can be stored at 4 °C for several days (up to 7 days) until embedding. For long-time storage, add 0.1% sodium azide to limit bacterial growth, however fast embedding is strongly advised since tissue can degrade with prolonged storage in sucrose.

3.1.3.2 Embedding and cryosectioning

- Transfer the samples into tissue freezing medium contained inside a mold. Be careful to transfer as little sucrose solution as possible. The mold can be prefabricated using aluminum foil (or use commercially available plastic molds). Align the samples and orientate them in the desired sectioning position along one side of the mold. Snap-freeze the sample using liquid nitrogen (most easily performed by placing the mold containing the sample on a metal plate inside a Styrofoam box that is filled at the bottom with liquid nitrogen to cool down the metal plate). Frozen sample blocks can be stored at ≤−20 °C until further use.

- Cut sections of 6–14 μm thickness using a cryostat at −20 °C (thinner sections allow improved penetration of antibodies while the structure may be better preserved in thicker sections).
- Mount sections on Superfrost slides. Store slides at ≤ −20 °C until further use.

3.1.3.3 Immunohistochemistry staining

- Cryosection slides are thawed at RT for 30 min.

  Note: The outer segment layer of the zebrafish retina is partly covered by retinal pigment epithelium pigments. For better visualization of the photoreceptor outer segments, the pigments can be bleached after sectioning. Consider that some epitopes may be affected by the bleaching step and that certain antibodies might not work after this step. Alternatively, the larvae/fish can be dark-adapted and fixed in the dark to induce pigment retraction, but this may be incomplete. Raising larvae in PTU (0.003% 1-phenyl-2-thiourea in E3) is another alternative, which does however somewhat impact retinal function and may interfere with the structure.

Optional bleaching step: Place slides in a vertical glass staining jar filled with pre-heated 10% H_2O_2/PBS at 65 °C (water bath). Incubate slides in bleaching solution until pigments are gone. Check every 10 min for remaining pigments (~20 min, time depends on slide thickness), remove the slides that are completely bleached and let them cool down to RT. Continue with staining protocol.

- Wash the slides two times for 5 min in 1× PBS to remove freezing medium or bleaching solution traces.
- Wash the slides an additional two times for 5 min in 0.5% Triton X-100/PBS to permeabilize the sections.
- Block sections in blocking solution (1× PBS, 1% BSA, 1% DMSO 0.5%, Triton X-100 plus 10% goat serum—or serum appropriate to the antibodies used) at RT for 30 min.
- Dilute primary antibodies in PBDT (1× PBS, 1% BSA, 1% DMSO, 0.5% Triton X-100) containing 2% goat serum (or a serum appropriate to the antibodies used) and incubate sections overnight at 4 °C in a humidified chamber. Primary antibody can be spread evenly over slides by covering the slides with a transparent plastic foil. A selection of antibodies that we have used successfully in zebrafish, as well as established transgenic lines, can be found in Table 1.

  Note: Triton X-100 can have a detrimental effect on structure preservation, especially for super-resolution microscopy. Consider applying a range of Triton-X concentrations (commonly between 0.1% and 0.5%) or other permeabilization detergents, such as saponin, to preserve the structure of the sample.
- Wash slides four times in PBDT for increasing lengths of time (5, 10, 15, 30 min)
- Dilute appropriate fluorescent-conjugated secondary antibodies in PBDT and incubate slides for 2 h at RT. Keep slides from now on in the dark to minimize light-induced bleaching.
- Wash slides four times in PDT (1× PBS, 0.1% Triton X-100, 1% DMSO) at RT for 5 min.

Table 1 Selection of antibodies and transgenic lines commonly used to study photoreceptors in zebrafish.

Antibody Name	Localization	Fixative[a]	Primary dilution[a]	Host	Source
Ab-opn1sw1	UV-opsin of cone PR	FA	1:300	Rabbit	generated by D. Hyde (Vihtelic, Doro, & Hyde, 1999) [http://zfin.org/ZDB-ATB-090508-3]
Ab-opn1sw2	Blue-opsin of cone PR	FA	1:250	Rabbit	generated by D. Hyde (Vihtelic et al., 1999) [http://zfin.org/ZDB-ATB-090508-2]
Ab-opn1mw1	Green-opsins of cone PR	FA	1:400	Rabbit	generated by D. Hyde (Vihtelic et al., 1999) [http://zfin.org/ZDB-ATB-090508-4]
Ab-Rho/ 1D1	Rod OS	FA	1:100	Mouse	generated by R. Molday (Hicks & Molday, 1986) [http://zfin.org/ZDB-ATB-081229-13]
Ab-4D2	Rod and some cone OS	FA	1:100	Mouse	generated by R. Molday (Hicks & Molday, 1986) [http://zfin.org/ZDB-ATB-111206-4]
Ab-4C12	Rod cell bodies	FA	1:100	Mouse	generated by J. Fadool (Morris, Schroeter, Bilotta, Wong, & Fadool, 2005) [http://zfin.org/ZDB-ATB-090506-2]
Ab-SV2	Outer plexiform layer	FA	1:100	Mouse	Developmental Studies Hybridoma Bank [http://zfin.org/ZDB-ATB-081201-1]
Ab-ZPR1	Arr3a-like protein of red/green-cone bodies	FA	1:200	Mouse	generated by R. Bremiller (Larison & Bremiller, 1990) [http://zfin.org/ZDB-ATB-081002-43]
Ab-Syntaxin-3	OS/PR synaptic terminal	FA	1:200	Rabbit	Alomone, Cat#ARN-005 [http://zfin.org/ZDB-ATB-180601-1]
Ab-Ctbp2l/ ribeye	PR ribbon synapses	FA	1:100	Rabbit	generated by T. Nicholson (Obholzer et al., 2008) [http://zfin.org/ZDB-ATB-091117-2]
Ab-Tuba	acetyl-alpha tubulin of ciliary axoneme	FA/TCA/ MeOH	1:500	Mouse G2b	Sigma-Aldrich, Clone 611B-1, Cat#:MABT868 [http://zfin.org/ZDB-ATB-081003-6]
Ab-Cc2d2a	Transition Zone	FA/MeOH	1:20	Mouse G1	generated by R. Bachmann-Gagescu (Bachmann-Gagescu et al., 2011) [http://zfin.org/ZDB-ATB-111130-3]
Ab-Ift88	Ciliary axoneme	FA	1:5000	Rabbit	generated by B. Perkins [https://zfin.org/ZDB-ATB-100730-1]

Dyes					
BODIPY	Membranes (strong signal in the OS)	FA/TCA/ MeOH	1:300	NA	Thermo Scientific, Cat#: C34556
DAPI	DNA/Nucleus	FA/TCA/ MeOH	final conc. 1 µg/mL	NA	Thermo Scientific, Cat#: D1306
DIO	Membranes (strong signal in the OS)	FA	1:200	NA	Thermo Scientific, Cat#: V22886
Phalloidin	F-Actin	FA/TCA/ MeOH	1:200	NA	Thermo Scientific, Cat#: A12379

Transgenes	Construct	Location	Source
Rod GFP	*Tg(zfRH1-3.7B:EGFP)*	Rod cell bodies	Hamaoka, Takechi, Chinen, Nishiwaki, and Kawamura (2002)
Cone GFP	*Tg(TaCP-EGFP-CAAX)*	Cone outer segments	unpublished (in house)
Cone Rab8a	*Tg(gnat2:mCherry-rab8a)zh403*	Cone OS-bound vesicles	Ojeda Naharros et al. (2017)
Rod Rab8a	*Tg(rho:mCherry-rab8a)zh401*	Rod OS-bound vesicles	Ojeda Naharros et al. (2017)
Blue opsin	*Tg(zfSWS2-3.5A:EGFP)*	Blue sensitive cone bodies	Takechi, Seno, and Kawamura (2008)
Red opsin	*Tg(LWS1up2.6kb:GFP)*	Red sensitive cone bodies	Tsujimura, Hosoya, and Kawamura (2010)
UV-Opsin	*Tg(zfSWS1-5.5A:EGFP)*	UV-sensitive cone bodies	Takechi, Hamaoka, and Kawamura (2003)

[a]*Condition tested successfully, Abbreviation: FA=4% formaldehyde prepared from paraformaldehyde solution, TCA=2% trichloroacetic acid/Water, MeOH=80% Methanol/DMSO. Abbreviations:* Ab antibody, FA Formaldehyde, MeOH Methanol, NA not applicable, PR photoreceptor, OS outer segment, TCA Trichloroacetic acid, Tg transgenic, UV ultra-violet.

Optional: Stain the outer segments with the membrane-interacting dyes BODIPY TR methyl ester or DiO for 20 min at RT and/or co-stain nuclei using DAPI for 3 min at RT. After co-staining, wash slides three times in 1× PBS for 5 min.

3.1.3.4 Mounting and imaging

- Keep slides dry in the dark, add antifade mounting medium such as Mowiol/DABCO on the slide and cover it with a microscope cover glass. Let Mowiol dry at RT in the dark for 1 h and store the stained slides at 4 °C. Image the slides within the next (few) weeks, long-term storage will reduce signal intensity.

3.1.3.5 Optimization and alternative staining strategies

The paucity of zebrafish-specific antibodies represents a challenge for immunohistochemistry in this organism. The selection of a well-conserved epitope increases the chance of cross-reactivity when purchasing an antibody that has not been specifically validated in zebrafish. Different fixation and staining methods should be tested to find the optimal staining conditions. Different fixation times (short fixation ≈30 min at RT, medium fixation ≈2 h at RT or long time fixation ≈ overnight at 4 °C) and fixatives (4% FA/phosphate buffer; 2% TCA/Water; 80% Methanol/DMSO) may result in improved staining.

3.2 Plastic-embedding and sections for histology

3.2.1 Key resource table

Name	Company	Catalog #
MS222/Tricaine[a]	Sigma-Aldrich	#886-86-2
Paraformaldehyde[a]	Sigma-Aldrich	#158127
Ethanol	VWR	#20821.296
Technovit 7100 kit	Kulzer	#64709003
Methylene-blue	Sigma-Aldrich	#M9140
Azure II	Sigma-Aldrich	#861065
Borax	Sigma-Aldrich	#B9876
Entellan	Merck	#107961
Material		
27G Needle	Braun	#4665406
BEEM plastic mold	Polysciences	#23257-1
Instrument		
Microtome RM2145	Leica	NA
Knife maker EM KMR2	Leica	NA
Heat plate	MEDAX Nagel	NA

[a]See basic buffers section for details on preparation.

3.2.2 Before you begin

- *Technovit 7100 Solution A*:
  - Dissolve 250 mg Hardener I in 25 mL Technovit 7100 solution. Mix well and store at 4 °C for up to 4 weeks.
- *Technovit 7100 Solution B*:
  - Mix 15:1 Solution A with Hardener II (barbiturate acid derivative). Use solution immediately.
- *Richardson (Romeis) staining solution*:
  - Dissolve 1 g Borax in 100 mL ddH_2O and filter the solution to remove undissolved particles.
  - Add 1 g Methylene Blue, stir solution until dye is completely dissolve.
  - Dissolve 1 g of Azur II in ddH_2O, mix Azure II solution 1:1 with methylene blue solution and filter the mixture.
  - Solution can be stored (light-protected) for up to 1 year at room temperature.

3.2.3 Step-by-step method details

3.2.3.1 Fixation and embedding

- Eyes of zebrafish larvae or adult fish are fixed in 4% FA at 4 °C overnight as described above.
- Wash off formaldehyde with two washes of 1× PBS for 5 min at RT.
- Dehydrate sample through washes with increasing ethanol (EtOH) gradient (50%, 70%, 80%, 90%, 95%, 100% Ethanol) for 10 min each.
- Transfer tissue into 1:1 Solution A/EtOH for pre-infiltration of the tissue. Incubate for 30 min and repeat this step a second time with fresh solution.
- Remove solution and incubate tissue in pure solution A for 60 min two times or overnight to complete rehydration of tissue.
- Embed the sample into the molds and fill them carefully with Solution B. Orientate them in the desired sectioning position along one side of the mold and cover the chamber with Parafilm to limit air access. Wait 20 min until the plastic starts to polymerize before placing the molds in an incubator at 37 °C to achieve complete polymerization.

  Note: Quick handling of Solution B is required. Slight movement of the chamber can affect the orientation since Solution B is not viscous. Orientation works better as soon the polymerization has started and Solution B becomes more gel-like.
- Polymerized blocks can be stored for years at room temperature.

3.2.3.2 Sectioning, staining, and mounting of sections

- Cut microtome sections of 1–3 μm thickness using glass blades.
- Transfer plastic sections to a glass slide in a drop of water. Flatten and orientate section and let the water evaporate on a warm surface at 57 °C.

Optional bleaching: place dry slides in a vertical staining jar filled with preheated 10% H_2O_2/PBS at 65 °C inside a water bath. Incubate slides in bleaching solution until pigments disappear. Check every 10 min for remaining pigments (~20 min, time depends on slide thickness), remove the slides that are completely bleached and let them cool down to RT. Wash slides twice with water to remove traces of bleaching solution.

- Stain slides with Richardson (Romeis) solution for exactly 15 s and wash the staining solution off immediately with water: dip the slide for 5 min in water while constantly moving the slide and changing the water in the staining jar three times.
- Let the slide dry on the warm surface at 57 °C, then add Entellan and cover the slide with a microscope glass cover slide. Slides can be stored for years at RT until they are imaged using a bright field microscope.

4 Ultrastructural visualization of photoreceptors using transmission electron microscopy and correlative light and electron microscopy

Transmission electron microscopy is used to study ultrastructural changes at subcellular resolution from ultra-thin sections imaged with an electron beam. In contrast to classical histology the improved resolution helps to identify subtle changes unnoticeable by previously described methods. However, the ultrathin sections are more laborious to prepare and finding a good section plane that perfectly shows the desired structure can be challenging.

4.1 Transmission electron microscope (TEM)

4.1.1 Key resource table

Name	Company	Catalog #
Solutions		
MS222/Tricaine[a]	Sigma-Aldrich	#886-86-2
Formaldehyde 16% (as aqueous Paraformaldehyde solution)	Electron microscopy Science	#15700
Glutaraldehyde EM Grade	Electron microscopy Science	#16220
Cacodylate	Merck	#8.20670
Epon	Sigma-Aldrich	#45345
Durcupan ACM	Sigma-Aldrich	#44611

—cont'd

Name	Company	Catalog #
Osmium tetroxide	Electron microscopy Science	Cat. 19152
Uranyl acetate	Merck	#8473
Propylene oxide	Sigma-Aldrich	#82320
Lead (II) nitrate	Sigma-Aldrich	#228621
Lead (II) citrate	Sigma-Aldrich	#15326
Toluidine blue O	Sigma-Aldrich	#198161
Borax	Sigma-Aldrich	#B9876
Material		
27G Needle	Braun	#4665406
Flat embedding molds	BEEM Flat	
TEM grids one hole	Agar Scientific	G2500C
Instrument		
Diamond Knife	Diatome	
Cryo-ultramicrotome	Leica Microsystems	Leica EM FCS

[a]*See basic buffers section for details on preparation.*

4.1.2 Before you begin

- *0.2 M Sodium Cacodylate Buffer:*
  Dissolve 21.4 g Sodium-cacodylate trihydrate in approximately 450 mL ddH_2O. Adjust pH = 7.35 and fill volume up to 500 mL with ddH_2O.
- *EM-Fixation solution*:
  - Dilute glutaraldehyde and formaldehyde in 0.2 M sodium Cacodylate buffer to a final concentration of 2.5% glutaraldehyde and 2% formaldehyde in 0.1 M Cacodylate buffer, adjust with ddH_2O. Prepare the fixation solution freshly the same day as fixing the samples.
- *Toluidine blue 1% + borax 1% pH 9.0 working solution*
  - Dissolve 0.4 g Toluidine blue and 0.4 g Borax (Borax sodium tetraborate decahydrate) in 40 mL ddH_2O.

4.1.3 Step-by-step method details

4.1.3.1 Fixation and embedding

- Zebrafish are euthanized (according to animal experiment license requirements) in ice cold MS222 and additionally decapitated to ensure better access of the fixative to the sample. This improves tissue preservation.
  - For larvae ($\leq$10 days post fertilization (dpf)), decapitate them and continue with the next step.

- For juvenile/adult fish, isolate the eyes and cut open the sclera using a microdissection scissor or a G27 needle to ensure fast fixation of the tissue. The eyes are immediately transferred into freshly prepared fixation solution.
- Fix tissue in EM-Fixation solution at 4 °C overnight.
- Rinse specimen in cacodylate buffer and transfer the eyes into 1% osmium tetroxide in 0.1 M sodium cacodylate buffer at room temperature for 1 h.
- Transfer the eyes into 1% aqueous uranyl acetate solution and incubate them at 4 °C for 2 h.
- Dehydrate samples in an alcohol series (70%, 80%, 96%, 100% ethanol) for 20 min each.
- Wash sample twice for 15 min in propylene oxide, to ensure a complete transition from alcohol to resin.
- Tissue is preinfiltrated overnight in a mixture of propylene oxide: Epon/Araldite (1:1) and subsequently in 100% Epon/Araldite for another 1.5 h.
- Place the eyes in a mold and cover them with fresh Epon/Araldite. Position the eye such that the lens is facing toward the bottom and the remnant of the optic nerve facing to the side parallel to the cutting plane (Fig. 3A). Let the Epon/Araldite polymerize at 60 °C for 28 h.

4.1.3.2 Cutting and imaging

- Resin-embedded tissue is trimmed and semi-thin (200 nm) sections are obtained with an ultramicrotome until region of interest is reached.

  To monitor the trimming process, sections are transferred to a glass slide and stained with toluidine blue solution for 1–2 min at room temperature. Wash the stained sections three times in ddH_2O and image section (Fig. 3B). The optic nerve is used as a landmark to verify the middle region of the eye.
- Within the region of interest, collect ultrathin (70 nm) sections on carbon-coated one hole TEM grids.
- The samples are post-stained with Reynold's lead citrate and examined with a Talos 120 transmission electron microscope at an acceleration voltage of 120 KV using a Ceta digital camera. Automatic image acquisition of large areas is provided by the MAPS software package (Thermo Fisher Scientific). Representative resulting images are seen in Fig. 3C and D.

4.2 Correlative light and electron microscopy (CLEM)

Correlative light and electron microscopy is a useful technique combining the advantages of both imaging modalities. There are many possibilities to obtain this correlation and each method has its own specific applications (for further review of these methods: Kopek et al., 2017; Oorschot, Lindsey, Kaslin, & Ramm, 2021). We will refer here to a CLEM method that we have established and is particularly applicable for zebrafish retina (Mateos et al., 2017). Photoreceptor subcellular details are gained through electron microscopy and protein expression is detected via light microscopy (Fig. 3E).

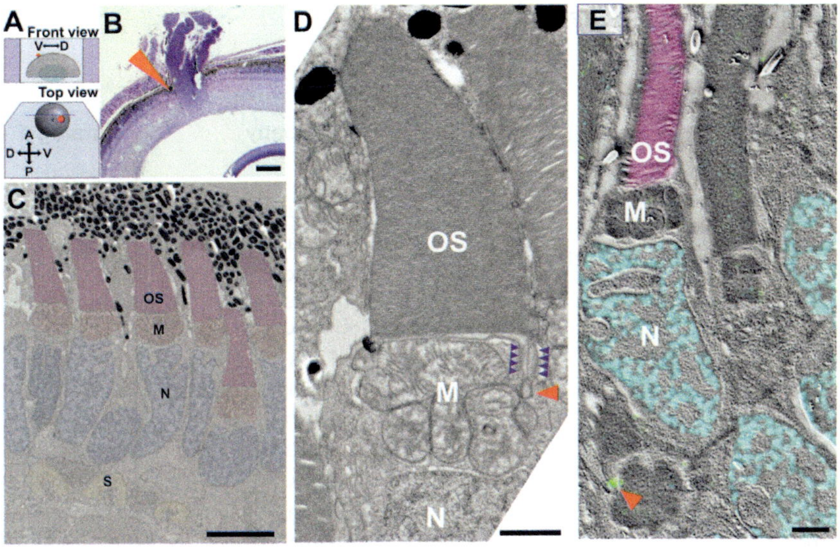

FIG. 3

Transmission electron microscopy (TEM) and Correlative Light and Electron Microscopy (CLEM) of zebrafish retina. (A) shows a schematic illustrating how to embed and align the adult eye inside the mold. The optic nerve (red) is not positioned in the center of the eye but slightly ventral. To obtain comparable section planes between different eyes, each eye is aligned such that the optic nerve faces to the right of the mold on the top view (parallel to the sectioning plane (blue line)). The cut retinae on the slides are all orientated in ventral/dorsal direction. (B) shows a retinal section stained with toluidine blue during the trimming process. The block can be trimmed until the optic nerve (red arrow) is visible. Sample ultrathin sections from this plane onwards for TEM imaging. (C) Overview of a larval retina at 5 dpf with the different structures pseudo-colored: Outer Segments (OS) in magenta, Mitochondrial clusters (M) in brown, nucleus (N) in blue and ribbon synapses (S) in yellow. Higher resolution image of a cone photoreceptor is shown in (D). Note the basal body marked by a red arrow, the connecting cilium indicated by the blue arrows and the organized membrane stacking of the outer segment. (E) Correlative Light and Electron Microscopy (CLEM) image on the same ultrathin section (100 nm) where a SEM image is overlaid with a widefield fluorescent image showing both the localization of green opsin (magenta), the basal body highlighted by GFP expression of the *tg(TaCP:GFP-centrin)*^{zh405} transgene (green, marked by the red arrowhead) and the nucleus stained with DAPI (blue). Besides the localization of the stained proteins, the ultrastructure can be observed in the electron microscopy image. Abbreviation: days post fertilization (dpf). Scale bar: B = 100 μm; C = 1.5 μm; D, E = 1 μm.

4.2.1 Key resource table

Name	Company	Catalog #
Solutions		
MS222/Tricaine[a]	Sigma-Aldrich	#886-86-2
Paraformaldehyde	Sigma-Aldrich	#158127
Glutaraldehyde EM Grade	Electron microscopy Science	#16220
Cacodylate	Merck	#8.20670
Agarose, peqGOLD Universal	VWR International GmbH	#35-1020
Local food brand gelatin	Dr. Oetker	Extra Gold
Sucrose	Merck	#1.07687
Methylcellulose	Sigma	#M-6385
Glycine	Sigma	#G-7126
Gelatin type B	Sigma	#G-6650
BSA	Applichem	#A6588.0050
Glucose	Sigma-Aldrich	#G8270
Glucoseoxidase	Sigma-Aldrich	#G7141
Catalase	Sigma-Aldrich	#C40
Beta-Mercaptoethylamine hydrochloride	Sigma-Aldrich	#M6500
Glycerol solution	Sigma-Aldrich	#49782
Material		
VANNAS micro Scissors	Electron microscopy Science	#72932-01
Silicone stripes	Picodent	Twinsil 22
Glass bottom petri dish: u-Dish 35 mm, high glass bottom	Ibidi, Germany	#81158
SEM aluminum stub	Agar Scientific	#G301F
Conducting carbon cement Leit-C	Plano, Germany	#AG3300
Silicon wafer Type: P/Boron; Orientation ⟨111⟩ ON; Growth method: CZ; Resistivity: 1–30 Ω/cm; Surface: polished; Laser cut at 7 × 7 mm	Si-Mat Silicon Materials	
Cryo-pin	Baltic Präparation, Germany	#16701950
Flat embedding molds	BEEM Flat	
Instrument		
Diamond Knife (Cryo Immuno)	Diatome	DCIMM3520
Cryo-ultramicrotome	Leica Microsystems	Leica EM FCS

[a]*see basic buffers section for details on preparation.*

4.2.2 Before you begin

- *CLEM Imaging Buffer*:
  - Prepare a 200 mM Phosphate buffer containing 10% glucose, 0.5 mg/mL glucoseoxidase, 40 µg/mL catalase and 15 mM beta-mercaptoethylamine hydrochloride. Adjust the pH to 8.0.
- *0.2 M Sodium Cacodylate Buffer*:
  - 0.2 M Sodium Cacodylate Buffer:
  - Dissolve 21.4 g Sodium-cacodylate trihydrate in approximately 450 mL ddH2O. Adjust pH = 7.35 and fill volume up to 500 mL with ddH2O.
  - EM-Fixation solution:
  - Dilute glutaraldehyde and formaldehyde in 0.2 M sodium Cacodylate buffer to a final concentration of 2.5% glutaraldehyde and 2% formaldehyde in 0.1 M Cacodylate buffer, adjust with ddH2O. Prepare the fixation solution freshly the day of fixing the samples.

4.2.3 Step-by-step method details

4.2.3.1 Fixation and mounting

- Prepare the fixative solution containing 0.1% glutaraldehyde, 4% formaldehyde in 0.1 M cacodylate buffer.
- Euthanize larva in ice-cold MS222 euthanization solution and immediately place tissue in pre-cooled fixative solution on ice and incubate it overnight at 4 °C on a shaker. To ensure complete penetration of the fixative, cut off the head prior to fixation.
- Dissect the eyes using micro scissors keeping the eye in fixative all the time.
- Wash away fixation solution in 1× PBS two times for 5 min at RT.
- Incubate the eyes in a pre-warmed 12% *w/v* gelatin solution in phosphate buffer for 10–30 min at 40 °C. Softly flick the tube to ensure that the solution can infiltrate the eyes.
- Embed two eyes per single silicon or polyethylene flat embedding mold using gelatin. Align the eyes and let the gelatin harden for 20 min at 4 °C.
- Trim the gelatin block under the binocular and separate the two eyes from each other.
- Transfer the embedded eye into 2.3 M Sucrose/Phosphate buffer. Incubate overnight at 4 °C and exchange the sucrose solution with fresh solution and store at −20 °C until further use.

4.2.3.2 Cryosectioning and immunolabeling

- The gelatin block is further trimmed to remove most of the gelatin around the eyes and transferred onto a cryo-pin. Freeze the sample in liquid nitrogen.
- Trim the eyes on a cryo-ultramicrotome using a diamond knife until region of interest is reached and start sampling 110 nm thin-sections at −120 °C. Transfer a section to a silicon wafer using a wired loop containing a film of 2% methylcellulose/Water and 2.3 M sucrose (1:1).

- Wash wafers by placing them upside down on a drop of ice-cold 1× PB for 5 min. Repeat washing step three times at 0 °C and twice at RT for 2 min.
- Incubate the wafers three times in 0.15% glycine/PBS for 1 min and wash three times in 1× PBS for 1 min.
- Incubate the wafer in PBG (1× PBS with 0.5% Bovine Serum Albumin BSA and 0.2% Gelatin type B) for 5 min.
- Dilute the primary antibody in PBG and incubate the wafer in the primary AB solution for 30 min at RT.
- Wash wafer six times 1 min in PBG and pre-incubate wafer in PBG for 5 min at RT.
- The secondary antibody, which must be suited for advanced resolution microscopy, is diluted in PBG to its working concentration (empirically tested) and incubated for 30 min at RT.
- Wash six times for 1 min in PBG followed by two washes of 1× PBS for 2 min.
- Counter stain nuclei with DAPI by placing wafer in DAPI (4 μg/mL) in PBS for 10 s, followed by two washes in 1× PBS for 2 min each.

4.2.3.3 Light microscopy

- Incubate the wafer for 10 s on a droplet of a solution of glycerol (40%) in PBS for standard light microscopy or in case of super-resolution in a dedicated imaging buffer (Mateos et al., 2017).
- Place the wafer (section facing down) in a glass bottom dish, onto a fresh drop of the 1:1 glycerol (80%) and CLEM imaging buffer solution. Use a pipette to remove most of the liquid underneath the wafer and tape it to the dish using silicon strips.
- Image section using an oil immersion objective with a high numerical aperture. Image the section with a fluorescent light microscope or even better with a super-resolution microscope such as STORM (Rust, Bates, & Zhuang, 2006).

Note: Find the area of interest, center it and acquire a reference widefield epifluorescence image. Change to the advanced resolution mode of the microscope. Find the right settings to ensure that the single molecule blinks are well separated in each frame and reduce laser power to minimize bleaching artifacts. Record the raw image in epifluorescence mode by acquiring a minimum of 30,000 frames.

4.2.3.4 Platinum coating and scanning electron microscopy

- Remove the wafer from the glass bottom petri dish by adding a drop of PBS to the edges until wafer is lifted. Wash two times in 1× PBS for 2 min.
- Post-fix section with 0.1% glutaraldehyde/PBS for 5 min at RT. Wash two times in 1× PBS for 2 min.
- Place the wafer upside down in a drop of ice-cold 2% methylcellulose/water for 5 min. Repeat this step once and remove all the liquid subsequently by centrifugation at 14′100 × *g* for 90 s.

- Mount the wafer on an SEM aluminum stub using conducting carbon cement. Coat the section with a 2–10 nm layer of platinum/carbon by rotary shadowing at 8° using an electron beam evaporation device settings: 1.55 kV, 55 mA, at 0.3 nm/s, and an angle of 8°, rotation level 4.
- Image the sections on a scanning electron microscope with an In-lens secondary electron detector.

4.2.3.5 Alignment of the light microscopy and SEM images

- This is achieved via the TrakEM2 (Cardona et al., 2012) package of the open source software imageJ (Schindelin et al., 2012).

  Note: Canvas size of both images (SEM and super resolution) must be adjusted and an image stack should be created. The light microscopy image is manually aligned to the electron microscopy image using landmarks such as recognizable edges of nuclei using "Align layer manually with landmarks" of the TrakEM2 interface of ImageJ.

For a detailed visual protocol please see Mateos et al. (2017).

5 Electrophysiological assessment of retinal function

Electroretinography (ERG) probes the signal transduction and overall functional integrity of the retina. Upon a light stimulus the photoreceptors hyperpolarize (a-wave) and glutamate release decreases at photoreceptor synapses. Downstream ON-bipolar cells depolarize upon glutamate reduction, generating the ERG b-wave. In contrast to the mouse ERG, the zebrafish a-wave is largely masked by the emerging larger b-wave. Hence, measurement and interpretation of the a-wave requires pharmacological blockage of the b-wave. Recording ERG from 5 dpf zebrafish larvae is a well established technique (Makhankov, Rinner, & Neuhauss, 2004); at this stage, rods are not mature but cones are fully functional, such that the ERG recordings at 5 dpf are mainly cone driven (Bilotta, Saszik, & Sutherland, 2001).

5.1 Key resource table

Name	Company	Catalog #
Solutions		
MS222/Tricaine[a]	Sigma-Aldrich	#886-86-2
Sodium Chloride/NaCl	Sigma-Aldrich	#S9888
Potassium Chloride/KCl	Sigma-Aldrich	#P9541
Calcium Chloride/CaCl	Sigma-Aldrich	#C4901
Magnesium sulfate/$MgSO_4$	Sigma-Aldrich	#M7506
Universal-Agarose, peqGOLD	VWR	#35-1020

Continued

—cont'd

Name	Company	Catalog #
Material		
Glass capillary: 0.58[a]1.00[a]100 mm with filament	SCIENCE PRODUCTS GMBH	#GB100F-10
Instrument		
Needle puller P-87	Sutter Instrument Co.	
HPX-2000 Xenon light source	Ocean Optics	
White LED light source	Ocean Optics, LSM series	
P55 A.C. Preamplifier	Astro-Med. Inc., Grass Technology	
DAQ Board, SCC-68	National Instruments	
Software		
LabView	National Instruments	
Matlab vers. R2020b	MathWorks	

[a]*See basic buffers section for details on preparation.*

5.2 Before you begin

- Pull the ERG recording glass capillary using the needle puller instrument.
  - Adjust the following program parameters: "heat," "pull," "velocity," "time," and "pressure" in order to generate a needle with a diameter of 15–30 μm (see Fig. 4A for an example needle tip); these parameters must be readjusted every time the filament on the needle puller is changed. Make sure no cracks are visible in your needle.
- Fill the needle with E3 and remove all bubbles inside the needle. Place needle in the pipette holder and make sure the pipette tip is completely filled.
- Prepare the recording chamber using 1.0% *w/v* agarose in E3 medium.

5.3 Step-by-step method details

5.3.1 Eye dissection

Perform the following steps inside the Faraday cage:

- Dark-adapt 5 dpf larvae for at least 30 min.
- Use dim red light during dissection.
- Transfer a single larva onto the recording chamber in which the reference electrode is inserted.
- Remove one eye with a looped tungsten wire to cut the optic nerve.
- Position the eye to face the light source; example of light spectrum information is provided in (Zang et al., 2021). According to the light source used, the light spectrum may vary, thereby influencing the relative contributions of different cone types to the ERG response.

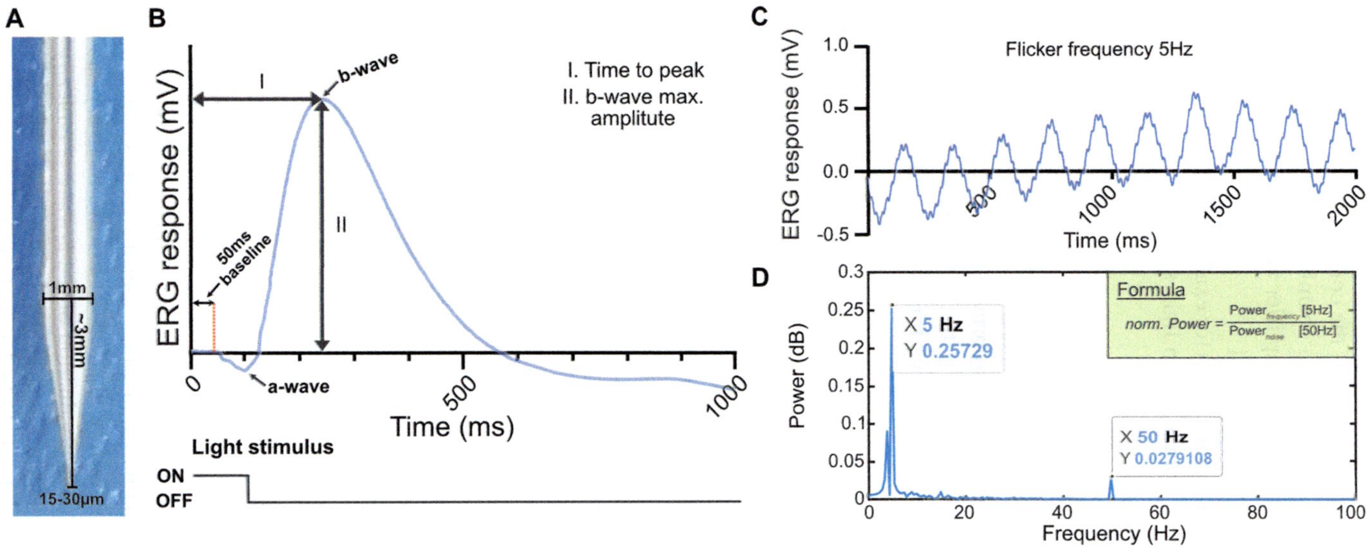

FIG. 4

Electroretinogram measurements in zebrafish. (A) Glass capillary used to measure the ERG responses with the approximate diameter and sizes in mm. (B) Illustration of a typical ERG response curve. Upon delivery of a 100 ms light stimulus (line below the graph) the photoreceptors hyperpolarize and generate the a-wave which is partly and sometimes fully covered by the emerging b-wave. Several parameters can be assessed form this curve such as the time to peak (I) and the b-wave maximum amplitude (II). The response is corrected by subtracting the average response of the first 50 ms (red line—baseline). (C) For the flicker ERG, the response to a flickering light stimulus (here 5 Hz) is measured. (D) Tansformation to power spectrum of the response shown in (C) by applying FFT in MATLAB. The power of the tested frequency (here 5 Hz) is normalized by the power of the line noise at 50 Hz (Formula is shown in the green box).

5.3.2 Measurement

- Place the recording pipette in the center of the cornea using a micromanipulator.
- Place a metal net in front of the recording chamber to reduce electrical noise.
- Electric signal is amplified by the pre-amplifier and digitized by DAQ Board.

5.3.2.1 Standard b-wave measurement

- Apply light flashes to stimulate the retina (for example, using LabView program to trigger flash stimulations from a Xenon light source):
    - Flash duration: 100ms;
    - Flash interval: 12s;
    - Flash intensities: 0.01%, 0.1%, 1%, 10% and 100% of maximal light intensity. Neutral density filters can be inserted over the light source to regulate light intensity. Alternatively, with certain light sources, it is possible to directly control the light intensity using an appropriate software. Measure the response with increasing light intensities (start with 0.01%) to minimize bleaching effects. A representative response curve for one light intensity is shown in Fig. 4B.

5.3.2.2 Flicker ERG measurement

To determine the temporal resolution of the retina, which gives an indication of the recovery time of the phototransduction cascade, the ERG in response to a flickering light stimulus can be recorded (Zang et al., 2021) (Fig. 4C and D). Flicker-ERG response fluctuates when the flickering light frequency is low but becomes continuous when the flickering frequency reaches around 20Hz in our set up. Note that if the amplitude of the b-wave is strongly decreased, in a mutant for example, it may become irrelevant to perform a flicker ERG, as the interpretation of the results may be limited.

- Program the white LED light source to produce different frequencies. Start the flickering frequency from 5Hz, increase frequency toward 20Hz (8, 10, 12, 15, and 18Hz) and assess the flicker ERG responses at various frequencies. An example of the light spectrum and the different flickering frequencies can be found in Zang et al. (2021).
- The response is recorded as above for the b-wave measurements.

5.3.3 Quantification and analysis

- The standard ERG measurement is based on the b-wave amplitude which represents the main readout. The baseline is defined as the average level 50ms after the flash is applied. The b-wave amplitude is quantified as the difference between the baseline and the maximal value after the flash. Other parameters such as b-wave time-to-peak or time for 50% of b-wave recovery can also be evaluated (Niklaus et al., 2017). Evaluation of the a-wave is typically not performed in zebrafish as the a-wave is hidden by the large b-wave and its evaluation requires pharmacological blockage of the b-wave.

- The flicker ERG response to a given frequency is transformed to power spectrum by Fast Fourier transform algorithm (FFT) in MATLAB: Normalize the power at the given frequency against the power of the background noise detected at 50 Hz. Compare the normalized power between (among) the different conditions.

6 Molecular examination of isolated photoreceptor outer segments

The primary cilia of photoreceptors, the outer segments, can be mechanically isolated and analyzed to decipher their protein or lipid composition. This untargeted approach allows an unbiased identification of the protein/lipid landscape of outer segments in healthy and diseased tissue. Mechanical isolation of the outer segments is followed by purification and enrichment using a non-continuous sucrose gradient.

6.1 Key resource table

Name	Company	Catalog #
Solutions		
MS222/Tricaine[a]	Sigma-Aldrich	#886-86-2
SYLGARD 184 Silicone Elastomer kit	Dow	#1673921
Protease inhibitor cOmplete Tablets	Roche	#04693159001
Sucrose	Merck	#1.07687
Material		
Protein LoBinding tubes	Eppendorf	#022431081
Meyhoefer Chalazion Curette: scoop shaped tip of 1.5 mm (#65-3170) or 1.8 mm (#65-3171) work best for adult zebrafish eyes	Sklar	#65-3170
VANNAS Scissors	E.M.S.	#72932-01
DumoStar fine tweezer	Dumont	#0209-55-PS-1
Paint brush: a single-hair brush is made by cutting all hairs off beside a single strand	Local supplier	NA
Sterile PS-Tube, 14 mL	Greiner Bio-One	#191161
Equipment		
Centrifuge 5810R	Eppendorf	
Swing rotor A-4-81	Eppendorf	
Centrifuge 5424R	Eppendorf	

[a]*see basic buffers section for details on preparation.*

6.2 Before you begin

6.2.1 Creation of silicon elastomer plates

- Mix SYLGARD® 184 base solution 10:1 with hardener and mix well
- Quickly spin the mixture at maximum speed in an appropriately sized centrifuge to remove bubbles in the viscous mixture.
- Pour Elastomer into Petri dish and let polymerize at 37 °C overnight.

6.3 Step-by-step method details

6.3.1 Retina dissection

- Dark-adapt adult fish for >3 h and work under red light during the entire dissection to ensure minimal RPE/outer segment interconnection.
- Clean a paintbrush in 1 M NaOH followed by a 5 min sonication step in water to remove residual dirt. Rinse the brush three times in ddH_2O and let it dry.
- Euthanize adult fish in ice cold MS222, remove both eyes and transfer them into a dish of ice cold 1× PBS on ice.
- Shorten the optic nerve as much as possible using a micro-dissecting spring scissor. Cut open the eye around the cornea and remove the lens together with the cornea (Fig. 5A). Transfer the sclera/retina into a new dish containing ice cold PBS. With two extra-fine tweezers pull softly at the edges of the white retina while pressing the sclera backwards. Remove remaining RPE cells from the retina and trash the sclera.

  Note: A scoop-headed curette with a 1.5 or 1.8 mm diameter can be used to hold the eye in place during the dissection. Remaining RPE on the retina can be removed by pushing and plucking at the edges of the RPE. The RPE will slowly start to detach in large chunks while most of the OSs remain attached to the retina (Fig. 5A).
- Transfer the retina into a 40 μL PBS droplet on the silicone elastomer plate and flatten the retina (outer segments facing upwards).
- Gently swipe over the retina to mechanically detach the OSs. Ensure that the whole retina is swiped to maximize the amount of isolated outer segments, while avoiding patches where RPE is still attached (Fig. 5A).
- Carefully remove the remaining retina now containing mainly neurons and inner segments; transfer the PBS droplet into a protein low-binding tube and store PBS containing the isolated OSs on ice. The amount and quality of the isolated outer segments can be inspected using light microscopy (Fig. 5B).

  Note: Differential interference contrast (DIC) microscopy gives the best result to look at the quality of the OSs.

6.3.2 Sucrose gradient

- Prepare 50% *w*/*v* Sucrose in PBS, sterile filtration.
- Prepare sucrose gradient:

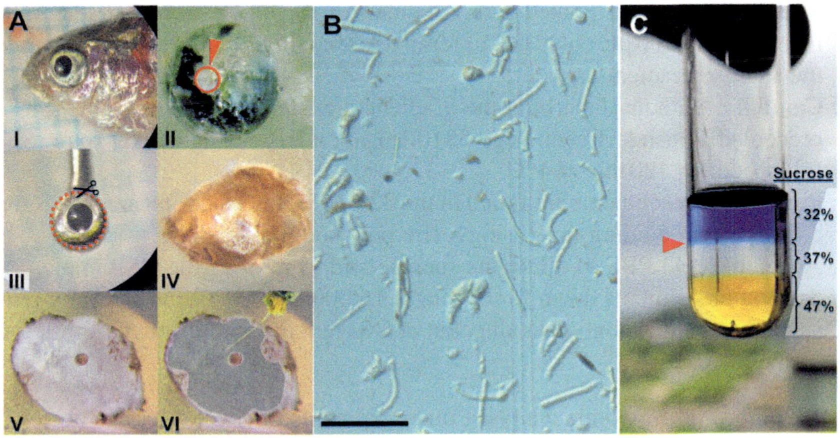

FIG. 5

Isolation of outer segments from adult zebrafish eyes. (A) shows an image series (I–VI) of the dissection steps of a zebrafish retina for mechanical isolation of OSs. The eye of an adult fish (I) is enucleated and stored in ice cold PBS. Next (II), the optic nerve (red arrow and circle) is cut as closely as possible to the eye cup. Using a scoop-headed curette the eye is positioned in place and cut open (III). The retina is removed using tweezers; Retinae where too much RPE is still attached as shown in (IV) are excluded from further OS isolation. Clean retinae, as shown in (V), are further processed and the OSs are mechanically removed by brushing over the clean, RPE pigment-free (gray) area, using a single-hair paintbrush (VI). The isolated OSs are collected and visually inspected using a bright field microscope as in (B). Subsequently, the OSs are purified by sucrose gradient centrifugation. (C) shows a pseudo-colored discontinuous sucrose gradient. The OS fraction is found at the interface between 32% and 37% sucrose (red arrow). Abbreviation: Outer segment (OS). Scale bar: B = 50 μm.

47% Sucrose	37% Sucrose	32% Sucrose
3 mL PBS 47 mL 50% Sucrose	13 mL PBS 37 mL 50% Sucrose	18 mL PBS 32 mL 50% Sucrose

- Create a sucrose gradient in a 15 mL centrifugation tube.

  Note: Using a fire polished glass pipette and carefully building the gradient starting from the top (32% sucrose) to the bottom (47% sucrose) layer has proven to be the most reliable way. Holding the tube against a strong light source allows to clearly visualize all three layers. Nonetheless, it is advised to mark the edges of the layers on the tube. For Fig. 5C the different layers were pseudo-colored to visualize the gradient.

- Slowly pipette the extracted outer segments on top of the sucrose gradient. Wash the tube with 100 μL cold PBS to remove remaining OSs and transfer them onto the sucrose gradient as well.
- Carefully (without disturbing the gradient) transfer the column to the 4 °C precooled Eppendorf centrifuge 5810R using a free swing bucket rotor. Spin at full speed for 120 min @4 °C
- Purified outer segments are found in the 32%/37% interface and are transferred to a low protein binding tube using a fire polished glass pipette.
- Add 1 mL cold PBS to dilute the sucrose and to retrieve the OS sample.
- Spin outer segments down for 60 min at 16′000rcf @ 4 °C and carefully remove supernatant so that only a small pellet of OSs remains. Store pellet at −80 °C until further use.

Note: For proteomic analysis, outer segments of several retinae need to be pooled. We achieved a sufficiently strong signal for LC-MS/MS analysis by pooling 7 retinae for the proteomic analysis and by pooling 2 retinae for lipidomics (Masek et al., 2022).

6.4 Quantification and statistical analysis

In general, the readout of an LC-MS/MS experiment contains protein/lipid identifiers and respective abundance values. Since the analysis of -omics data highly depends on the research question, we will not discuss the analysis steps in detail but rather provide useful tools and links to annotate and analyze the data. For proteomic data, additional annotation terms (such as symbol, description etc.) can be found based on the unique identifier. Be aware that the latest zebrafish genome annotation is less complete than for human and mouse, such that you may need to perform some extra work to identify the corresponding orthologue name. Web-based tools such as BioMart or the R-Package "biomaRt" from Ensembl are useful sources that link many different databases together and can assist in linking hits from zebrafish proteomic datasets to mouse and human orthologues. Fast translation of a unique identifier to another (e.g.: Ensembl to Entrez ID) can be done using the "*bitr*" command (Biological Id Translator) of the clusterProfiler R Package. Depending on the research question, a targeted analysis of the top hits from the differential abundance analysis is the easiest. To find novel and potentially unexpected information in the dataset, over representation analysis (ORA) is an easy start point. The web-based tool WEBgestalt allows users to find over-represented gene ontology terms or pathways within a list of significantly different proteins. Other popular sources for ORA are PantherDB or DAVID (see Table 2). For lipidomic data, the web-tool MetaboAnalyst is a powerful tool to run differential abundance analysis, enrichment or network analysis (see Table 2).

Table 2 Useful web resources for -omic hit annotations, gene ontology analysis and identifier transformation.

Databases	
Zebrafish Information Network (ZFIN)	https://zfin.org/
Ensembl BioMart to annotate hits found in omics analysis	https://www.ensembl.org/biomart/martview
WEB-based **GE**ne **S**e**T** **A**na**L**ysis **T**oolkit—Translating gene lists into biological insights (WEBGESTALT)	http://www.webgestalt.org/
Protein **AN**alysis **TH**rough **E**volutionary **R**elationships (PANTHER)	http://www.pantherdb.org/
Database for **A**nnotation, **V**isualization and **I**ntegrated **D**iscovery (DAVID)	https://david.ncifcrf.gov/
MetaboAnalyst	https://www.metaboanalyst.ca/

7 Conclusion

The protocols described here allow to evaluate the morphology, ultrastructure, protein, and lipid composition, as well as the function of zebrafish retinal photoreceptor outer segments. The comprehensive picture of this highly specialized ciliary compartment achieved by this combination of methods provides an excellent model for understanding human cilia-related retinal disease.

Acknowledgment

We would like to thank R. Molday, D. Hyde, J. Fadool, B. Perkins and T. Nicholson for sharing their antibodies with us. We also want to thank the Kawamura lab for sharing their transgenic fish lines. In addition, we would like to thank the technical staff of the Center for Microscopy and Image Analysis, in especial to Carmen Kaiser for sample preparation and sectioning.

Fundings

MM and RBG are supported by Swiss National Science Foundation grant # PP00P3_198895. SCFN, MG and JZ are supported by Swiss National Science Foundation grant #310030_204648. JM and UZ are supported by Swiss National Science Foundation grant #316030_145037.

References

Ahnelt, P. K., & Kolb, H. (2000). The mammalian photoreceptor mosaic-adaptive design. *Progress in Retinal and Eye Research*, *19*(6), 711–777. https://doi.org/10.1016/S1350-9462(00)00012-4.

Allison, W. T., Barthel, L. K., Skebo, K. M., Takechi, M., Kawamura, S., & Raymond, P. A. (2010). Ontogeny of cone photoreceptor mosaics in zebrafish. *Journal of Comparative Neurology*, *518*(20), 4182–4195. https://doi.org/10.1002/CNE.22447.

Bachmann-Gagescu, R., & Neuhauss, S. C. (2019). The photoreceptor cilium and its diseases. *Current Opinion in Genetics and Development*, *56*, 22–33. https://doi.org/10.1016/j.gde.2019.05.004. Figure 1.

Bachmann-Gagescu, R., Phelps, I. G., Stearns, G., Link, B. A., Brockerhoff, S. E., Moens, C. B., et al. (2011). The ciliopathy gene cc2d2a controls zebrafish photoreceptor outer segment development through a role in Rab8-dependent vesicle trafficking. *Human Molecular Genetics*, *20*(20), 4041–4055. https://doi.org/10.1093/hmg/ddr332.

Bilotta, J., Saszik, S., & Sutherland, S. E. (2001). Rod contributions to the electroretinogram of the dark-adapted developing zebrafish. *Developmental Dynamics*, *222*(4), 564–570. https://doi.org/10.1002/dvdy.1188.

Branchek, T., & Bremiller, R. (1984). The development of photoreceptors in the zebrafish, Brachydanio rerio. I. Structure. *Journal of Comparative Neurology*, *224*(1), 107–115. https://doi.org/10.1002/CNE.902240109.

Cardona, A., Saalfeld, S., Schindelin, J., Arganda-Carreras, I., Preibisch, S., Longair, M., et al. (2012). TrakEM2 software for neural circuit reconstruction. *PLoS One*, *7*(6), e38011. https://doi.org/10.1371/JOURNAL.PONE.0038011.

Carter-Dawson, L. D., & Lavail, M. M. (1979). Rods and cones in the mouse retina. I. Structural analysis using light and electron microscopy. *Journal of Comparative Neurology*, *188*(2), 245–262. https://doi.org/10.1002/cne.901880204.

Fadool, J. M. (2003). Development of a rod photoreceptor mosaic revealed in transgenic zebrafish. *Developmental Biology*, *258*(2), 277–290. https://doi.org/10.1016/S0012-1606(03)00125-8.

Hamaoka, T., Takechi, M., Chinen, A., Nishiwaki, Y., & Kawamura, S. (2002). Visualization of rod photoreceptor development using GFP-transgenic zebrafish. *Genesis*, *34*(3), 215–220. https://doi.org/10.1002/GENE.10155.

Hicks, D., & Molday, R. S. (1986). Differential immunogold-dextran labeling of bovine and frog rod and cone cells using monoclonal antibodies against bovine rhodopsin. *Experimental Eye Research*, *42*(1), 55–71. https://doi.org/10.1016/0014-4835(86)90017-5.

Kopek, B. G., Paez-Segala, M. G., Shtengel, G., Sochacki, K. A., Sun, M. G., Wang, Y., et al. (2017). Diverse protocols for correlative super-resolution fluorescence imaging and electron microscopy of chemically fixed samples. *Nature Protocols*, *12*(5), 916–946. https://doi.org/10.1038/nprot.2017.017.

Larison, K. D., & Bremiller, R. (1990). Early onset of phenotype and cell patterning in the embryonic zebrafish retina. *Development (Cambridge, England)*, *109*(3), 567–576. https://doi.org/10.1242/DEV.109.3.567.

Makhankov, Y. V., Rinner, O., & Neuhauss, S. C. F. (2004). An inexpensive device for non-invasive electroretinography in small aquatic vertebrates. *Journal of Neuroscience Methods*, *135*(1–2), 205–210. https://doi.org/10.1016/j.jneumeth.2003.12.015.

Masek, M., Etard, C., Hofmann, C., Hülsmeier, A. J., Zang, J., Takamiya, M., et al. (2022). Loss of the Bardet-Biedl protein Bbs1 alters photoreceptor outer segment protein and lipid composition. *Nature Communications*, *13*(1), 1–17. https://doi.org/10.1038/s41467-022-28982-6.

Mateos, J. M., Barmettler, G., Doehner, J., Naharros, I. O., Guhl, B., Neuhauss, S. C. F., et al. (2017). Correlative super-resolution and Electron microscopy to resolve protein localization in zebrafish retina. *JoVE (Journal of Visualized Experiments)*, *2017*(129), e56113. https://doi.org/10.3791/56113.

Morris, A. C., Schroeter, E. H., Bilotta, J., Wong, R. O. L., & Fadool, J. M. (2005). Cone survival despite rod degeneration in XOPS-mCFP transgenic zebrafish. *Investigative Ophthalmology & Visual Science*, *46*(12), 4762–4771. https://doi.org/10.1167/IOVS.05-0797.

Niklaus, S., Cadetti, L., Vom Berg-Maurer, C. M., Lehnherr, A., Hotz, A. L., Forster, I. C., et al. (2017). Shaping of signal transmission at the photoreceptor synapse by EAAT2 glutamate transporters. *ENeuro*, *4*(3). https://doi.org/10.1523/ENEURO.0339-16.2017.

Obholzer, N., Wolfson, S., Trapani, J. G., Mo, W., Nechiporuk, A., Busch-Nentwich, E., et al. (2008). Vesicular glutamate transporter 3 is required for synaptic transmission in zebrafish hair cells. *Journal of Neuroscience*, *28*(9), 2110–2118. https://doi.org/10.1523/JNEUROSCI.5230-07.2008.

Ojeda Naharros, I., Gesemann, M., Mateos, J. M., Barmettler, G., Forbes, A., Ziegler, U., et al. (2017). Loss-of-function of the ciliopathy protein Cc2d2a disorganizes the vesicle fusion machinery at the periciliary membrane and indirectly affects Rab8-trafficking in zebrafish photoreceptors. *PLoS Genetics*, *13*(12), 1–32. https://doi.org/10.1371/journal.pgen.1007150.

Oorschot, V., Lindsey, B. W., Kaslin, J., & Ramm, G. (2021). TEM, SEM, and STEM-based immuno-CLEM workflows offer complementary advantages. *Scientific Reports*, *11*(1), 1–16. https://doi.org/10.1038/s41598-020-79637-9.

Raymond, P. A., Barthel, L. K., & Curran, G. A. (1995). Developmental patterning of rod and cone photoreceptors in embryonic zebrafish. *Journal of Comparative Neurology*, *359*(4), 537–550. https://doi.org/10.1002/CNE.903590403.

Rust, M. J., Bates, M., & Zhuang, X. (2006). Sub-diffraction-limit imaging by stochastic optical reconstruction microscopy (STORM). *Nature Methods*, *3*(10), 793–796. https://doi.org/10.1038/nmeth929.

Schindelin, J., Arganda-Carreras, I., Frise, E., Kaynig, V., Longair, M., Pietzsch, T., et al. (2012). Fiji—an open source platform for biological image analysis. *Nature Methods*, *9*(7), 676–682. https://doi.org/10.1038/NMETH.2019.

Takechi, M., Hamaoka, T., & Kawamura, S. (2003). Fluorescence visualization of ultraviolet-sensitive cone photoreceptor development in living zebrafish. *FEBS Letters*, *553*(1–2), 90–94. https://doi.org/10.1016/S0014-5793(03)00977-3.

Takechi, M., Seno, S., & Kawamura, S. (2008). Identification of cis-acting elements repressing blue opsin expression in zebrafish UV cones and pineal cells. *Journal of Biological Chemistry*, *283*(46), 31625–31632. https://doi.org/10.1074/JBC.M806226200.

Tarboush, R., Chapman, G. B., & Connaughton, V. P. (2012). Ultrastructure of the distal retina of the adult zebrafish, Danio rerio. *Tissue and Cell*, *44*(4), 264–279. https://doi.org/10.1016/J.TICE.2012.04.004.

Tsujimura, T., Hosoya, T., & Kawamura, S. (2010). A single enhancer regulating the differential expression of duplicated red-sensitive opsin genes in zebrafish. *PLoS Genetics*, *6*(12), e1001245. https://doi.org/10.1371/JOURNAL.PGEN.1001245.

Vihtelic, T. S., Doro, C. J., & Hyde, D. R. (1999). Cloning and characterization of six zebrafish photoreceptor opsin cDNAs and immunolocalization of their corresponding proteins. *Visual Neuroscience*, *16*(3), 571–585. https://doi.org/10.1017/S0952523899163168.

Willoughby, J. J., & Jensen, A. M. (2012). Generation of a genetically encoded marker of rod photoreceptor outer segment growth and renewal. *Biology Open*, *1*(1), 30–36. https://doi.org/10.1242/BIO.2011016.

Young, R. W. (1967). The renewal of photoreceptor cell outer segments. *The Journal of Cell Biology*, *33*(1), 61–72. https://doi.org/10.1083/jcb.33.1.61.

Zang, J., Gesemann, M., Keim, J., Samardzija, M., Grimm, C., & Neuhauss, S. C. F. (2021). Circadian regulation of vertebrate cone photoreceptor function. *eLife*, *10*, e68903. https://doi.org/10.7554/ELIFE.68903.

Zimmermann, M. J. Y., Nevala, N. E., Yoshimatsu, T., Osorio, D., Nilsson, D. E., Berens, P., et al. (2018). Zebrafish differentially process color across visual space to match natural scenes. *Current Biology*, *28*(13), 2018–2032.e5. https://doi.org/10.1016/J.CUB.2018.04.075.

CHAPTER

Visualizing multiciliated cells in the zebrafish

8

Hannah M. Wesselman*, Allison E. Gatz, and Rebecca A. Wingert*

Department of Biological Sciences, Center for Stem Cells and Regenerative Medicine, Center for Zebrafish Research, Boler-Parseghian Center for Rare and Neglected Diseases, University of Notre Dame, Notre Dame, IN, United States

**Corresponding authors: e-mail address: hwesselm@nd.edu; rwingert@nd.edu*

Chapter outline

Methods in Cell Biology, Volume 175, ISSN 0091-679X, https://doi.org/10.1016/bs.mcb.2022.12.001
Copyright © 2023 Elsevier Inc. All rights reserved.

Abstract

Ciliated cells serve vital functions in the body ranging from mechano- and chemo-sensing to fluid propulsion. Specialized cells with bundles dozens to hundreds of motile cilia known as multiciliated cells (MCCs) are essential as well, where they direct fluid movement in locations such as the respiratory, central nervous and reproductive systems. Intriguingly, the appearance of MCCs has been noted in the kidney in several disease conditions, but knowledge about their contributions to the pathobiology of these states has remained a mystery. As the mechanisms contributing to ciliopathic diseases are not yet fully understood, animal models serve as valuable tools for studying cilia development and how alterations in ciliated cell function impacts disease progression. Like other vertebrates, the zebrafish, *Danio rerio*, has numerous ciliated tissues. Among these, the embryonic kidney (or pronephros) is comprised of both monociliated cells and MCCs and therefore provides a setting to investigate both ciliated cell fate choice and ciliogenesis. Considering the zebrafish nephron resembles the segmentation and function of human nephrons, the zebrafish provide a tractable model for studying conserved ciliogenesis pathways in vivo. In this chapter, we provide an overview of ciliated cells with a special focus on MCCs, and present a suite of methods that can be used to visualize ciliated cells and their features in the developing zebrafish. Further, these methods enable precise quantification of ciliated cell number and various cilia-related characteristics.

1 Introduction

Cilia are organelles present on the apical surface of most vertebrate cells. These hair-like projections are comprised of >500 proteins that are arranged into several distinct structural zones known as the ciliary tip, microtubule-based axoneme, transition zone, and at their base, a centriole-derived basal body (Fliegauf, Benzing, & Omran, 2007; Ishikawa, Ueno, Omori, & Kikuchi, 2021). Primary cilia, which exist as a single immotile cilium, are nearly ubiquitous on body cells and perform roles including mechano-sensing, chemo-sensing, and signal transduction (Satir & Christensen, 2007). Motile cilia, however, are less common, and exist as either a single cilium or in bundles comprised of dozens to hundreds of cilia on so-called multiciliated cells (MCCs) (Boutin & Kodjabachian, 2019; Lewis & Stracker, 2021). The motility of these organelles is derived from dynein arms that periodically bind to additional neighboring doublets. Many motile cilia are also associated with an additional central doublet of microtubules within the axoneme, commonly referred to as a "9+2" configuration. Dynein arms from the outer tubules can bind to this inner doublet. While motile cilia can also contribute to cell signaling and signal transduction like their primary counterparts, the added motility enables fluid propulsion and drives the local movement of secreted signaling molecules (Davis, Brueckner, & Katsanis, 2006; Marra, Li, & Wingert, 2016; Vincensini, Blisnick, & Bastin, 2011).

These functionalities make cilia key players in developmental processes and the homeostasis of multiple tissues (Oh & Katsanis, 2012; Sharma, Berbari, & Yoder, 2008; Sreekumar & Norris, 2019). For example, MCCs are present in the central nervous system where they drive the movement of cerebrospinal fluid, in the respiratory system to aid in movement of mucous, and in reproductive organs where they facilitate the movement of gametes (Choksi, Lauter, Swoboda, & Roy, 2014; Spassky & Meunier, 2017). As cilia are essential in various functions, aberrant cilia formation or function can have catastrophic outcomes. Motile cilia facilitate left-right patterning, and altered ciliogenesis results in laterality defects (Fliegauf et al., 2007). Similarly, loss of MCC function can lead to infertility, hydrocephalus, and chronic respiratory disease (Fliegauf et al., 2007). Kidney epithelial cells possess primary cilia, and in the case of their disruption, have been extensively linked to disease conditions with characteristic cystic phenotypes (Bagnat & Gray, 2020; McConnachie, Stow, & Mallett, 2021; Pazour, Quarmby, Smith, Desai, & Schmidts, 2020; Waters & Beales, 2011). While typically absent from healthy kidney tissue, MCCs have been found in some disease states (Datsis & Boman, 1974; Duffy & Suzuki, 1968; Katz & Morgan, 1984; Larsen & Ghadially, 1974; Lungarella, de Santi, & Tosi, 1984; Ong & Wagner, 2005; Pazour et al., 2020).

Recent research has made tremendous strides in understanding the mechanisms of ciliopathic disease progression. Results of such studies have identified key cilia proteins as well as the components of ciliogenic transcriptional programs that contribute to proper cilia formation and function (Reiter & Leroux, 2017). Mammalian ciliated cells are often difficult to access for analyses as they occupy deeper tissues, like the central nervous system, and researchers have capitalized on several models for studying cilia formation and function. Such models include, but are not limited to, planarian epidermis, cultured rodent cells, derivation from induced pluripotent stem cells, *Xenopus* embryonic epidermis, and the zebrafish (Basquin, Orfila, & Azimzadeh, 2015; Corkins, Krneta-Stankic, Kloc, & Miller, 2021; Firth et al., 2014; Grondona, Granados-Durán, Fernández-Llebrez, & López-Ávalos, 2013; Lee & Ostrowski, 2021; Legendre, Zaragosi, & Mitchison, 2021; Morales & Wingert, 2017; Poureetezadi & Wingert, 2016). These models allow for live imaging of cilia beating patterns and in vivo models specifically provide the additional benefit of studying cilia in the context of entire tissues.

Here, we outline a suite of methods tailored for the study of MCCs in the zebrafish model. Work by many labs has revealed that zebrafish embryos exhibit conserved function and localization of MCCs with other vertebrates including mammals. Due to such characteristics like high fecundity, transparent development through early embryonic stages, and conservation of approximately 70% of genes with humans, the zebrafish offer an appealing animal model for studying topics such as MCC specification, ciliogenesis, ciliary function, and ciliopathies (Corkins et al., 2021; Howe et al., 2013; Kimmel, Ballard, Kimmel, Ullmann, & Schilling, 1995; Kramer-Zucker et al., 2005; Li, Xu, Jerman, & Sun, 2019; Malicki, Avanesov, Li, Yuan, & Sun, 2011; Marra et al., 2016).

2 Zebrafish models of ciliopathic disease

Similar to cilia found in mammals, cilia found in zebrafish exhibit either the "9+0" or "9+2" configuration. Motile cilia are found in the Kupffer's Vesicle (KV), which is analogous to the mammalian node, the pronephros, and the eye (Corkins et al., 2021). More specifically, MCCs are found in the central nervous system, ear structures, lateral line system, olfactory placode, and the embryonic kidney (pronephros) (Leventea, Hazime, Zhao, & Malicki, 2016). The pronephros is of particular interest, as MCCs are interspersed among monociliated cells within several tubule segments and monociliated cells alone comprise the distal-most segment and duct (Fig. 1A) (Liu, Pathak, Kramer-Zucker, & Drummond, 2007; Ma & Jiang, 2007; Marra & Wingert, 2016). This offers a particularly interesting model to study the development of monociliated cells and MCCs simultaneously in a single organ, and furthermore their collective functions related to fluid clearance can be monitored with several dynamic physiological assays (Bolten, Pratsinis, Alter, Fricker, & Huwyler, 2022).

Similar to ciliopathies in humans, aberrant ciliogenesis often affects multiple tissues in the zebrafish. Deficiency in early patterning cilia often results in *situs inversus* of the heart as well as overall body curvature phenotypes (Bisgrove, Makova, Yost, & Brueckner, 2012; Fliegauf et al., 2007). Interestingly, few ciliopathies are linked to "curl up" tail phenotypes, like *pkd* knockdowns, while majority consist of "curl down" phenotypes, like those characteristically found in *pgc1a* mutants (Bagnat & Gray, 2020; Chambers, Addiego, Flores-Mireles, & Wingert, 2020;

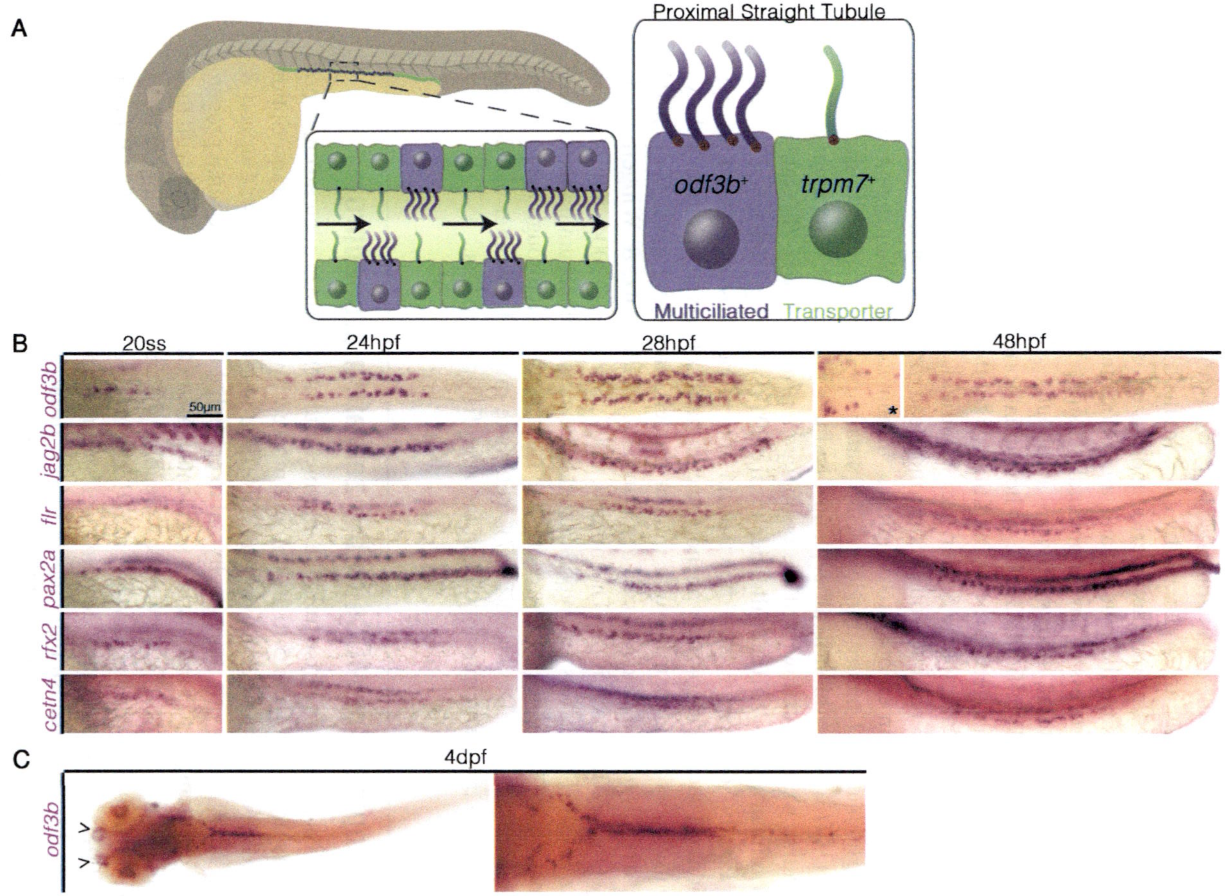

FIG. 1

MCCs visualized using WISH. (A) Schematic of a zebrafish embryo at 24 hpf. Green cells illustrate those of transporter cell fate, and blue/purple cells illustrate those of a multiciliated cell fate. The inset shows the salt-and-pepper like distribution of MCCs among primary epithelium. For example, in the proximal straight tubule (boxed to the right) is comprised of *odf3b* expressing MCCs and *trpm7* expressing transporter cells. (B) MCC markers, stained via WISH, are shown in purple at various developmental stages. Animals marked with *odf3b* were imaged dorsally, as this gene is not expressed in any part of the dorsal tissues, allowing for clear visualization. Asterisk denotes more proximal image of *odf3b* at 48 hpf in which MCCs exist in the curved portion of the proximal convoluted tubule. All other markers were imaged at a tilt, to allow for visualization of the ciliated cells while avoiding interference from other tissues. The number of MCCs increases over time. (C) 4 dpf WT animal stained with *odf3b* via WISH. MCCs are visible in the nasal placodes (arrowheads) and throughout the nephron tubule (right panel).

Chang et al., 2017; Obara et al., 2006; Omori et al., 2008; Shi, Su, Lipschutz, & Lobo, 2017; Sullivan-Brown et al., 2008; Sun et al., 2004; van Rooijen et al., 2008). Ciliopathic models may also exhibit aberrant structures in the ear-like structure or otic vesicle. The otic vesicle is comprised of two otoliths, which are protein and mineral aggregates, that are organized by cilia. In this tissue, cilia first arise around 18 h post fertilization (hpf) in the anterior and posterior region of the otic vesicle. These structures become the macula and are associated with the otoliths, which are essential in the process of environmental sensing. The cristae appear in three patches (anterior, lateral, and posterior) at about 3 dpf and further assist in mechano-sensation (Whitfield, 2020). When cilia formation is altered, otoliths can be absent, fused together, or an additional otolith may be present (Stooke-Vaughan, Huang, Hammond, Schier, & Whitfield, 2012; Whitfield, 2020). Aberrant cilia in the central nervous system may lead to hydrocephaly, a phenotype commonly observed in ciliopathic conditions in humans and other mammals. For example, humans and mouse models with mutations in the critical transcription factor, *FOXJ1*, often exhibit hydrocephalus, and this phenotype is recapitulated in the zebrafish with knockdowns of the orthologs *foxj1a* and *foxj1b* (D'Gama et al., 2021; Hellman et al., 2010; Wallmeier et al., 2019; Yu, Ng, Habacher, & Roy, 2008). Finally, decreased cilia function in the pronephros can result in cyst formation, yet another ciliopathic phenotype commonly observed in human kidneys. Pronephric cysts develop in several mutant models, including *ift88*, *fleer*, *pkd2*, *ruvbl1*, *lrcce6l*, *arl13b*, and *ppargc1a* (Chambers, Clark, Gatz, & Wingert, 2020; Pathak, Obara, Mangos, Liu, & Drummond, 2007; Sun et al., 2004; Tsujikawa & Malicki, 2004). These cysts are hypothesized to be caused by accumulation of fluid and expansion of the tubules, due to defective fluid propulsion through the pronephros (Naylor, Qubisi, & Davidson, 2017). Yet, there may still be other mechanisms contributing to ciliogenic phenotypes in the zebrafish pronephros.

3 Development of MCCs in the zebrafish pronephros

The zebrafish pronephros arise from the intermediate mesoderm, a bilateral cluster of progenitors organized from 5 to 10 hpf and marked by expression of transcription factor *pax2a* (Drummond & Davidson, 2016; Drummond et al., 1998; Gerlach & Wingert, 2013). Following a mesenchymal to epithelial transition, the renal progenitor cells differentiate to form distinct segments that are highly conserved with their mammalian counterparts (Desgrange & Cereghini, 2015; Gerlach & Wingert, 2014; McKee, Gerlach, Jou, Cheng, & Wingert, 2014; Wingert & Davidson, 2008; Wingert et al., 2007). These segment domains—glomerulus, neck, proximal convoluted tubule (PCT), proximal straight tubule (PST), distal early (DE), and distal late (DL)—are established and partly differentiated by 24 hpf (Chambers, Clark, et al., 2020; Chambers et al., 2019; Chambers, Poureetezadi, Addiego, Lahne, & Wingert, 2018; Cheng & Wingert, 2015; Drummond, Li, Marra, Cheng, & Wingert, 2016;

Kroeger et al., 2017; Morales et al., 2018; Naylor, Chang, Qubisi, & Davidson, 2018; Naylor, Przepiorski, Ren, Yu, & Davidson, 2013; Naylor et al., 2016; O'Brien et al., 2011; Outtandy, Russell, Kleta, & Bockenhauer, 2019; Poureetezadi, Cheng, Chambers, Drummond, & Wingert, 2016; Vasilyev et al., 2009; Wingert & Davidson, 2008, 2011; Wingert et al., 2007). MCCs are distributed in a "salt and pepper" fashion among the mono-ciliated transporter cells of the PCT, PST, and DE by 24 hpf. Notch signaling is critical in achieving this unique distribution (Li, Cheng, Verdun, & Wingert, 2014; Liu et al., 2007; Ma & Jiang, 2007). As early as 18 somite stage (ss), *jag2b* (a Notch ligand) is expressed in the intermediate mesoderm in a similar speckled pattern (Liu et al., 2007; Ma & Jiang, 2007; Marra et al., 2019). This ligand is excreted to the surrounding cells, activating downstream Notch pathways, resulting in lateral inhibition of MCC fate. Inhibition of *jag2b* or *notch3* results in trans-fating of the pronephros, where the majority of cells adopt an MCC fate (Liu et al., 2007; Ma & Jiang, 2007). Interestingly, *jag2b* and *odf3b* are also co-expressed with *pax2a* as it restricts caudally, further supporting that MCCs are gradually differentiated through development (Marra, Adeeb, et al., 2019). Also, *jag2b* expression persists through 48 hpf, as MCC development continues even after the pronephros are fully patterned at 24 hpf (Fig. 1B, Liu et al., 2007; Ma & Jiang, 2007; Marra et al., 2019).

Development of MCCs is further controlled by activity of essential transcription factors and co-activators. Highly conserved factors such as *rfx2, e2f5, etv5a, irx2a, mecom,* and *foxj1a* have been identified as positive regulators of MCC fate in the pronephros (Chong, Zhang, Zhou, & Roy, 2018; Dubruille et al., 2002; Li et al., 2014; Liu et al., 2007; Marra, Cheng, et al., 2019; Marra et al., 2016; Xie et al., 2020; Yu et al., 2008). These transcription factors are accompanied by the expression of other key factors, including *gmnc, multicilin, myb, and mcidas* (Stubbs, Vladar, Axelrod, & Kitner, 2012; Tan et al., 2013; Zhou et al., 2015, 2020). Gmnc has even been noted as a master regulator of MCC fate, and works upstream of *multicilin* and *myb* (Tan et al., 2013; Zhou et al., 2015). It is hypothesized that these regulators are required for cells to activate expression of more organizational and structural components of ciliogenesis (Lewis & Stracker, 2021). In the zebrafish, these structural markers include *odf3b*, *cetn4*, and *flr,* among others. Interestingly, while many ciliogenic factors are conserved across vertebrates, some factors exhibit divergent functions. For example, Mcidas/MCDIS is obligatory for MCC development in *Xenopus laevis* and mice, but is not essential in zebrafish (Zhou et al., 2020). Conversely, Gmnc has even been noted as a master regulator of MCC fate, and works upstream of *multicilin* and *myb* (Tan et al., 2013; Zhou et al., 2015). Additional factors not explicitly expressed in MCCs at 24 hpf also play a role in MCC specification and cilia outgrowth. This includes regulators required for prostaglandin signaling (*ptgs1, ptgs2*), *ppargc1a*, and *esrrga*, suggesting that there may be more dynamic signaling mechanisms at play that have yet to be elucidated (Chambers, Addiego, et al., 2020; Jin et al., 2014; Marra, Adeeb, et al., 2019; Wesselman, Flores-Mireles, & Wingert, 2021; Wesselman, Nguyen, Chambers, Drummond, & Wingert, 2023).

4 Methods to assess MCC development

To study novel mechanisms contributing to MCC formation, it is imperative to visualize cilia and ciliated cells. Here, we describe several methods that can be implemented as powerful approaches for studying cilia development and ciliopathic disease progression. Whole mount in situ hybridization (WISH) allows for detection of gene transcripts and analysis of cell identity. In the following text we have included WISH protocols conducted for MCC markers, which allows for quantification of MCC number (Fig. 1). To further explore cell identity, we outline a protocol for fluorescent in situ hybridization (FISH). This allows for visualization of transcripts that are co-expressed in the same cell (e.g., *odf3b* and *pax2a*) and to resolve when transcripts are expressed in adjacent cell types (e.g., *odf3b* and *trpm7*) transcripts that would be difficult to discern using WISH (Fig. 2). We also define the steps of whole mount immunofluorescence (IF), which allows for visualization of ciliary proteins and their structural arrangements into cilia and basal bodies in various tissues, including the pronephros and otic vesicle (Fig. 3). While whole mount IF is useful to detect and quantify basal bodies as well as cilia length, it is less suitable for viewing nephron tubule dilation or apical-basal polarity. For this reason, we also describe the methods to cryosection animals and visualize MCCs using IF on cross-sections of the embryo (Fig. 4). Together, with the quantification methods we describe, these methods allow researchers to study the development, structure, and function of multiciliated cells. Techniques to assess ciliary ultrastructure in the zebrafish have been nicely described elsewhere (Jaffe, Thiberge, Bisher, & Burdine, 2010).

4.1 Embryo sample preparation and fixation

4.1.1 Materials and reagents

(1) E3 embryo media: 50× solution consists of 250 mM NaCl, 8.5 mM KCl, 16.5 mM $CaCl_2$, 16.5 mM $MgSO_4$ in distilled water. Dilute this to 1× for working solution, and add 5 mL of 0.1% methylene blue (Sigma-Aldrich, M9140). Store at room temperature.

(2) Pronase: From *Streptomyces griseus* (Roche, 11459643001). Dilute in E3 without methylene blue to make 50 mg/mL stock solution. Store at −20 °C.

(3) PBST: Dilute 10× PBS (American Bioanalytical, AB11072) to 1× for working solution. Add Tween-20 to 0.1%. Store at room temperature.

(4) 4% Paraformaldehyde (PFA): Dissolve powder to make 4% PFA (wt/vol) (Electron Microscopy Services, 19210) in 1× PBS, and bring to a boil under a fume hood. Cool to room temperature, then filter sterilize using a 0.45 μm filter system (Corning, 430516). Aliquot and store at −20 °C. Thaw fresh and chill on ice for fixation. Do not refreeze.

(5) 100% Methanol (MeOH, BDH 1135).

(6) Embryo dishes, 10,020 mm (Greiner Bio-One, 82050-918).

(7) Tricaine®-S (pharmaceutical grade tricaine methanesulfonate (TMS) also known as MS-222) is an FDA approved (ANADA 200-226) fish anesthetic.

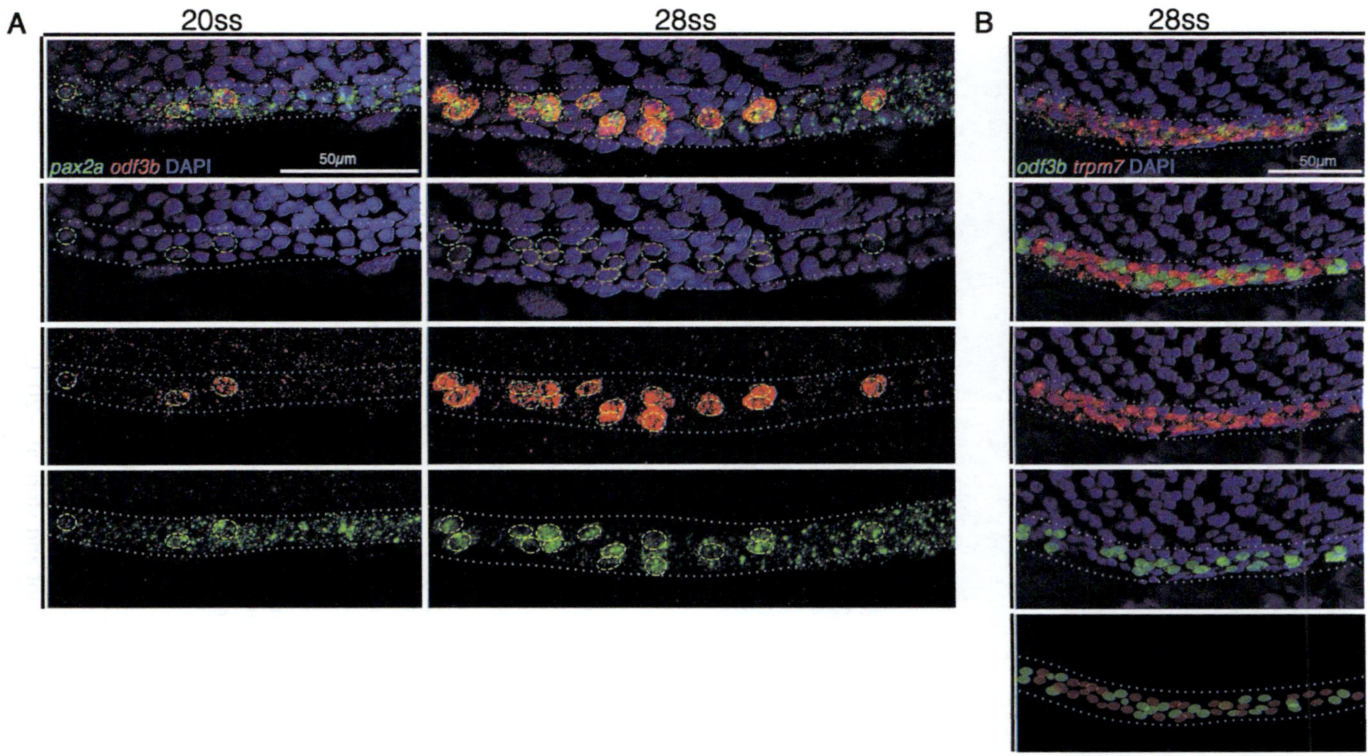

FIG. 2

MCCs and transporter cells analyzed using FISH. (A) FISH conducted with green (*pax2a*) and red (*odf3b*) staining at the 20 ss (left) and 28 ss (right). Cells that are co-expressing these markers are denoted with a dashed yellow circle. The white dashed line outlines the nephron. The top row is the merged image, while the second row is DAPI (blue) alone, the third row is *odf3b* (red) alone, and the bottom row is *pax2a* (green) alone. (B) MCCs are marked in green (*odf3b*) and transporter cells are marked in red (*trpm7*), stained via FISH. The top panel includes the merged channels with nuclei in blue (DAPI). The second panel includes the merged channel with a space filling overlay made in Adobe Illustrator. Each oval represents a DAPI stain, while the associated color denotes its expression of either *odf3b* or *trpm7*, as these markers are not co-expressed. The second and third panels are the separated channels with the space filling overlay for the corresponding color. The bottom panel is the space filling model overlay alone. This method of creating separate color-coded overlays can facilitate cell count quantifications.

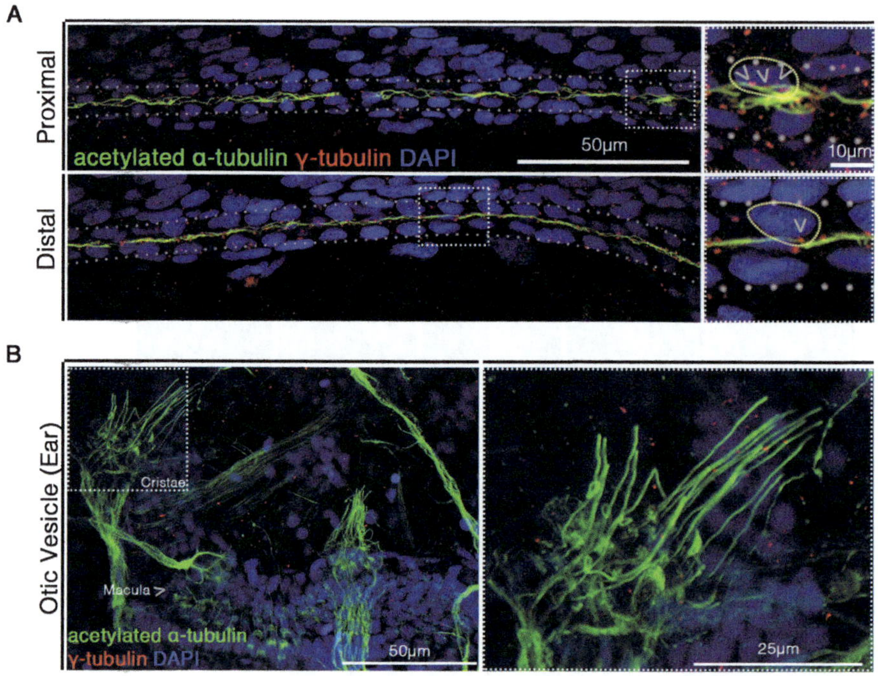

FIG. 3

Cilia of MCCs in the pronephros and ear visualized with IHC. (A) IHC performed on a 28 hpf embryo with anti-acetylated alpha tubulin (green) to mark cilia, anti-gamma tubulin (red) to mark basal bodies, and DAPI (blue) to mark nuclei. MCCs are interspersed among primary epithelium in the proximal segments (top) and can be identified by the presence of cilia bundles and multiple basal bodies per nuclei stain (top, inset). Conversely, transporter cells occupy the distal portion of the nephron and are monociliated (bottom). These cells can be identified by the decreased fluorescent intensity of anti-acetylated alpha tubulin and the presence of only one basal body per nuclei (bottom, inset). The dotted box denotes the region magnified in the inset. The yellow dotted line circles a single DAPI cell, and the arrowheads denote the basal bodies associated with the single nuclei. (B). IHC performed on a 4 dpf embryo with anti-acetylated alpha tubulin (green) to mark cilia, anti-gamma tubulin (red) to mark basal bodies, and DAPI (blue) to mark nuclei. The ear structure may be identified by the characteristic cristae and macula. The anterior macula is labeled with an arrowhead, and the anterior crista are surrounded by a dotted boxed line. The anterior crista is magnified in the inset (right).

(8) Flat-bottom microcentrifuge tube (VWR; 87003-300; 87003-298).
(9) Glass vials (VWR, 225012).
(10) Proteinase K (Roche, 3115879001): Dissolve in distilled water to make a 10 mg/mL stock; aliquot and store at −20 °C.
(11) Transfer pipet (Samco, 204) or another pipet with wide tip.

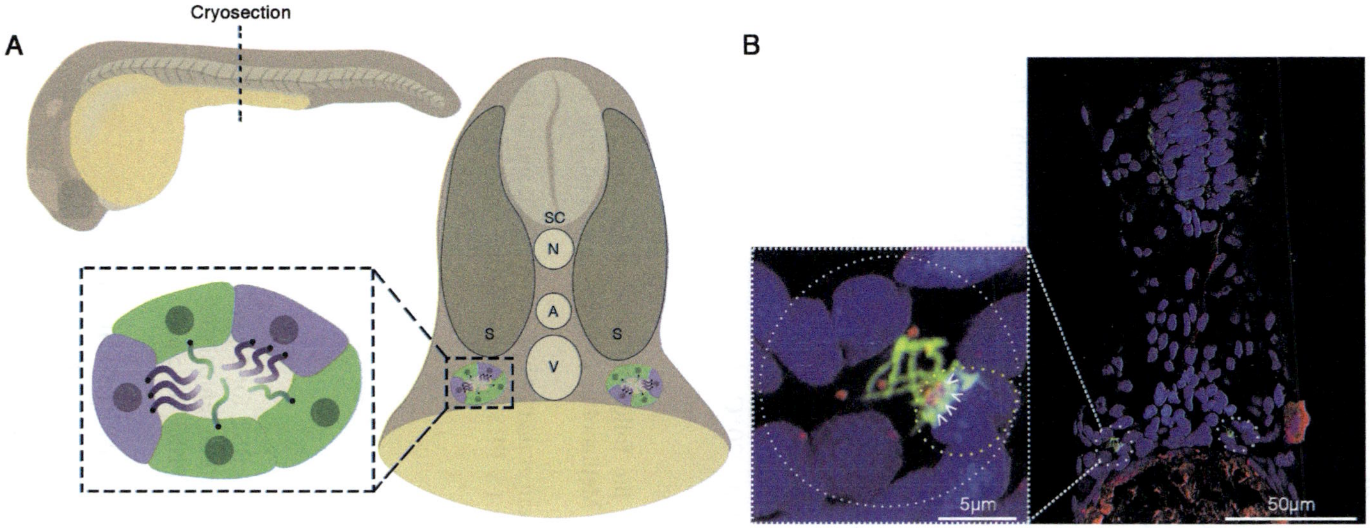

FIG. 4

MCCs within the pronephric tubule visualized by performing cryosectioning followed by IHC. (A) Schematic illustrating a 24 hpf zebrafish embryo and approximate section location. The sectioned view of the animal denotes the spinal cord (SC), somites (S), neural tube (N), aorta (A), vein (V), and pronephros (purple and green cells). The inset contains a magnified view of the intermediate pronephros, which are comprised of MCCs (purple), and monociliated transporter cells (green). (B) IHC of a cryosection from the intermediate nephron using with anti-acetylated alpha tubulin (green) to mark cilia, anti-gamma tubulin (red) to mark basal bodies, and DAPI (blue) to mark nuclei. The nephrons are circled with a dotted white line, and the inset shows a magnified view. The yellow dotted line denotes a single DAPI cell, and the arrowheads show the basal bodies associated with the single multiciliated DAPI cell.

4.1.2 Collection and fixation

(1) Collect embryos and store in E3 at 28.5 °C until they reach the developmental stage of interest.
(2) Remove the embryo chorion using forceps (suitable for animals younger that the 20 ss) or pronase. Using the latter method, add 100 μL of pronase per 5 mL of E3. Incubate at room temperature for approximately 10 min, while intermittently swirling the dish to gently agitate the samples.
(3) Rinse embryos twice with fresh E3 media.
(4) Add 1 mL of 0.2% tricaine per 5 mL of E3 to euthanize animals.
(5) Transfer sample to flat bottom glass vial or microcentrifuge tube.
(6) Remove E3-tricaine mixture, and replace with freshly thawed ice cold 4% PFA.
(7) Fix for a minimum of 2 h at room temperature or overnight at 4 °C.
(8) If using these samples for immunohistochemistry, samples can proceed directly to acetone permeabilization. If using these samples for WISH, FISH, or long-term storage, samples must be moved to MeOH.
(9) Remove fixative from samples.
(10) Wash embryos 2 times in 1× PBST for 5 min, followed by 2 washes in 100% MeOH for 5 min.
(11) Incubate embryos at −20 °C in 100% MeOH for at least 20 min (overnight incubation is preferred). Embryos can be stored for days to several months in 100% MeOH at −20 °C.

4.1.3 Permeabilization

(1) Rehydrate embryos with the following series of washes: 50% Methanol in 1× PBST for 5 min, 30% Methanol in 1× PBST for 5 min, and 100% 1× PBST twice for 5 min.
(2) Permeabilize embryos younger that 24 hpf with 5 μg/mL and 24 hpf or older with 10 μg/mL proteinase K in 1× PBST. The digestion times are as follows:
   1. 5 ss: 1 min
   2. 10–12 ss: 1.5 min
   3. 15 ss: 2 min
   4. 20 ss: 3 min
   5. 24 hpf: 4 min
   6. 36 hpf: 5–7 min
   7. 48 hpf: 10 min
(3) Wash embryos twice in 1× PBST for 5 min.
(4) Fix embryos in 4% PFA at room temperature for 20 min.
(5) Wash embryos twice in 1× PBST for 5 min.

4.2 Whole mount in situ hybridization (WISH)

WISH allows for the spatial visualization of gene transcripts at various developmental stages (Cheng, Li, Marra, Verdun, & Wingert, 2014; Femino, Fay, Fogarty, & Singer, 1998; Hauptmann & Gerster, 2000; Marra et al., 2019; Marra, Ulrich, White, Springer, & Wingert, 2017; Thisse & Thisse, 2008). As seen in Fig. 1, this method allows for visualization of MCCs in the embryonic kidney. The zebrafish pronephros arises from intermediate mesoderm precursor cells and is fully segmented into discrete areas of unique cell types by 24h post fertilization (hpf). The embryonic kidney begins filtering the blood around 48 hpf, as the podocyte precursors of each nephron migrate to the midline and undergo morphogenesis events to form a single glomerulus (Gerlach & Wingert, 2013). These structures continue to change overtime, as the proximal cells migrate rostrally, resulting a hook-like structure. MCCs are interspersed among primary epithelial transporter cells, which are monociliated (Fig. 1A). Interestingly, MCCs appear throughout development, as expression of MCC markers initiates around the 20 ss (Fig. 1B). This pattern readily allows for quantification of MCC number, and we have found that MCC number increases as the animal ages (Marra, Cheng, et al., 2019). Furthermore, MCCs also inhabit more proximal segments at later stages, and can be detected throughout the coiled proximal tubule segment beginning at approximately 48 hpf and persisting through 4 dpf (Fig. 1C). Here, we describe the methods used to visualize various MCC markers within the zebrafish embryo.

4.2.1 Materials and reagents

(1) HYB+: 50% formamide, 5× SSC, 0.1% Tween-20, 5 mg/mL yeast torula RNA, 50 μg/μL heparin.
(2) MAB: Into 1.5 L of distilled water, combine 23.2 g maleic acid, 17.5 g NaCl, 55.0 g Trisma base, 8 mL of 1 M Tris-HCl (pH 9.5), and fill to 2 L.
(3) MABT: MAB + 0.1% Tween-20.
(4) Fetal Calf Serum (FCS): Thaw at 37 °C (even if already heat-inactivated) then incubate at 55 °C for 4 h before aliquoting.
(5) Bovine Serum Albumin (BSA): Dissolve powder to make a 10% weight/volume stock in MAB.
(6) Antibodies (anti-DIG [Fab fragments]; anti-FLU [Fab fragments]) (Sigma-Aldrich).
(7) Staining Buffer: Freshly combine 100 mM Tris pH 9.5, 50 mM MgC12, 100 mM NaCl, 0.1% Tween.
(8) Dimethylformamide (DMF).
(9) NBT: Dilute powder at 50 mg/mL in 70% DMF, 30% water.
(10) BCIP: Dilute powder at 50 mg/mL in 100% DMF.
(11) INT: Dilute powder at 55 mg/mL in 70% DMF, 35% water.

4.2.2 Riboprobe production

(1) Combine on ice: 1.5 μL DNA template (linearized plasmid) or 100–200 ng PCR product, 12.5 μL distilled, molecular biology grade water, 2.0 μL RNA labeling mix, either DIG or FLUOR. Note: for experiments that aim to visualize two markers, use two different epitope labels.
(2) Vortex to mix the components.
(3) Incubate RNA synthesis for 2 h at 37 °C.

(4) To destroy the DNA template, add 35 μL of molecular biology grade water, 5 μL of 10× DNaseI buffer (thaw first at 37 °C to get into solution), 2 μL of DNaseI enzyme.
(5) Vortex to mix the components.
(6) Incubate for 30 min at 37 °C.
(7) To purify RNA, add 5 μL (1/10th volume) of 3 M NaAcetate, 110 μL (2 volumes) of ice-cold 100% Ethanol.
(8) Incubate tube in −20 °C for 1 h or overnight.
(9) Pellet RNA by centrifuging at 4 °C for 20 min at maximum speed.
(10) Remove supernatant.
(11) Wash with 200 μL of ice-cold 70% EtOH, then spin tube again at 4 °C for 5 min.
(12) Remove supernatant, avoiding RNA pellet. Dry briefly at room temperature.
(13) Add 100 μL of molecular-grade water to resuspend the RNA and store at −20 °C.

4.2.3 Probe hybridization

(1) Remove the 1× PBST from embryo samples. Add 500–1000 μL of HYB+. Reminder: samples should be in flat bottom tubes. This is essential for even distribution of probe and wash solutions.
(2) Incubate embryos in HYB+ to prehybridize the samples for at least 4 h at 70 °C.
(3) While samples are incubating, prepare RNA probes. Add 500 ng of RNA antisense stock in 500 μL of HYB+. This mixture should contain all probes needed for detection.
(4) Remove the prehybridization HYB+ solution from the sample tubes.
(5) Add enough probe/HYB+ so that all embryos are covered.
(6) Incubate samples overnight at 70 °C.

4.2.4 Hot washes and antibody labeling

(1) Remove the riboprobe and save in microcentrifuge tube in the −20 °C. Riboprobes can be reused 2–4 times. After several uses, consider adding an additional 500 ng of probe to improve signal.
(2) Wash embryos 2 times each for 30 min with 50% formamide/2× SSCT at 70 °C.
(3) Wash embryos 1 time for 15 min with 2× SSCT at 70 °C.
(4) Wash embryos 2 times each for 30 min with 0.2× SSCT at 70 °C.
(5) Wash embryos 2 times in MABT for 5 min each at room temperature.
(6) Prepare fresh block solution by thawing BSA and FCS at 37 °C, and mix ingredients with MABT to final volume.
(7) Remove the MABT and replace with block solution. Incubate embryos in block solution for a minimum of 2 h at room temperature. For multiciliated cell markers, we recommend blocking more than 4 h at room temperature or overnight at 4 °C.
(8) Add corresponding antibody (we recommend labeling DIG first as it typically lasts longer) diluted 1:5000 in block and incubate for 4 h at room temperature or overnight at 4 °C. Cover the tubes in foil to protect from light.

4.2.5 Probe detection

(1) Remove antibody/block solution.
(2) Wash embryos between 8 and 10 times for 15 min each in MABT. Keep embryos shielded from light during all washes. For multiciliated cell markers, we recommend completing 2–4 washes at room temperature, followed by an overnight incubation at 4 °C.
(3) Prepare fresh staining buffer. Wash embryos 1 time in staining buffer.
(4) Using transfer pipet, transfer embryos in the staining buffer to 12-well staining dish so the stain reaction can be monitored over time.
(5) Remove the staining buffer and add staining solution. Note: for double staining, complete purple stain first.
   1. Staining solution for purple substrate: To 10 mL of buffer, add 45 μL of NBT and 35 μL of BCIP.
   2. Staining solution for red substrate: To 10 mL of buffer, add 31.5 μL of INT and 35 μL of BCIP.

(6) Check the progress of the color reaction every 15–20 min at a stereomicroscope with lighting for reflected illumination.
(7) When the reaction is complete, wash embryos 2 times in 1× PBST.
(8) If you are not going to detect any other probes, fix the samples in ice-cold 4% PFA, and store at 4 °C. To prevent faded stain, image samples within a few days of color reaction.

4.2.6 Second color detection

(1) For detection of another labeled transcript, wash embryos 2 times for 15 min in 0.1 M glycine, pH 2.2.
(2) Wash embryos 2 times for a minimum of 15 min in MABT.
(3) Incubate embryos in block for 30 min at room temperature.
(4) Add appropriate antibody diluted in block and incubate for 2 h at room temperature, or overnight at 4 °C.
(5) Repeat steps 1–8 of initial probe detection.
(6) Fix samples in ice-cold 4% PFA, and store at 4 °C (long-term), shielded from light.

4.3 Fluorescent in situ hybridization (FISH)

FISH is similar to WISH, as it allows for the visualization of transcripts, as seen in Fig. 2 (Brend & Holley, 2009; Marra et al., 2017). However, FISH exhibits improved spatial resolution, allowing for investigations of colocalized transcripts (Fig. 2A) or those which are directly adjacent (Fig. 2B). This is particularly useful in the case of imaging MCCs, as these cells are interspersed among primary transporter cells. FISH allows for simultaneous visualization of primary cells and MCCs to determine cell ratios and other analyses.

4.3.1 Materials and reagents

(1) 4′,6-Diamidino-2-phenylindole, dihydrochloride (DAPI) Invitrogen Life Sciences, D1306; Stock solution 5 mg/mL.
(2) Blocking reagent (Roche).
(3) 1 maleic acid buffer: 150 mM maleic acid, 100 mM NaCl (pH 7.5).
(4) Antibodies (anti-DIG[POD]; anti-FLU[POD]) (Sigma-Aldrich).

4.3.2 Riboprobe production and probe hybridization

Preparation of riboprobes and initial hybridization steps are identical to those found in the above WISH protocol. For FISH, however, we suggest increasing the antisense probe added to the HYB+ mix to decrease stain time. For example, we recommend adding 750 ng of RNA antisense stock in 500 μL of HYB+.

4.3.3 Hot washes and antibody labeling

(1) Remove HYB+/riboprobe mixes and store in microcentrifuge tube in −20 °C.
(2) Wash embryos 2 times for 30 min in 50% formamide/2× SSC at 70 °C. (Note: these salt washes do not contain Tween-20.)
(3) Wash embryos 1 time for 15 min in 2× SSC at 70 °C.
(4) Wash embryos 1 time for 30 min in 0.2× SSC at 70 °C.
(5) Block for 4 h at room temperature in 1× maleic acid buffer plus 2% blocking reagent. Add enough solution to cover embryos (about 500 μL).
(6) Prepare the antibody/blocking reagent mix. For DIG probes make a 1:1000 dilution of Anti-Dig POD Fab fragments. For FLU probes make a 1:500 dilution of Anti-Flu POD Fab fragments.
(7) Incubate overnight in the 4 °C. From this point on, protect the tubes from light.

4.3.4 Fluorescent staining

(1) Wash samples 4 times for 20 min in 1× maleic acid buffer at room temperature. (Note: incubation overnight at 4 °C may decrease probe background. We recommend this long incubation for *trpm7*.)
(2) Wash embryos 2 times for 5 min in 1× PBS at room temperature.
(3) Prepare staining solution at room temperature. Use a 1:50 dilution of tyramide reagent in Perkin Elmer amplification dilution buffer. Select from TSA Plus Cy3 Solution, TSA Plus Fluorescein Solution, or TSA Plus Cy5 Solution. Each staining kit can be used in conjunction with the Anti-Dig-POD or Anti-Flu-POD. (Note: Follow kit instructions for proper resuspension of reagents.)
(4) Apply staining solution to tubes, assuring all embryos are submerged in solution.
(5) Staining time may vary for each reaction, depending on the expression pattern of the gene and the amount of antisense probe in the probe-HYB+ mixture. We recommend testing stain times with trial and error, using previously published staining times as a framework to work within (Marra et al., 2017).

1. Staining may also be monitored by removing 2–3 embryos with a transfer pipet from the clutch. Place these embryos on a glass slide, and rinse 1–2 times with PBS. The reaction can be evaluated with a fluorescent microscope.
2. Note: the fluorescent signal may be faint at this time, but will likely become cleaner after additional washes.

(6) If detecting a single riboprobe, wash embryos in 1× PBST and fix with 4% PFA. If detecting a second probe, wash embryos in the following series, each 5 min: 30% MeOH in 1× PBS, 50% MeOH in 1× PBS, 75% MeOH in 1× PBS, and then 100% MeOH.
(7) Incubate in a solution of 1% H_2O_2 in MeOH for 30 min at room temperature. This solution is made fresh.
(8) Wash embryos for 5 min each in the following series of rehydration solutions: 75% MeOH in 1× PBS, 50% MeOH in 1× PBS, and finally 30% MeOH in 1× PBS.
(9) Wash embryos 2 times for 5 min in 1× PBS.
(10) Block the embryos again for at least 1 h at room temperature.
(11) Apply the appropriate block/antibody solution and incubate at 4 °C overnight.
(12) Wash embryos 4 times for 20 min in 1× maleic acid buffer at room temperature. (Optional: Incubate the embryos overnight at 4 °C to decrease probe background.)
(13) Wash embryos 2 times for 5 min in 1× PBS.
(14) Prepare staining solution from Perkin Elmer kit for a different fluorescent signal of choice.
(15) Apply staining solution to tubes with enough solution to cover embryos.
(16) Again, determine staining time via trial and error, keeping in mind previously reported stain times.
(17) Wash 3 times for 10 min in 1× PBS.
(18) Perform DAPI staining to label nuclei: Prepare working solution by diluting 1 μL of DAPI 5 mg/mL stock into 1500 μL of PBS. DAPI is light sensitive, so thaw in the dark and keep tubes covered. Add the working DAPI solution to the samples, and put on rocker for 15 min at room temperature. Rinse 2 times in PBST, then fix in 4% PFA.
(19) Store samples in PFA at 4 °C until ready for imaging.

4.4 Whole mount immunofluorescence (IF)

While the previous methods have focused on transcript visualization, IF allows for visualization of proteins. This serves as a valuable tool to evaluate the structure of cilia of both transporter cells and MCCs (Chambers, Addiego, et al., 2020; Chambers et al., 2019; Marra, Adeeb, et al., 2019; Marra et al., 2017; Wesselman et al., 2021). As seen in Fig. 3, cilia are visualized using anti-acetylated alpha tubulin, and basal bodies are marked using anti-gamma tubulin. This method can mark cilia in multiple tissues. MCCs can be imaged in the proximal segments of the pronephros while monociliated cells are evaluated in the distal segment alone at 28 hpf (Fig. 3A). MCCs can also be visualized in the ear at 4 dpf with this same protocol (Fig. 3B). IF may also be combined with FISH, allowing for additional analyses of cilia in distinct cell populations (Marra et al., 2017).

4.4.1 Materials and reagents

(1) PBDT: 1× PBST with 1% DMSO

4.4.2 Permeabilization and primary antibody

(1) Note: embryos do not need to be incubated in MeOH for any period of time for IHC. However, embryos that have been stored in MeOH must be rehydrated with PBST (see step 1 under "permeabilization") before proceeding.
(2) Remove PBST, and wash embryos for 5 min in distilled water.
(3) Wash embryos for 7 min in chilled acetone (kept at −20 °C).
(4) Wash 5 min in PBDT.
(5) Block embryos with 10% FCS on a rocker for 2 h at room temperature.
(6) Dilute primary antibody in PBDT with 1% FCS. To visualize MCCs, use the following antibodies:

Primary antibody	Source	ID
Anti-tubulin acetylated (1:400)—cilia	Sigma	T6793; RRID:AB_477585
Anti-γ-tubulin (1:400)—basal bodies	Sigma	T5192; RRID:AB_261690
Anti-PKC (1:250)—apical surface	Santa Cruz	SC216; RRID:AB_2300359
Anti-$Na^{+}K^{+}$ ATPase (1:35)—basolateral surface	DSHB	AB_528092 RRID:AB_528092

(7) After 2 h, remove block and add enough primary antibody solution to cover all animals. Incubate overnight at 4 °C. Be sure to protect samples from light from this point forward.

4.4.3 Secondary antibody

(1) Wash embryos for 5 min in PBDT with 1% FCS and 0.1 M NaCl.
(2) Wash embryos 5 times for 30 min in PBDT with 1% FCS and 0.1 M NaCl.
(3) Wash 1 time for 30 min in PBDT with 1% FCS.

(4) Add appropriate secondary antibody (1:500) in PBDT. Incubate overnight at 4 °C. Be sure to protect samples from light. We recommend using the following secondary antibodies:

Secondary antibody	Source	ID
Goat anti-Mouse IgG (H+L) Highly Cross-Adsorbed Secondary Antibody, Alexa Fluor 568	Invitrogen	A11031; RRID:AB_144696
Goat anti-Rabbit IgG (H+L) Highly Cross-Adsorbed Secondary Antibody, Alexa Fluor 594	Invitrogen	A11037; RRID:AB_2534095
Goat anti-Mouse IgG (H+L) Highly Cross-Adsorbed Secondary Antibody, Alexa Fluor 488	Invitrogen	A11029; RRID:AB_138404
Goat anti-Rabbit IgG (H+L) Highly Cross-Adsorbed Secondary Antibody, Alexa Fluor 488	Invitrogen	A11034; RRID:AB_2576217

4.4.4 Nuclear staining and final washes

(1) Wash embryos for 5 min in PBDT.
(2) Dilute DAPI 1:1500 in PBS. Add DAPI solution to embryos, and incubate on rocker for 15 min.
(3) Wash embryos for 5 min in PBDT with 1% FCS and 0.1 M NaCl.
(4) Wash embryos 5 times for 30 min in PBDT with 1% FCS and 0.1 M NaCl.
(5) Wash 1 time for 30 min in PBDT.
(6) Replace wash with fresh PBDT. Samples can now be stored in PBDT at 4 °C until ready for mounting and imaging.

4.5 Cryosectioning

The previous sections have described methods for whole mount analyses of MCCs. While these methods provide valuable insight, it is difficult to evaluate tubule polarity and basal body docking without the use of 3D rendering software. As a supplement to whole mount methods, cryosectioning and IF have been well established for zebrafish tissues, including the kidney (McCampbell, Springer, & Wingert, 2014; McCampbell, Springer, & Wingert, 2015). These approaches can be implemented to allow for a different view of cilia in various locations. As MCCs are interspersed among monociliated cells in multiple segments, sections from some segments will yield visualizations of both cell types (Fig. 4A). This method may also be useful to investigate co-localization of cilia associated proteins (Fig. 4B) as well as apical-basal polarity (Wesselman et al., 2021) or morphological analyses of tubule dilation (Gerlach & Wingert, 2014). Here, we describe a technique for preparation and cryosectioning of samples, followed by a modified IF protocol tailored for cryosections.

4.5.1 Materials and reagents

(1) Cryostat machine, capable of 10 μm sections
(2) Ethanol
(3) Formaldehyde
(4) Tissue-Tek® Cryomold® Molds (VWR, Cat #: 4565)
(5) Tissue freezing media (TFM) (VWR, Cat #: 15148-031)
(6) Dissection needle or fine forceps
(7) Charged glass slides (VWR, Cat #: 16004-406)
(8) Slide warmer

4.5.2 Fixation and sucrose washes

(1) Remove chorion and fix animals as described in the above WISH protocol: anesthetize embryos with tricaine followed by fixation in 4% PFA in PBS. Fix at 4 °C overnight or room temperature for 4 h.
(2) After fixation, wash 3 times for 15 min in PBS on a shaker.
(3) Wash 1 time for 15 min in 5% sucrose in PBS on a shaker.
(4) Incubate overnight at 4 °C in 30% sucrose in PBS.
(5) Incubate overnight in a 1:1 solution of tissue freezing medium (TFM) and 30% sucrose in PBS. Note: Samples can be stored in this solution for a few days, but will disintegrate over time. We recommend mounting embryos the next day.

4.5.3 Mounting

(1) Add a thin layer of TFM to the bottom of a cryomold.
(2) Transfer embryos from 1:1 TFM/Sucrose to mold using a glass pipet. Transfer as little liquid as possible. We recommend mounting 3–5 embryos per mold.
(3) Orient embryos using metal needle or forceps, assuring that the sample is not angled. Any tilt to the animal will result in uneven sections.
    (a) Note: Direction of mount can change depending on the region of interest. For beginners, we recommend mounting head first (head toward the closest side of the mold). This allows for the location in the animal to be approximated using the presence and dimensions of the yolk.
(4) Once embryos are arrayed, add more TFM to the mold, trying not to disrupt the sample. Fill to the top of the mold, but do not overflow. An even edge is essential for mounting to the cryostat.
(5) Check that samples are still even, and adjust embryos with needle as needed.
(6) Freeze samples at −80 °C.
(7) We recommend putting the mold directly onto the metal of the −80 °C freezer shelf. This will facilitate quicker freezing.

4.5.4 Sectioning

(1) General notes about cryosectioning:
    (a) Make sure the brakes are on, and the guard is covering the blade.

(b) The cryosection chamber temperature should be between −19 °C and −25 °C, depending on humidity.
   (i) If sections curl or fly away, try increasing temperature.
   (ii) If sections are easily squished or slide, try decreasing temperature.
(c) Antiroll plate position must be close enough to allow sections to slide underneath, but far enough away to avoid touching the TFM block.
(d) Clean the blade/antiroll plate after picking up sections.

(2) Once completely frozen, move samples from the −80 °C freezer into the cryostat to prewarm the samples for about 20–30 min.
(3) Remove TFM block from the mold. Note: before removing TFM block from the mold, make a mark on the block to indicate the direction of the sample. This allows for proper orientation of the block during mounting.
(4) Move the mounting chuck into the cryostat, and add a small amount of TFM to the center. Immediately push the sample into the TFM, assuring that the sample is facing up. Note: Add enough TFM to assure the block is securely attached to the chuck.
(5) Allow TFM to freeze completely, and transfer the metal chuck onto the cryostat holder and tighten it. Note: Pre-cooling the chuck for a short amount of time will help the new TFM to freeze quickly.
(6) Adjust the angle of the chuck, so the sample block is parallel to the blade.
(7) The goal is to cut evenly through the block, but this may take a few sections to adjust.
(8) Select the depth of the sample. We recommend 12–14 μm as a starting point.
(9) Remove guard, turn off brake, and take sections as desired.
(10) Test adjustments using rough sections (30 μm) until arriving at the desired location in the animal.
   (a) If mounted head first, slides can be monitored under a stereo microscope to detect yolk ball, which can act as a key to the location in the animal.
(11) When the sample is at the desired location, be sure to adjust the depth (e.g., 12–14 μm). Take a maximum of 9 sections, and proceed to transfer.
(12) Turn on the brakes, pull back the antiroll plate.
(13) Transfer sections onto slides by moving the slide downward at an angle. Rest a hand on blade guard as needed, and stabilize the slide on the back of the blade holder. Move the slide slowly until all sections are collected.
(14) Double check sections under the microscope to check the section location and angle.
(15) Transfer slides to the slide warmer (50 °C) for at least 30 min (maximum 4 h), then store slides at −80 °C.
(16) Clean the blade and antiroll plate with a small paint brush or cotton swab. Wait 10 s for the blade/plate to cool down before sectioning again.

4.6 Cryosection IF

4.6.1 Materials and reagents

(1) Coplin jars
(2) Hydrophobic pap pen (EMS, Cat #: 71310)

(3) Water bath
(4) Optional: sodium citrate buffer (pH 6)
 (a) 400 mL distilled water
 (b) 1.47 g sodium citrate dihydrate
 (c) Mix. Fill to 500 with water
 (d) pH to 6 using HCl
 (e) Add 2.5 mL 10% Tween
(5) Optional: EDTA buffer (pH 8)
 (a) 400 mL distilled water
 (b) 0.185 g EDTA
 (c) Mix. Fill to 500 mL with water
 (d) pH to 8 using 1 N NaOH (2 g NaOH per 50 mL water)
(6) Vectashield (VWR Cat #: 101098-042)
(7) Clear nail polish or another sealant
(8) Glass cover slips

4.6.2 Primary antibody

(1) Retrieve samples from the freezer (−80 °C) and allow to briefly dry on warmer (50 °C).
(2) Outline samples with hydrophobic pap pen, repeating the outline if needed to form a discernable ridge. Note: avoid touching the samples.
(3) Allow sections to dry on the warmer for 20 min, or until pap pen has dried.
(4) OPTIONAL: Heat Induced Antigen Retrieval (HIAR). This may be required for other antibodies, but it is not necessary for those we have tested thus far.
 (a) Preheat water bath to 95–100 °C.
 (b) Preheat either sodium citrate (pH 6) or EDTA (pH 8) buffer in Coplin jars in the water bath.
  (i) The buffer may differ, depending on the antibody. We recommend determining the best option using trial and error.
 (c) Transfer slides to buffer, and return Coplin jars with the slides to the water bath for 30 min.
 (d) Remove from water bath and let cool for 30 min.
 (e) Quick rinse with PBS in Coplin jars. To do so, pour liquid down the side of the jar to avoid washing away samples.
 (f) Proceed with the remainder of the protocol.
(5) Wash samples for a minimum of 30 min in PBS in the Coplin jars.
(6) Prepare humidity chamber by adding water to the bottom. The water level should be sufficient to cover the bottom half of the chamber but low enough to avoid touching the glass slides.
(7) Prepare block solution: 1× PBS, 0.1% Tween, 0.1% Triton X 100, 1% DMSO, 10% FBS

(8) Remove samples from the Coplin jar, gently tap each slide on its side to remove excess liquid, and place in humidity chamber. Immediately begin adding block solution to avoid slides drying out. Note: 200 μL of block is usually sufficient to cover the sample, depending on the pap pen border.
(9) Block for 2 h at room temperature. Note: If the humidity chamber does not have a tight seal, we recommend avoiding evaporation by adding weight to the top of the humidity chamber or sealing with parafilm.
(10) Prepare primary antibody dilutions in block to desired concentration.
  (a) We recommend anti-acetylated alpha-tubulin monoclonal (1:1000), anti-gamma tubulin monoclonal (1:400).
(11) Tip slides on to a paper towel to remove block. Quickly add primary antibody directly to slides to avoid drying the samples.
(12) Incubate in primary overnight at 4 °C. Keep samples protected from light from this point forward. Note: for other antibodies, consider incubating at room temperature overnight.

4.6.3 Washes and secondary antibody

(1) Move samples from chamber to Coplin jar with 0.1% PBST. Remember to keep samples protected from light.
(2) Wash 3 times for 30 min in 0.1% PBST.
(3) Prepare secondary antibody by diluting 1:500 in PBST.
(4) Remove from Coplin jar, gently tap off excess liquid, and place in humid slide chamber.
(5) Immediately begin adding secondary antibody to avoid samples from drying out.
(6) Incubate 1 h at room temperature.
(7) Tap off excess liquid, and place back in humidity chamber.
(8) Add DAPI diluted 1:15,000 in PBS. Incubate for 30 min.
(9) Tap of extra liquid, and transfer to Coplin jar with 0.1% PBST.
(10) Wash 3 times for 15 min in 0.1% PBST.
(11) Wash 1 time for at least 10 min in PBS.

4.6.4 Mounting samples for imaging and storage

(1) Remove samples from PBS, gently tapping the slide on its side to remove excess liquid.
(2) Add 1–2 drops of vectashield to appropriately sized cover slip.
(3) Slowly lower the cover slip (vectashield facing toward the sample) at an angle until flat.
(4) Use clear nail polish or other adhesive to seal the sides of the cover slip.
(5) Let sealant dry at room temperature, and store slides at 4 °C shielded from light.

4.7 Mounting and imaging samples

The previous sections have described methods to stain for MCC transcripts or cilia. The next section will describe our recommendations for properly mounting and imaging of these samples. All of our imaging has been completed on a Nikon A1R or C2 confocal microscope.

4.7.1 Materials and reagents

(1) Glycerol
(2) Aqua polymount
(3) Glass slides
(4) Coverslips

4.7.2 Whole mount samples for light microscopy

(1) Transfer embryos from PFA to a dish containing 1× PBST.
(2) Transfer to a dish containing glycerol, and use pipet tip to submerge animals.
(3) Transfer animals in a small drop of glycerol to a glass depression or bridge slide.
(4) Gently position the animal (now in a drop of glycerol). We recommend using a thin plastic pipette tip or dissection needle. For imaging multiciliated cells, we recommend dorsal images (for clean probes like *odf3b*) or tilted images (for probes with spinal cord expression like *pax2a* or high background, like *cent4*).
(5) Once the animal is in position, cover with a glass coverslip. Animals may be further adjusted by moving the coverslip.
(6) Once imaged, samples can be returned to 1× PBST for a brief wash, and then returned to 4% PFA for storage.

4.7.3 Whole mount samples for fluorescent microscopy

(1) Place 1–10 embryos on a glass slide in a drop of 1× PBST.
(2) Remove the yolk ball with fine forceps, but keep the yok sac extension intact as much as possible.
(3) Using fine forceps, move the animal to a clean drop of 1× PBST. Be sure to secure the animal at the head or very tip of the tail, as clamping anywhere else may damage the tissue of interest. The goal of this step is to clean the animal of any residual yolk or debris that could cause interference.
(4) Prepare a mounting slide by adding a drop of aqua polymount to a glass slide. Note: we have found that we are able to mount 2 animals per glass slide.
(5) Transfer the de-yolked animal from PBST to the aqua polymount using fine forceps. Use the forceps to gently submerge the animal.
(6) Place a coverslip over the sample.
(7) Mounted animals can now be stored at 4 °C protected from light.

4.7.4 General imaging notes

(1) All cilia images were taken with the 60× objective.
(2) We advise using a slower imaging technique to capture the fine structures (e.g., 1 z-stack per 8 s).
(3) Maximum z-stack should be 0.5 μm, though this is dependent on the confocal microscope being used. We recommend using the optical section range set by the microscope.
(4) FISH images can be imaged at a quicker pace (e.g., 1 z-stack per 4 s).

4.8 Scoring cilia: Using fluorescent images to extrapolate quantitative measurements

Once samples have been processed and imaged, accurate data analysis is key in understanding the distribution and nature of MCCs. Here, we describe our methods of extrapolating quantitative data from images (Wesselman et al., 2021). It is worth noting that other labs have used in depth coding to quantify cilia, like CiliaQ (Hansen, Rassmann, Stüven, Jurisch-Yaksi, & Wachten, 2021). However, this may be difficult for individuals without a background in computational science, and the overlaid MCC bundles in the pronephros may be difficult to discern, depending on the microscopy technique used. These described methods require more hands-on manipulations, but may be useful for quantifying cilia phenotypes.

4.8.1 Materials

(1) ImageJ (Fiji: https://imagej.nih.gov/ij/)
(2) Optional: Graphpad Prism v 9 for statistical analysis (Graphpad: https://www.graphpad.com/scientific-software/prism/)

4.8.2 Fluorescent intensity plots

(1) Open file using ImageJ.
(2) Create a maximum z-stack with the cilia channel, reduce background (recommend by 50.0).
(3) Using this maximum projection, use the segmented line tool to follow along the path of the cilia from proximal to distal. Note: be sure to note the direction of the line, as this will need to be consistent across treatments.
(4) Set the line width to include all cilia of interest. Note: this must be kept consistent across treatments.
(5) Use the plot profile tool to generate a list of fluorescent intensity across this line.
(6) Right click to generate a "list." Import this list to a spreadsheet for fluorescent analysis.
(7) These values can be used for various analyses.

4.8.3 Cilia length measurements

(1) Open file using ImageJ.

(2) Create a maximum z-stack with the cilia channel, reduce background (recommend by rolling ball radius of 50.0 pixels).
(3) Compare maximum projection to the z-stack to identify individual cilia.
  (a) Sometimes easier to measure directly on the z stack, and trace while scrolling through the stack.
  (b) Only measure those that are confidently in the beginning/end.
    (i) This can be particularly challenging on wildtype images because the cilia are overlaid. If all cilia are not discernable, try reducing the z-stack step-size in the original image.
(4) Measure length of each identified cilia using the segmented line tool with a width of 1.

4.8.4 DAPI measurements

(1) Open file using ImageJ.
(2) While scrolling through the z-stack, tubule (two layers of cells with the lumen in the middle) should become visible.
(3) Only create maximum projection of the center 2–3 stacks.
(4) Create maximum projections of the cilia channel.
(5) Merge the cilia channel with the DAPI channel.
(6) Using the counter tool, count all DAPI that come into contact with cilia/make up the tubule.
  (a) If the sample was mounted at an angle, consider counting DAPI while scrolling through the z-stack.

4.8.5 Basal body measurements

(1) Open file using ImageJ.
(2) Create a maximum projection of cilia and basal body channels. Reduce background as needed and merge with DAPI channel (see above).
(3) Before counting, determine inclusion/exclusion criteria for basal bodies. As antibodies like gamma-tubulin are associated with high background, it is imperative to be consistent when scoring these organelles.
  (a) We recommend including:
    (i) Basal bodies within the radius of the tubule. For example, any basal body greater than ~4 μm away from the center of the tubule is excluded, as this is likely background or a basal body in another nearby tissue.
    (ii) All ciliated basal bodies.
  (b) We recommend excluding:
    (i) Basal bodies outside of the cilia path, beyond that of the radius of the tubule.
    (ii) Background signal. This may be avoided by reducing the background in ImageJ or adjusting the contrast.
(4) Using the counter tool, count each basal body within the tubule.

(5) Of these identified basal bodies, determined which are "ciliated" or "unciliated."
   (a) Ciliated: A ciliated basal body will either have a cilium directly projecting from it or are covered in cilia.
   (b) Unciliated basal bodies either do not come into contact with cilia.

5 Conclusion

This chapter has established the zebrafish as a useful model organism to study ciliated cell development and ciliopathies due to its rapid development, transparency through early stages, and recapitulation of mammalian phenotypes. These methods can be utilized to visualize cilia of both monociliated cells and MCCs and study the ontogeny of these cell types in the developing zebrafish. Additionally, these methods may be applied to visualizing cilia in other tissues beyond those described, including the central nervous system and lateral line. Taken together, these approaches provide a powerful arsenal for ciliated cell research and can be implemented to study ciliopathic zebrafish disease models and assess new therapeutic approaches to treat ciliary conditions.

Acknowledgments

This work was supported in part by start-up funds from the University of Notre Dame (to R.A.W), the Sean Cocchia Rare Disease Grant from the Warren Center for Drug Discovery at the University of Notre Dame (to R.A.W), a Graduate Women in Science National Fellowship (to H.M.W), a Warren Center Drug Development Welter Family Fellowship (to H.M.W), the Notre Dame Center for Stem Cells and Regenerative Medicine Fellowship (to H.M.W), and a College of Science Summer Undergraduate Research Fellowship (to A.E.G). We thank the Department of Biological Sciences for administrative support and the Center for Zebrafish Research at the University of Notre Dame for their world-class animal care. Imaging seen in this manuscript was carried out in part in the Notre Dame Integrated Imaging Facility (AR1, C2 confocal microscopes), and we especially thank S.C. for her knowledge and expertise. Thank you to M.L. for providing H.M.W. training on the cryostat. Finally, we thank the members of our lab and express our deep gratitude to those members who have helped develop and/or refine these protocols and then shared their incredible skillset with those around them, especially A.M., J.C., and B.C.

References

Bagnat, M., & Gray, R. S. (2020). Development of a straight vertebrate body axis. *Development*, *146*, dev175794.

Basquin, C., Orfila, A. M., & Azimzadeh, J. (2015). The planarian *Schmidtea mediterranea* as a model for studying motile cilia and multiciliated cells. *Methods in Cell Biology*, *127*, 243–262.

Bisgrove, B. W., Makova, S., Yost, H. J., & Brueckner, M. (2012). RFX2 is essential in the ciliated organ of asymmetry and an RFX2 transgene identifies a population of ciliated cells sufficient for fluid flow. *Developmental Biology*, *363*, 166–178.

Bolten, J. S., Pratsinis, A., Alter, C. L., Fricker, G., & Huwyler, J. (2022). Zebrafish (*Danio rerio*) larva as an in vivo vertebrate model to study renal function. *American Journal of Physiology. Renal Physiology*, *322*, F280–F294.

Boutin, C., & Kodjabachian, L. (2019). Biology of multiciliated cells. *Current Opinion in Genetics & Development*, *56*, 1–7.

Brend, T., & Holley, S. A. (2009). Zebrafish whole mount high-resolution double fluorescent in situ hybridization. *Journal of Visualized Experiments*, *25*, 1229.

Chambers, J. M., Addiego, A., Flores-Mireles, A. L., & Wingert, R. A. (2020). *ppargc1a* controls ciliated cell development by regulating prostaglandin biosynthesis. *Cell Reports*, *33*, 108370.

Chambers, B. E., Clark, E. G., Gatz, A. E., & Wingert, R. A. (2020). Kctd15 regulates nephron segment development by repressing Tfap2a activity. *Development*, *147*. dev191973.

Chambers, B. E., Gerlach, G. F., Clark, E. G., Chen, K. H., Levesque, A. E., Leshchiner, I., et al. (2019). Tfap2a is a novel gatekeeper of nephron differentiation during kidney development. *Development*, *146*, dev172387.

Chambers, J. M., Poureetezadi, S. J., Addiego, A., Lahne, M., & Wingert, R. A. (2018). *ppargc1a* controls nephron segmentation during zebrafish embryonic kidney ontogeny. *eLife*, *7*, e40266.

Chang, M. Y., Ma, T. L., Hung, C. C., Tian, Y. C., Chen, Y. C., Yang, C. W., et al. (2017). Metformin inhibits cyst formation in a zebrafish model of polycystin-2 deficiency. *Scientific Reports*, *7*, 7161.

Cheng, C. N., Li, Y., Marra, A. N., Verdun, V., & Wingert, R. A. (2014). Flat mount preparation for observation and analysis of zebrafish embryo specimens stained by whole mount in situ hybridization. *Journal of Visualized Experiments*, *89*, e51604.

Cheng, C. N., & Wingert, R. A. (2015). Nephron proximal tubule patterning and corpuscles of Stannius formation are regulated by the *sim1a* transcription factor and retinoic acid in zebrafish. *Developmental Biology*, *399*, 100–116.

Choksi, S. P., Lauter, G., Swoboda, P., & Roy, S. (2014). Switching on cilia: Transcriptional networks regulating ciliogenesis. *Development*, *141*, 1427–1441.

Chong, Y. L., Zhang, Y., Zhou, F., & Roy, S. (2018). Distinct requirements of E2f4 versus E2f5 activity for multiciliated cell development in the zebrafish embryo. *Developmental Biology*, *443*, 165–172.

Corkins, M. E., Krneta-Stankic, V., Kloc, M., & Miller, R. K. (2021). Aquatic models of human ciliary diseases. *Genesis*, *59*, e23410.

Datsis, S. A., & Boman, I. A. (1974). Ciliated renal tubular epithelium in congenital nephrosis. *Beiträge zur Pathologie*, *151*, 297–303.

Davis, E. E., Brueckner, M., & Katsanis, N. (2006). The emerging complexity of the vertebrate cilium: New functional roles for an ancient organelle. *Developmental Cell*, *11*, 9–19.

Desgrange, A., & Cereghini, S. (2015). Nephron patterning: Lessons from *Xenopus*, zebrafish and mouse studies. *Cell*, *4*, 483–499.

D'Gama, P., Qiu, T., Cosacak, M. I., Rayamajhi, D., Konac, A., Hansen, J. N., et al. (2021). Diversity and function of motile ciliated cell types within ependymal lineages of the zebrafish brain. *Cell Reports*, *37*, 109775.

Drummond, I. A., & Davidson, A. J. (2016). Zebrafish kidney development. *Methods in Cell Biology*, *134*, 391–429.

Drummond, B. E., Li, Y., Marra, A. N., Cheng, C. N., & Wingert, R. A. (2016). The *tbx2a/b* transcription factors direct pronephros segmentation and corpuscle of Stannius formation in zebrafish. *Developmental Biology*, *421*, 52–66.

Drummond, I. A., Majumdar, A., Hentschel, H., Elger, M., Solnica-Krezel, L., Schier, A. F., et al. (1998). Early development of the zebrafish pronephros and analysis of mutations affecting pronephric function. *Development*, *125*, 4655–4667.

Dubruille, R., Laurencon, A., Vandaele, C., Shishido, E., Coulon-Bublex, M., Swoboda, P., et al. (2002). Drosophila regulatory factor X is necessary for ciliated sensory neuron differentiation. *Development*, *129*, 5487–5498.

Duffy, J. L., & Suzuki, Y. (1968). Ciliated human renal proximal tubular cells. Observations in three cases of hypercalcemia. *The American Journal of Pathology*, *53*, 609–616.

Femino, A. M., Fay, F. S., Fogarty, K., & Singer, R. H. (1998). Visualization of single RNA transcripts in situ. *Science*, *280*, 585–590.

Firth, A. L., Dargitz, C. T., Qualls, S. J., Menon, T., Wright, R., Singer, O., et al. (2014). Generation of multiciliated cells in functional airway epithelia from human induced pluripotent stem cells. *Proceedings of the National Academy of Sciences of the United States of America*, *111*, E1723–E1730.

Fliegauf, M., Benzing, T., & Omran, H. (2007). When cilia go bad: Cilia defects and ciliopathies. *Nature Reviews Molecular Cell Biology*, *8*, 880–893.

Gerlach, G. F., & Wingert, R. A. (2013). Kidney organogenesis in the zebrafish: Insights into vertebrate nephrogenesis and regeneration. *Wiley Interdisciplinary Reviews: Developmental Biology*, *2*, 559–585.

Gerlach, G. F., & Wingert, R. A. (2014). Zebrafish pronephros tubulogenesis and epithelial identity maintenance are reliant on the polarity proteins Prkc iota and zeta. *Developmental Biology*, *396*, 183–200.

Grondona, J. M., Granados-Durán, P., Fernández-Llebrez, P., & López-Ávalos, M. D. (2013). A simple method to obtain pure cultures of multiciliated ependymal cells from adult rodents. *Histochemistry and Cell Biology*, *139*, 205–220.

Hansen, J. N., Rassmann, S., Stüven, B., Jurisch-Yaksi, N., & Wachten, D. (2021). CiliaQ: A simple, open-source software for automated quantification of ciliary morphology and fluorescence in 2D, 3D, and 4D images. *The European Physical Journal E: Soft Matter and Biological Physics*, *44*, 18.

Hauptmann, G., & Gerster, T. (2000). Multicolor whole-mount in situ hybridization. *Methods in Molecular Biology*, *137*, 139–148.

Hellman, N. E., Liu, Y., Merkel, E., Austin, C., Le Corre, S., Beier, D. R., et al. (2010). The zebrafish *foxj1a* transcription factor regulates cilia function in response to injury and epithelial stretch. *Proceedings of the National Academy of Sciences of the United States of America*, *107*, 18499–18504.

Howe, K., Clark, M. D., Torroja, C. F., Torrance, J., Berthelot, C., Muffato, M., et al. (2013). The zebrafish reference genome sequence and its relationship to the human genome. *Nature*, *496*, 498–503.

Ishikawa, T., Ueno, H., Omori, T., & Kikuchi, K. (2021). Cilia and centrosomes: Ultrastructural and mechanical perspectives. *Seminars in Cell and Developmental Biology*, *110*, 61–69.

Jaffe, K. M., Thiberge, S. Y., Bisher, M. E., & Burdine, R. D. (2010). Imaging cilia in zebrafish. *Methods in Cell Biology*, *97*, 415–435.

Jin, D., Ni, T. T., Sun, J., Wan, H., Amack, J. D., Yu, G., et al. (2014). Prostaglandin signalling regulates ciliogenesis by modulating intraflagellar transport. *Nature Cell Biology*, *16*, 841–851.

Katz, S. M., & Morgan, J. J. (1984). Cilia in the human kidney. *Ultrastructural Pathology*, *6*, 285–294.

Kimmel, C. B., Ballard, W. W., Kimmel, S. R., Ullmann, B., & Schilling, T. F. (1995). Stages of embryonic development of the zebrafish. *Developmental Dynamics*, *203*, 253–310.

Kramer-Zucker, A. G., Olale, F., Haycraft, C. J., Yoder, B. K., Schier, A. F., & Drummond, I. A. (2005). Cilia-driven fluid flow in the zebrafish pronephros, brain and Kupffer's vesicle is required for normal organogenesis. *Development*, *132*, 1907–1921.

Kroeger, P. T., Jr., Drummond, B. E., Miceli, R., McKernan, M., Gerlach, G. F., Marra, A. N., et al. (2017). The zebrafish kidney mutant zeppelin reveals that *brca2/fancd1* is essential for pronephros development. *Developmental Biology*, *428*, 148–163.

Larsen, T. E., & Ghadially, F. N. (1974). Cilia in lupus nephritis. *The Journal of Pathology*, *114*, 69–73.

Lee, L., & Ostrowski, L. E. (2021). Motile cilia genetics and cell biology: Big results from little mice. *Cellular and Molecular Life Sciences*, *78*, 769–797.

Legendre, M., Zaragosi, L. E., & Mitchison, H. M. (2021). Motile cilia and airway disease. *Seminars in Cell and Developmental Biology*, *110*, 19–33.

Leventea, E., Hazime, K., Zhao, C., & Malicki, J. (2016). Analysis of cilia structure and function in zebrafish. *Methods in Cell Biology*, *133*, 179–227.

Lewis, M., & Stracker, T. H. (2021). Transcriptional regulation of multiciliated cell differentiation. *Seminars in Cell & Developmental Biology*, *110*, 51–60.

Li, Y., Cheng, C. N., Verdun, V. A., & Wingert, R. A. (2014). Zebrafish nephrogenesis is regulated by interactions between retinoic acid, *mecom*, and notch signaling. *Developmental Biology*, *386*, 111–122.

Li, Y., Xu, W., Jerman, S., & Sun, Z. (2019). In vivo analysis of renal epithelial cells in zebrafish. *Methods in Cell Biology*, *154*, 163–181.

Liu, Y., Pathak, N., Kramer-Zucker, A., & Drummond, I. A. (2007). Notch signaling controls the differentiation of transporting epithelia and multiciliated cells in the zebrafish pronephros. *Development*, *134*, 1111–1122.

Lungarella, G., de Santi, M. M., & Tosi, P. (1984). Ultrastructural study of the ciliated cells from renal tubular epithelium in acute progressive glomerulonephritis. *Ultrastructural Pathology*, *6*, 1–7.

Ma, M., & Jiang, Y. J. (2007). Jagged2a-notch signaling mediates cell fate choice in the zebrafish pronephric duct. *PLoS Genetics*, *3*, e18.

Malicki, J., Avanesov, A., Li, J., Yuan, S., & Sun, Z. (2011). Analysis of cilia structure and function in zebrafish. *Methods in Cell Biology*, *101*, 39–74.

Marra, A. N., Adeeb, B. D., Chambers, B. E., Drummond, B. E., Ulrich, M., Addiego, A., et al. (2019). Prostaglandin signaling regulates renal multiciliated cell specification and maturation. *Proceedings of the National Academy of Sciences of the United States of America*, *116*, 8409–8418.

Marra, A. N., Chambers, B. E., Chambers, J. M., Drummond, B. E., Adeeb, B. D., Wesselman, H. M., et al. (2019). Visualizing gene expression during zebrafish pronephros development and regeneration. *Methods in Cell Biology*, *154*, 183–215.

Marra, A. N., Cheng, C. N., Adeeb, B., Addiego, A., Wesselman, H. M., Chambers, B. E., et al. (2019). Iroquois transcription factor *irx2a* is required for multiciliated and transporter cell fate decisions during zebrafish pronephros development. *Scientific Reports*, *9*, 6454.

Marra, A. N., Li, Y., & Wingert, R. A. (2016). Antennas of organ morphogenesis: The roles of cilia in vertebrate kidney development. *Genesis*, *54*, 457–469.

Marra, A. N., Ulrich, M., White, A., Springer, M., & Wingert, R. A. (2017). Visualizing multiciliated cells in the zebrafish through a combined protocol of whole mount fluorescent in situ hybridization and immunofluorescence. *Journal of Visualized Experiments*, *129*, e56261.

Marra, A. N., & Wingert, R. A. (2016). Epithelial cell fate in the nephron tubule is mediated by the ETS transcription factors *etv5a* and *etv4* during zebrafish kidney development. *Developmental Biology*, *411*, 231–245.

McCampbell, K. K., Springer, K. N., & Wingert, R. A. (2014). Analysis of nephron composition and function in the adult zebrafish kidney. *Journal of Visualized Experiments*, *90*, e51644.

McCampbell, K. K., Springer, K. N., & Wingert, R. A. (2015). Atlas of cellular dynamics during zebrafish adult kidney regeneration. *Stem Cells International*, *2015*, 547636.

McConnachie, D. J., Stow, J. L., & Mallett, A. J. (2021). Ciliopathies and the kidney: A review. *American Journal of Kidney Diseases*, *77*, 410–419.

McKee, R., Gerlach, G. F., Jou, J., Cheng, C. N., & Wingert, R. A. (2014). Temporal and spatial expression of tight junction genes during zebrafish pronephros development. *Gene Expression Patterns*, *16*, 104–113.

Morales, E. E., Handa, N., Drummond, B. E., Chambers, J. M., Marra, A. N., Addiego, A., et al. (2018). Homeogene *emx1* is required for nephron distal segment development in zebrafish. *Scientific Reports*, *8*, 18038.

Morales, E. E., & Wingert, R. A. (2017). Zebrafish as a model of kidney disease. *Results and Problems in Cell Differentiation*, *60*, 55–75.

Naylor, R. W., Chang, H. G., Qubisi, S., & Davidson, A. J. (2018). A novel mechanism of gland formation in zebrafish involving transdifferentiation of renal epithelial cells and live cell extrusion. *eLife*, *7*, e38911.

Naylor, R. W., Przepiorski, A., Ren, Q., Yu, J., & Davidson, A. J. (2013). HNF1β is essential for nephron segmentation during nephrogenesis. *Journal of the American Society of Nephrology*, *24*, 77–87.

Naylor, R. W., Qubisi, S. S., & Davidson, A. J. (2017). Zebrafish pronephros development. *Results and Problems in Cell Differentiation*, *60*, 27–53.

Naylor, R. W., Skvarca, L. B., Thisse, C., Thisse, B., Hukriede, N. A., & Davidson, A. J. (2016). BMP and retinoic acid regulate anterior-posterior patterning of the non-axial mesoderm across the dorsal-ventral axis. *Nature Communications*, *7*, 12197.

Obara, T., Mangos, S., Liu, Y., Zhao, J., Wiessner, S., Kramer-Zucker, A. G., et al. (2006). Polycystin-2 immunolocalization and function in zebrafish. *Journal of the American Society of Nephrology*, *17*, 2706–2718.

O'Brien, L. L., Grimaldi, M., Kostun, Z., Wingert, R. A., Selleck, R., & Davidson, A. J. (2011). Wt1a, Foxc1a, and the notch mediator Rbpj physically interact and regulate the formation of podocytes in zebrafish. *Developmental Biology*, *358*, 318. e330.

Oh, E. C., & Katsanis, N. (2012). Cilia in vertebrate development and disease. *Development*, *139*, 443–448.

Omori, Y., Zhao, C., Saras, A., Mukhopadhyay, S., Kim, W., Furukawa, T., et al. (2008). Elipsa is an early determinant of ciliogenesis that links the IFT particle to membrane-associated small GTPase Rab8. *Nature Cell Biology*, *10*, 437–444.

Ong, A. C. M., & Wagner, B. (2005). Detection of proximal tubular motile cilia in a patient with renal sarcoidosis associated with hypercalcemia. *American Journal of Kidney Diseases*, *45*, 1096–1099.

Outtandy, P., Russell, C., Kleta, R., & Bockenhauer, D. (2019). Zebrafish as a model for kidney function and disease. *Pediatric Nephrology*, *34*, 751–762.

Pathak, N., Obara, T., Mangos, S., Liu, Y., & Drummond, I. A. (2007). The zebrafish fleer gene encodes an essential regulator of cilia tubulin polyglutamylation. *Molecular Biology of the Cell*, *18*, 4353–4364.

Pazour, G. J., Quarmby, L., Smith, A. O., Desai, P. B., & Schmidts, M. (2020). Cilia in cystic kidney and other diseases. *Cellular Signaling*, *69*, 109519.

Poureetezadi, S. J., Cheng, C. N., Chambers, J. M., Drummond, B. E., & Wingert, R. A. (2016). Prostaglandin signaling regulates nephron segment patterning of renal progenitors during zebrafish kidney development. *eLife*, *5*. pii:e17551.

Poureetezadi, S. J., & Wingert, R. A. (2016). Little fish, big catch: Zebrafish as a model for kidney disease. *Kidney International*, *89*, 1204–1210.

Reiter, J., & Leroux, M. (2017). Genes and molecular pathways underpinning ciliopathies. *Nature Reviews Molecular Cell Biology*, *18*, 533–547.

Satir, P., & Christensen, S. T. (2007). Overview of structure and function of mammalian cilia. *Annual Review of Physiology*, *69*, 377–400.

Sharma, N., Berbari, N. F., & Yoder, B. K. (2008). Ciliary dysfunction in developmental abnormalities and diseases. *Current Topics in Developmental Biology*, *85*, 371–427.

Shi, Y., Su, Y., Lipschutz, J. H., & Lobo, G. P. (2017). Zebrafish as models to study ciliopathies of the eye and kidney. *Clinical Nephrology Research*, *1*, 6–9.

Spassky, N., & Meunier, A. (2017). The development and functions of multiciliated epithelia. *Nature Reviews Molecular Cell Biology*, *18*, 423–436.

Sreekumar, V., & Norris, D. P. (2019). Cilia and development. *Current Opinion in Genetics and Development*, *56*, 15–21.

Stooke-Vaughan, G. A., Huang, P., Hammond, K. L., Schier, A. F., & Whitfield, T. T. (2012). The role of hair cells, cilia and ciliary motility in otolith formation in the zebrafish otic vesicle. *Development*, *139*, 1777–1787.

Stubbs, J. L., Vladar, E. K., Axelrod, J. D., & Kitner, C. (2012). Multicilin promotes centriole assembly and ciliogenesis during multiciliated cell differentiation. *Nature Cell Biology*, *14*, 140–147.

Sullivan-Brown, J., Schottenfeld, J., Okabe, N., Hostetter, C. L., Serluca, F. C., Thiberge, S. Y., et al. (2008). Zebrafish mutations affecting cilia motility share similar cystic phenotypes and suggest a mechanism of cyst formation that differs from pkd2 morphants. *Developmental Biology*, *314*, 261–275.

Sun, Z., Amsterdam, A., Pazour, G. J., Cole, D. G., Miller, M. S., & Hopkins, N. (2004). A genetic screen in zebrafish identifies cilia genes as a principal cause of cystic kidney. *Development*, *131*, 4085–4093.

Tan, F. E., Vladar, E. K., Ma, L., Fuentealba, L. C., Hoh, R., Espinoza, F. H., et al. (2013). Myb promotes centriole amplification and later steps of the multiciliogenesis program. *Development*, *140*, 4277–4286.

Thisse, C., & Thisse, B. (2008). High-resolution in situ hybridization to whole-mount zebrafish embryos. *Nature Protocols*, *3*, 59–69.

Tsujikawa, M., & Malicki, J. (2004). Intraflagellar transport genes are essential for differentiation and survival of vertebrate sensory neurons. *Neuron*, *42*, 703–716.

van Rooijen, E., Giles, R. H., Voest, E. E., van Rooijen, C., Schulte-Merker, S., & van Eeden, F. J. (2008). LRRC50, a conserved ciliary protein implicated in polycystic kidney disease. *Journal of the American Society of Nephrology*, *19*, 1128–1138.

Vasilyev, A., Liu, Y., Mudumana, S., Mangos, S., Lam, P. Y., Majumdar, A., et al. (2009). Collective cell migration drives morphogenesis of the kidney nephron. *PLoS Biology*, *7*, e9.

Vincensini, L., Blisnick, T., & Bastin, P. (2011). 1001 model organisms to study cilia and flagella. *Biology of the Cell*, *103*, 109–130.

Wallmeier, J., Frank, D., Shoemark, A., Nöthe-Menchen, T., Cindric, S., Olbrich, H., et al. (2019). De novo mutations in FOXJ1 result in a motile ciliopathy with hydrocephalus and randomization of left/right body asymmetry. *The American Journal of Human Genetics*, *105*, 1030–1039.

Waters, A. M., & Beales, P. L. (2011). Ciliopathies: An expanding disease spectrum. *Pediatric Nephrology*, *26*, 1039–1056.

Wesselman, H. M., Flores-Mireles, A., & Wingert, R. A. (2021). Esrrγa regulates nephron development and ciliogenesis by controlling prostaglandin synthesis and cooperation with Ppargc1a. *bioRxiv*. https://doi.org/10.1101/2021.11.17.468871.

Wesselman, H. M., Nguyen, T. K., Chambers, J. M., Drummond, B. E., & Wingert, R. A. (2023). Advances in understanding the genetic mechanisms of zebrafish renal multiciliated cell development. *Journal of Developmental Biology*, *11*, 1.

Whitfield, T. T. (2020). Cilia in the developing zebrafish ear. *Philosophical Transactions of the Royal Society, B: Biological Sciences*, *375*, 20190163.

Wingert, R. A., & Davidson, A. J. (2008). The zebrafish pronephros: A model to study nephron segmentation. *Kidney International*, *73*, 1120–1127.

Wingert, R. A., & Davidson, A. J. (2011). Zebrafish nephrogenesis involves dynamic spatiotemporal expression changes in renal progenitors and essential signals from retinoic acid and *irx3b*. *Developmental Dynamics*, *240*, 2011–2027.

Wingert, R. A., Selleck, R., Yu, J., Song, H. D., Chen, Z., Song, A., et al. (2007). The *cdx* genes and retinoic acid control the positioning and segmentation of the zebrafish pronephros. *PLoS Genetics*, *3*, 1922–1938.

Xie, H., Kang, Y., Wang, S., Zheng, P., Chen, Z., Roy, S., et al. (2020). E2f5 is a versatile transcriptional activator required for spermatogenesis and multiciliated cell differentiation in zebrafish. *PLoS Genetics*, *16*, e1008655.

Yu, X., Ng, C., Habacher, H., & Roy, S. (2008). Foxj1 transcription factors are master regulators of the motile ciliogenic program. *Nature Genetics*, *40*, 1445–1453.

Zhou, F., Narasimhan, V., Shboul, M., Chong, Y. L., Reversade, B., & Roy, S. (2015). Gmnc is a master regulator of the multiciliated cell differentiation program. *Current Biology*, *25*, 3267–3273.

Zhou, F., Rayamajhi, D., Ravi, V., Narasimhan, V., Chong, Y. L., Lu, H., et al. (2020). Conservation as well as divergence in Mcidas function underlies the differentiation of multiciliated cells in vertebrates. *Developmental Biology*, *465*, 168–177.

CHAPTER

9

Isolation of ciliary ectosomes and analysis of amidated peptide-mediated chemotaxis in *Chlamydomonas*

Raj Luxmi* and Stephen M. King*

Department of Molecular Biology and Biophysics, University of Connecticut Health Center, Farmington, CT, United States

**Corresponding authors: e-mail address: luxmi@uchc.edu; king@uchc.edu*

Chapter outline

Abstract

Ciliary ectosomes are vesicles that bud from the ciliary membrane. Isolation and analysis of these structures can shed light on their bioactive cargoes and identify proteins and biomolecules involved in intercellular communication and various physiological processes. Most

Methods in Cell Biology, Volume 175, ISSN 0091-679X, https://doi.org/10.1016/bs.mcb.2022.09.009
Copyright © 2023 Elsevier Inc. All rights reserved.

published methods to isolate ciliary ectosomes are based on their size (100 nm to 1 μm) to separate cilia-derived vesicles from isolated cilia and/or intact cells. However, it is often difficult to determine the origin of extracellular vesicles and to distinguish ciliary ectosomes from ectosomes budded from the plasma membrane or from exosomes that derive from multivesicular bodies. Here, we describe procedures to isolate and purify ciliary ectosomes from the unicellular green alga, *Chlamydomonas reinhardtii*, through differential and iodixanol density gradient ultracentrifugation; in this organism, the ciliary membrane is the only membrane directly exposed to the environment and thus ectosomes are of known origin. Ciliary ectosomes contain enzymes and α-amidated peptide products required to mediate peptidergic-signaling cascades; one identified amidated peptide acts as a chemotactic modulator for *C. reinhardtii* gametes. Classical methods used to assess chemotaxis do not provide quantitative measurements of the chemotactic gradient or the real-time effects on the migration of fast moving cells. Consequently, we developed a chemotaxis assay protocol using microfluidic channel slides that provides quantitative and qualitative measurements of the chemotactic gradient and cell migration. Here, we describe how to establish a stable gradient of a bioactive substance in microfluidic channel slides and perform quantitative assays to assess chemotaxis of both individual cells and populations of *C. reinhardtii*.

1 Introduction

Cilia are membrane-bound microtubule-based structures that project into the extracellular environment and serve as motile, sensory and secretory organelles participating in a series of signaling cascades inside cells (Basto & Marshall, 2015; Wheway, Nazlamova, & Hancock, 2018). Extracellular vesicles are membrane-delimited structures that carry a variety of biologically active molecules and play a role in mediating intercellular communication (Abels & Breakefield, 2016; Amintas et al., 2021; Malkin & Bratman, 2020; Skotland, Sagini, Sandvig, & Llorente, 2020). These vesicles are named based on their origin. Exosomes derive from the fusion of multivesicular bodies (MVBs) with the plasma membrane such that the intraluminal vesicles of MVBs are released into the environment. In contrast, ectosomes form by budding directly from the plasma membrane into the extracellular space. Multiple studies in a variety of eukaryotes (including green algae, nematodes and mammals) have revealed that cilia also release extracellular vesicle-like structures, termed ciliary ectosomes or cilia-derived vesicles, by outward budding of the ciliary membrane (Cao et al., 2015; Luxmi, Kumar, Mains, King, & Eipper, 2019; Wang et al., 2014; Wood, Huang, Diener, & Rosenbaum, 2013; Ye, Nager, & Nachury, 2018; Zuo, Guo, & Lipschutz, 2009).

Ciliary ectosome content is subject to regulation at two distinct steps. First, proteins must be specifically trafficked to the ciliary base and allowed into the organelle by passage through the transition zone gate (Garcia-Gonzalo & Reiter, 2017). Subsequently, proteins destined for secretion must be moved to ciliary sites of ectosome release (Luxmi & King, 2022). Studies in the unicellular green alga *Chlamydomonas reinhardtii* have revealed that ectosomal content changes during development and that the ectosomal membrane has a composition distinct from that of the ciliary membrane (Long et al., 2016; Luxmi et al., 2019). This latter observation provides direct evidence for protein sorting at the level of the ciliary membrane although the trafficking mechanism remains to be established. Furthermore, biochemical and physiological analyses have made clear that ciliary ectosomes are bioactive and can act as a cell-autonomous source of degradative enzymes and/or mediate signals between individual cells and organisms (Kubo et al., 2009; Luxmi et al., 2019; Luxmi & King, 2022; Wang, Nikonorova, Gu, Sternberg, & Barr, 2020; Wood et al., 2013).

In *C. reinhardtii*, these cilia-derived vesicles are involved in peptidergic signaling and contain bioactive amidated peptide products and the peptide amidating enzyme—peptidylglycine α-amidating monooxygenase (PAM)—that converts C-terminal Gly residues into amides with the concomitant release of glyoxylate (Luxmi et al., 2019). Amidation of peptides is a key modification necessary for the bioactivity of numerous secreted peptide signals; it acts to increase the affinity of the peptide for its cognate receptor and to extend peptide half-life in the extracellular environment as amidated peptides are poorly recognized by carboxypeptidases. One identified amidated peptide product (GATI-*amide*) acts as a chemotactic modulator for *C. reinhardtii* gametes; *minus* gametes exhibit chemo-attraction, while *plus* gametes are repelled by the amidated peptide (Luxmi et al., 2019). As a traditional, agar plate-based, chemotaxis assay does not provide quantitative information on formation of a stable gradient of the chemical substance under test or a direct measure of the chemotactic migration of cells, we developed a microfluidic chemotaxis assay to follow the motile behavior of *C. reinhardtii* gametes in a GATI-*amide* gradient (Luxmi et al., 2019).

Here, we provide a detailed and reliable method to isolate ciliary ectosomes from *C. reinhardtii*. We also describe a detailed protocol for the microfluidic chemotaxis assay that yields quantitative and qualitative measurements of the migration of the *C. reinhardtii* gametes within a stable gradient of a bioactive substance.

2 Isolation of ciliary ectosomes from mating gametes

2.1 Cell culture and growth conditions

Chlamydomonas reinhardtii strains can be obtained from the *Chlamydomonas* Resource Center (https://www.chlamycollection.org/) for a nominal fee. The mating-type *minus* CC-124 and *plus* CC-125 strains are cultured in 100 mL of R medium* aerated with 95% air and 5% CO_2 under a 12 h/12 h light/dark cycle at 22 °C for a period of 2–3 days (Harris, 2009; Witman, 1986). To induce gametogenesis, vegetative cells of *minus* and *plus* mating-types are harvested by centrifugation at 1,500 x *g* for 5 min, washed and then resuspended in 100 mL of nitrogen-deficient minimal medium (M-N medium*). Cells are then incubated in M-N medium for 18–24 h with 95% air and 5% CO_2 under a 12 h/12 h light/dark cycle at 22 °C. The mating competency of gametes can be assessed by mixing equal numbers of mating-type *minus* and *plus* gametes for 20 min and directly observing the number of quadriciliate cells formed by phase contrast microscopy; it is often helpful to fix the sample prior to microscopy with 2% paraformaldehyde. Mating efficiency can be calculated as follows:

$$\begin{aligned}\%\ \text{QC formation} &= 100 \times (2 \times \text{QC})/[(2 \times \text{QC}) + \text{S}], \text{ where}\\ \text{QC} &= \text{number of quadriciliates counted, and}\\ \text{S} &= \text{number of single cells with only two ciliacounted.}\end{aligned}$$

Gametes with high mating efficiency (*i.e.* exhibiting >80% agglutination and quadriciliate formation upon mixing) are needed for the successful isolation of ciliary ectosomes in high yield. If gametes do not exhibit sufficient mating competency after 24 h, they can be further incubated in the M-N medium for an additional period of 12 h and their mating ability reassessed.

*The recipes for R and M-N media are available at https://www.chlamycollection.org/methods/media-recipes/minimal-or-m-medium-and-derivatives-sager-granick/.

2.2 Protocol for initial isolation of ciliary ectosomes

1. Gametes of both mating-types were centrifuged using a Thermo Scientific Sorvall ST8 benchtop centrifuge and an 8 x 50 mL swing-out rotor, washed, and gently re-suspended in 20 mL of fresh M-N medium at a density of $5–7 \times 10^6$ cells/mL. Resuspension of cells should be performed by gentle pipetting or swirling the flask to avoid deciliation; it is especially important to avoid using pipettes with a narrow bore, which impart a high shear force. Examine the cells by phase contrast microscopy to ensure they have full-length cilia, are not contaminated with cell debris, and have not lost their cilia during washing.

   Note—The volume of medium used for resuspension of cells should be adjusted depending upon the rotors and ultracentrifuge available as this volume of medium will be used to isolate the ectosomes. As a Beckman 70 Ti fixed-angle rotor can hold eight 26.3 mL polycarbonate tubes, a total volume of 210 mL can be processed *per* run. If this or equivalent rotors are available, gametes of both mating-types can be resuspended in a larger volume of 100 mL to avoid cell aggregation and minimize debris contamination.
2. Equal volumes (20 mL at $5–7 \times 10^6$ cells/mL) of gametes of both mating-types are mixed in a 125 mL conical flask, for 1 h with very gentle aeration at 22 °C in constant light. At the end of the incubation, examine the cells under the microscope to ensure that most are quadriciliates.
3. After 1 h incubation, cells are pelleted by centrifugation at 1,500 x *g* for 5 min at room temperature in a Thermo Scientific Sorvall ST8 benchtop centrifuge with an 8 x 50 mL swing-out rotor. Before centrifugation, an aliquot of cells can be saved to make a cell lysate if needed for subsequent analyses. The remaining supernatant that contains the ciliary ectosomes is then carefully removed without

disturbing the cell pellet and placed in a new 50 mL polypropylene conical tube. Examine the supernatant under the microscope to ensure there are no cilia or cells present. If cells or cellular debris are observed, it is best to re-centrifuge the supernatant at 1,500 x *g* for an additional 5 min.

4. To remove any cilia, the supernatant from step #3 is then further centrifuged at 20,000 × *g* for 30 min at 4 °C; we use a Thermo Scientific™ Fiberlite™ F21-8 x 50y Fixed-Angle rotor in a Thermo Scientific™ Sorvall™ RC 6 Plus high speed centrifuge. Remove the supernatant without touching the pellet and transfer to another tube. It is best to leave behind ~1 mL of the supernatant above the pellet to avoid any chance of contamination.
5. The sample is next centrifuged at 200,000 × g for 60 min at 4 °C using a TL-100.3 rotor in a Beckman TL-100 ultracentrifuge; 18 mL of supernatant can be processed at one time. Following centrifugation, the resulting ectosome-depleted supernatant is removed completely with a pipette; this represents the soluble mating secretome.
6. The pellets obtained from step #5 are enriched in ciliary ectosomes and are then further washed with 10 mM HEPES, pH 7.4 buffer. First, resuspend the pellets in 1–3 mL of buffer, pool into one tube and then centrifuge at 200,000 x *g* for 60 min at 4 °C. Remove the supernatant with a pipette. For biochemical analysis, the ectosome-enriched pellet is resuspended in 100–200 μL TMT buffer [20 mM 2-[tris(hydroxymethyl)-methylamino]-ethanesulfonic acid (TES), pH 7.4, 10 mM mannitol, 1% Triton X-100] containing a protease inhibitor cocktail (cOmplete ULTRA Tablets, # 05892791001, Roche, Basel, Switzerland) and 0.3 mg/mL phenylmethylsulfonyl fluoride (PMSF, Sigma Chemical Co., St. Louis, MO). Ectosome samples may be stored at −80 °C prior to further analysis. For electron microscopy (EM), ectosome pellets are resuspended in 10 mM HEPES, pH 7.4 buffer containing protease inhibitors and stored at 4 °C until use. Note that ciliary ectosome pellets are almost transparent and viscous, and thus care is needed during resuspension to avoid sample loss especially through adherence to plastic pipette tips.

2.3 Protocol for further purification of ciliary ectosomes

The above procedure yields pellets of partially purified washed ectosomes. However, these can be further purified using density gradient sedimentation. We routinely use iodixanol density step gradients for this purpose (Long et al., 2016).

1. Washed ectosome pellets are resuspended in 1 mL of 10 mM HEPES, pH 7.4 containing protease inhibitors and a sufficient volume of 60% (w/v) iodixanol (Sigma-Aldrich, D1556) added to the ectosome solution to achieve a final iodixanol concentration of 15% (w/v).
2. A solution of 34% (w/v) iodixanol (900 μL) made in 10 mM HEPES, pH 7.4 with protease inhibitor is added to the bottom of a 2.2 mL thin-wall polyallomer tube.

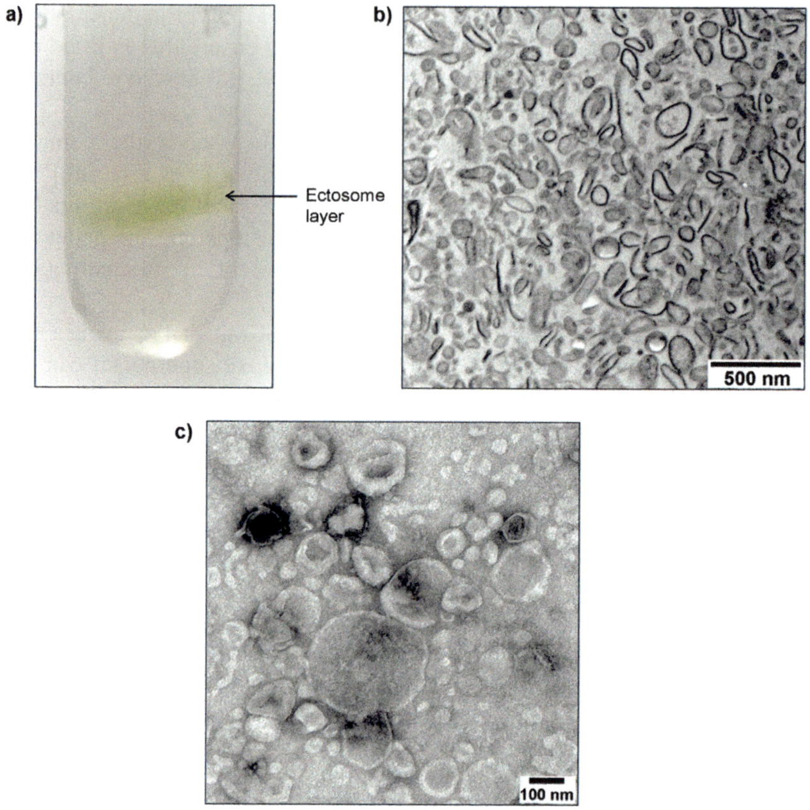

FIG. 1

Isolation of ciliary ectosomes. (A) Image of the ectosome layer (arrow) obtained after iodixanol density step gradient centrifugation; this particular sample has a slightly green tinge due to the presence of some chlorophyll—usually the layer is whiter. (B) Thin section electron micrograph (EM) of an agarose-embedded ectosome pellet; bar = 500 nm. (C) Negative stained EM image of purified ciliary ectosomes; bar = 100 nm.

3. The 15% ectosome-iodixanol solution (1.3 mL) is then slowly layered on top of the 34% iodixanol solution and the sample centrifuged at 259,000 x *g* for 2 h in a TLS-55 swinging-bucket rotor. Ectosomes become enriched in a white layer that accumulates at the interface between 15% and 34% iodixanol layers (Fig. 1A). Take off the upper layer with a pipette and carefully collected the white ectosome layer using a fresh pipette tip and dilute to a final volume of 3 mL in 10 mM HEPES, pH 7.4 containing protease inhibitors. This sample is then centrifuged at 431,000 x *g* for 30 min using a TL100.3 rotor. The final ectosome pellet is resuspended in 100 μL of 10 mM HEPES, pH 7.4 containing protease inhibitors, and can be used directly, stored at 4 °C for short periods, or frozen at −80 °C.

For thin section electron microscopy analysis, ectosome pellets can be embedded in low melting point agarose, fixed with glutaraldehyde and osmium tetroxide/ potassium ferricyanide, dehydrated and embedded in epoxy resin prior to ultrathin sectioning (Fig. 1B). Alternatively, ectosomes can be subject to negative staining with uranyl acetate and visualized directly (Fig. 1C).

3 Microfluidic assay to assess chemotaxis in *C. reinhardtii*

Chemotaxis is the directional migration of cells in response to a gradient of a chemical substance. Positive chemotaxis refers to the movement of cells up the gradient towards the source of the substance, while negative chemotaxis is the movement of cells down the gradient or away from the chemical substance. In *C. reinhardtii*, a qualitative assessment of chemotaxis can be made by watching the migration of cells on the moist surface of an agar plate towards the source of a chemoattractant; for example the chemoattractant can be embedded within an agarose block and placed directly on the agar surface (see Luxmi et al., 2019). However, such assays although useful for an initial assessment of chemotactic activity are not readily quantifiable. Therefore, to measure directly the chemotactic behavior of *C. reinhardtii* cells towards the bioactive GATI-*amide* peptide, we devised a microfluidic approach involving the generation of stable peptide gradients within a commercial microscope slide containing two access ports interconnected by a microfluidic channel.

3.1 Cell culture

C. reinhardtii strains (CC-124 mating-type *minus*) are cultured in 50 mL of R medium for a period of 2 days under the growth conditions described in Section 2.1. To prepare gametes, vegetative cells are washed and incubated in M-N medium for 24 h as described above. Harvest cells by centrifugation at 1,500 x *g* for 5 min and wash with M-N medium. Gametes are then further diluted to a density of 1.2×10^6 cells/mL with M-N medium prior to use in the chemotaxis assay. It may be necessary to optimize the cell density of other strains for use in the assay chamber.

3.2 Establishment of the chemotactic gradient

1. A gradient of the test chemotactic substance needs to be established in a microfluidic channel slide (μ slide I 0.2 Luer; ibidi GmbH, Gräfelfing, Germany, catalog # 80166). The microfluidic channel slide has two reservoirs, one serves as the cell reservoir and other as the chemo-effector reservoir. The two reservoirs are connected by an enclosed channel with dimensions of 50 mm × 5 mm × 0.2 mm. The channel volume is 50 μL and the volume of each reservoir is 60 μL.
2. Before setting up the gradient, microfluidic channel slides are placed into 10 cm Petri dishes humidified with M-N medium-soaked paper towels. The channel is filled with 50 μL M-N media using a pipette (Fig. 2A); note that the fluid is kept

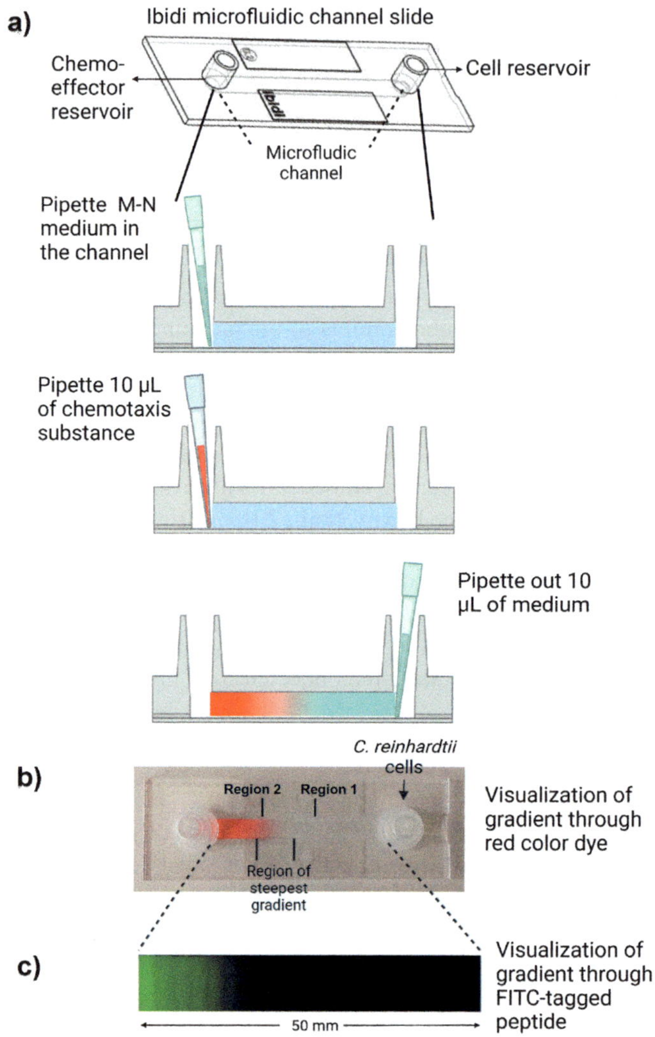

FIG. 2

Establishment of a peptide gradient in an ibidi microfluidic channel slide. (A) Schematic representation of pipetting the chemotactic agent and formation of the gradient in the ibidi microfluidic channel slide. (B) Image of a gradient visualized using red food color dye (provided by ibidi). The region of the steepest gradient is readily visible in the image. To calculate the chemotaxis index, the number of *C. reinhardtii* cells in regions I and 2 were counted at various time points. (C) Image of a gradient visualized using a FITC-tagged peptide in the microfluidic channel. Note that the region of steepest gradient and its stability can vary depending on the molecular composition and mass of the molecule/compound used.

Panel A: Illustration was created with BioRender.com

within the channel by surface tension. To set up the gradient, 10 μL of the test substance (diluted in M-N media) is first added to the bottom of the chemo-effector reservoir; in our experiments, the GATI-*amide* peptide was diluted to 10 μM and the vehicle control (dimethylsulfoxide) to 0.1% (v/v) using M-N medium prior to adding into the reservoir. The gradient is established by removing 10 μL of M-N medium from the channel at the bottom of the cell reservoir thereby drawing the test substance into the channel from the opposite reservoir. Both reservoirs are then covered with the supplied lids and kept at room temperature for 15 min to allow for the formation of a stable gradient of the test substance (Fig. 2A) (and see Application Note 01, https://ibidi.com/img/cms/support/AN/AN01_Gradient_SlideI.pdf). To avoid convective flow, the test substance and medium used to fill the microfluidic channel should be of the same density.

3. Gradient formation can be directly visualized using a dye (such as a commercial food coloring dye) (Fig. 2B), FITC-tagged fluorescent beads (Molecular Probes, FluoSpheres®, 10 μm F8836) or FITC tagged-peptides (here we used the FITC-tagged peptide FITC-Ahx-GPGDFSTYV, which has a molecular mass of 1542 Da—Ahx is 1,6-aminohexanoic acid) (Fig. 2C). Although the gradient is visible for 18–20 h, it flattens somewhat after 4–6 h. The FITC-tagged beads or the dye can be mixed directly with the test substance and applied to the microfluidic channel as described above. Low power imaging allows the shape of the gradient to be assessed (Figs. 2B and C).

3.3 Procedure for the chemotaxis assay

1. After establishment of the chemotactic gradient, remove the lid from the cell reservoir and add 10 μL of gametic cells (from a culture at 1.2×10^6 cells/mL) to the cell reservoir chamber in the dark or under dim red light, and then re-cover with the lid. Incubate the petri dish containing the microfluidic channel slides in the dark for periods of 1–4 h. Note that for *C. reinhardtii*, it is important to keep the slides in the dark (or under red light) so as to avoid phototactic effects which could have confounding effects on the assay. Thus, the entire experiment should be conducted in a darkened room and a red filter used to cover the microscope light source throughout the imaging process.
2. Images of cells (Fig. 3) within the channel and their movement are recorded at discrete time intervals using an inverted microscope; we employ a Nikon Eclipse TE300 (Nikon Instruments, Melville, NY) with a 4× objective lens. It is essential that the microfluidic channel slides be moved very carefully, to avoid disturbing the gradient.

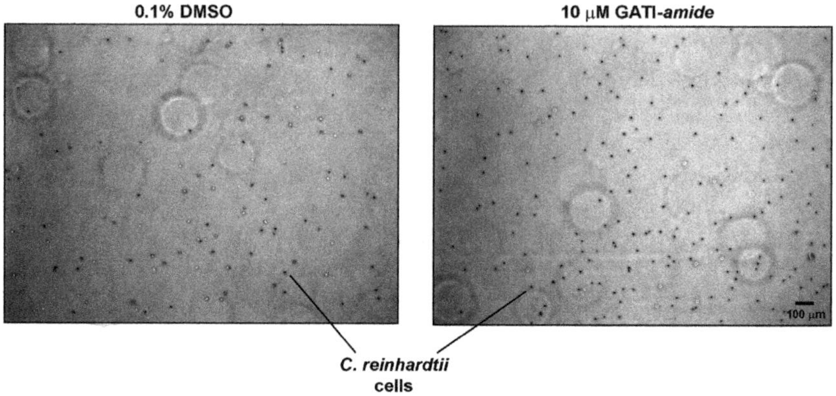

FIG. 3

Phase contrast images of *C. reinhardtii* cells in the presence of gradients of 0.1% DMSO (left panel) and 10 μM GATI-*amide* peptide (right panel) in the microfluidic channel is shown. The peptide/solvent vehicle source is from the left. Bar = 100 μm.

3.4 Data analysis

The number of cells in each image of specified channel regions can be counted manually or using the mosaic plugin (http://rsb.info.nih.gov/ij/) within the Image J software package. The chemotaxis index, which is a quantitative measure of the response of cells to the test chemotactic substance, is determined as a function of time and calculated using the formula:

Chemotaxis index (CI) = [(Number of cells in region 2/number of cells in region 1) for cells responding to the test substance/same ratio for cells responding to the vehicle control] (Fig. 4A). The regions to be counted should be the same for both test and control slides (Fig. 2B).

Cells are tracked using the manual tracking plugin of Image J (http://rsb.info.nih.gov/ij/plugins/track/track.html). The data of tracked cells are then analyzed using on-line chemotaxis and migration tool software (https://ibidi.com/chemotaxis-analysis/171-chemotaxis-and-migration-tool.html). These tools allow one to display individual cell trajectory plots within a reference frame (Fig. 4B), follow the center-of-mass of the entire population (Fig. 4C), and calculate the velocity of how fast cells are migrating (Fig. 4D). This allows for assessment of chemotactic behavior on the level of both individual cells and entire populations.

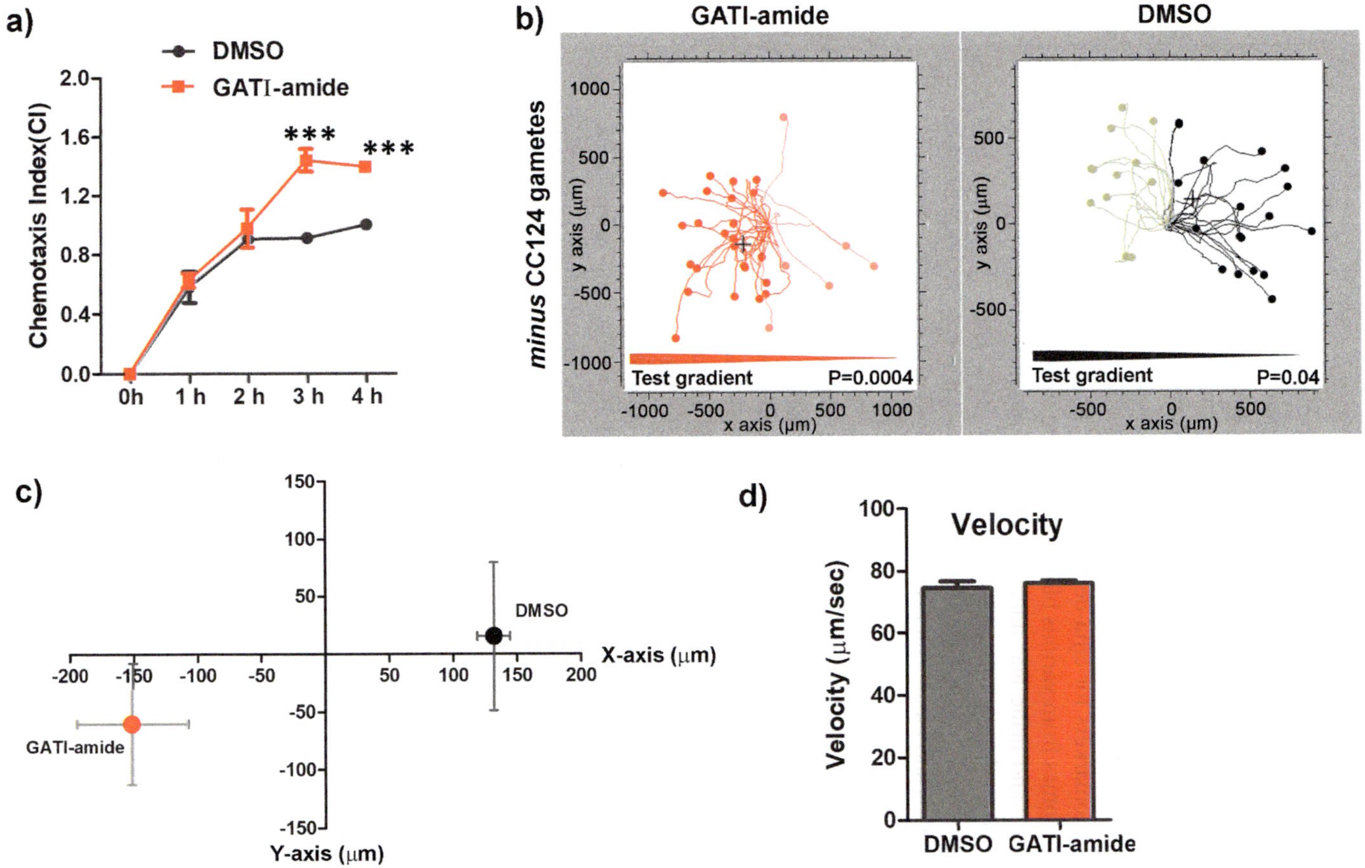

FIG. 4

Analysis of chemotactic migration of *minus* CC-124 gametes in a microchannel slide. (A) The chemotaxis index (CI) of *minus* CC-124 gametes was measured over a period of 4 h in response to 10 μM GATI-*amide* synthetic peptide and vehicle control (0.1% DMSO). Asterisks indicate a significant difference between the control and peptide gradients; *** = $P < 0.001$. Data are the average of three independent experiments ± SEM. (B) Trajectory plots within a single reference frame for individual *minus* CC-124 gametes in the presence of a gradient of 10 μM GATI-*amide* peptide (red elongated triangle) and 0.1% DMSO (black elongated triangle) are shown. Migration of cells was tracked for 10 s (0.5 s/frame). The p-values given by the Rayleigh test (this is a statistical test for the uniformity of a circular distribution of cell endpoints) are indicated. Darker and lighter tracks indicate the direction of cell movement within the gradient. (C) Center of mass (displacement of the total population of cells) and swimming velocity (D) for populations of cells under both conditions was calculated from the trajectory plots. Results are the average of three independent experiments ± SEM.

Data used to prepare this figure were obtained from Luxmi, R., Kumar, D., Mains, R.E., King, S.M., & Eipper, B.A. (2019). Cilia-based peptidergic signaling. PLoS Biology, *17, e3000566.*

4 Concluding remarks

The density gradient ultracentrifugation protocol for isolation of ciliary ectosomes from cilia of *C. reinhardtii* provides highly purified populations of cilia-derived vesicles. This method can be used to characterize ciliary ectosome cargo proteins. Furthermore, as *C. reinhardtii* vegetative cells and resting gametes are surrounded by a multi-layer cell wall, the ciliary membrane provides the only exposed source for the release of EVs into the surrounding environment. Thus, any vesicle isolated is of known ciliary origin. Potentially then, analysis of *C. reinhardtii* ectosomes may identify general biomarkers for cilia-derived vesicles. In mammalian systems, isolating ciliary ectosomes from various biological fluid sources such as blood plasma, cell culture media, and urine and under varying pathological conditions becomes more complex due to the multiple potential origins of EVs. Thus, determining the source of EVs is pivotal to understanding their biology. It is highly recommended to optimize the methods and include an extra step (immuno-pull down or fluorescent labeling) in combination with differential centrifugation to isolate EVs of ciliary origin.

Ciliary EVs isolated from mating *C. reinhardtii* gametes mediate peptidergic signaling through an amidated product that acts as a chemotactic modulator for *C. reinhardtii* gametes. In this report, we describe a protocol to accurately and quantitatively measure chemotaxis of *C. reinhardtii* cells using commercially available microfluidic Ibidi channel slides. This diffusion-based microfluidic chemotaxis assay protocol provides a long-term stable gradient of chemotactic substance that is easy to establish. The long microfluidic channel is useful for imaging and the quantitative analysis of chemotactic behavior of individual cells. Additionally, the effect on a cell population can be monitored at various points in the long microfluidic channel, which correspond to differing concentrations of the chemotactic substance. Finally, the microfluidic channel slide used in this chemotaxis assay is commercially available thereby making it easy and practical to perform this assay in other laboratories.

Acknowledgment

Our laboratory is supported by R35-GM140631 from the National Institutes of Health (to SMK).

References

Abels, E. R., & Breakefield, X. O. (2016). Introduction to extracellular vesicles: Biogenesis, RNA cargo selection, content, release, and uptake. *Cellular and Molecular Neurobiology*, *36*, 301–312.

Amintas, S., Vendrely, V., Dupin, C., Buscail, L., Laurent, C., Bournet, B., et al. (2021). Next-generation cancer biomarkers: Extracellular vesicle DNA as a circulating surrogate of tumor DNA. *Frontiers in Cell and Developmental Biology*, *8*, 622048.

Basto, R., & Marshall, W. F. (2015). Preface. *Methods in Cell Biology*, *127*, xxi–xxii.

Cao, M., Ning, J., Hernandez-Lara, C. I., Belzile, O., Wang, Q., Dutcher, S. K., et al. (2015). Uni-directional ciliary membrane protein trafficking by a cytoplasmic retrograde IFT motor and ciliary ectosome shedding. *Elife*, *4*, 05242.

Garcia-Gonzalo, F. R., & Reiter, J. F. (2017). Open sesame: how transition fibers and the transition zone control ciliary composition. *Cold Spring Harbor Perspectives in Biology*, *9*, a028134.

Harris, E. (2009). The Chlamydomonas sourcebook. *Introduction to Chlamydomonas and its laboratory use: Vol. 1* (2nd ed.). San Diego: Elsevier.

Kubo, T., Kaida, S., Abe, J., Saito, T., Fukuzawa, H., & Matsuda, Y. (2009). The *Chlamydomonas* hatching enzyme, sporangin, is expressed in specific phases of the cell cycle and is localized to the flagella of daughter cells within the sporangial cell wall. *Plant and Cell Physiology*, *50*, 572–583.

Long, H., Zhang, F., Xu, N., Liu, G., Diener, D. R., Rosenbaum, J. L., et al. (2016). Comparative analysis of ciliary membranes and ectosomes. *Current Biology*, *26*, 3327–3335.

Luxmi, R., & King, S. M. (2022). Cilia-derived vesicles: An ancient route for intercellular communication. *Seminars in Cell and Developmental Biology*, *129*, 82–92.

Luxmi, R., Kumar, D., Mains, R. E., King, S. M., & Eipper, B. A. (2019). Cilia-based peptidergic signaling. *PLoS Biology*, *17*, e3000566.

Malkin, E. Z., & Bratman, S. V. (2020). Bioactive DNA from extracellular vesicles and particles. *Cell Death & Disease*, *11*, 584.

Skotland, T., Sagini, K., Sandvig, K., & Llorente, A. (2020). An emerging focus on lipids in extracellular vesicles. *Advanced Drug Delivery Reviews*, *159*, 308–321.

Wang, J., Nikonorova, I. A., Gu, A., Sternberg, P. W., & Barr, M. M. (2020). Release and targeting of polycystin-2-carrying ciliary extracellular vesicles. *Current Biology*, *30*, R755–R756.

Wang, J., Silva, M., Haas, L. A., Morsci, N. S., Nguyen, K. C. Q., Hall, D. H., et al. (2014). *C. elegans* ciliated sensory neurons release extracellular vesicles that function in animal communication. *Current Biology*, *24*, 519–525.

Wheway, G., Nazlamova, L., & Hancock, J. T. (2018). Signaling through the primary cilium. *Frontiers in Cell and Developmental Biology*, *6*, 8.

Witman, G. B. (1986). Isolation of *Chlamydomonas* flagella and flagellar axonemes. *Methods in Enzymology*, *134*, 280–290.

Wood, C. R., Huang, K., Diener, D. R., & Rosenbaum, J. L. (2013). The cilium secretes bioactive ectosomes. *Current Biology*, *23*, 906–911.

Ye, F., Nager, A. R., & Nachury, M. V. (2018). BBSome trains remove activated GPCRs from cilia by enabling passage through the transition zone. *Journal of Cell Biology*, *217*, 1847–1868.

Zuo, X., Guo, W., & Lipschutz, J. H. (2009). The exocyst protein Sec10 is necessary for primary ciliogenesis and cystogenesis *in vitro*. *Molecular Biology of the Cell*, *20*, 2522–2529.

CHAPTER

Methods for *Paramecium tetraurelia* ciliary membrane protein identification and function

10

Megan Valentine[a], Junji Yano[b], Sukanya Lodh[c], Ashikun Nabi[d], Bin Deng[e], and Judith Van Houten[b,]*

[a]*State University of New York, Plattsburgh, NY, United States*
[b]*University of Vermont, Burlington, VT, United States*
[c]*Marquette University, Milwaukee, WI, United States*
[d]*Luminex, Austin, TX, United States*
[e]*Vermont Biomedical Research Network, University of Vermont, Burlington, VT, United States*
**Corresponding author: e-mail address: judith.vanhouten@uvm.edu*

Chapter outline

Methods in Cell Biology, Volume 175, ISSN 0091-679X, https://doi.org/10.1016/bs.mcb.2022.12.003
Copyright © 2023 Elsevier Inc. All rights reserved.

Abstract

In this chapter we provide some tools to study the ciliary proteins that make it possible for *Paramecium* cells to swim by beating their cilia. These proteins include many ion channels, accessory proteins, peripheral proteins, structural proteins, rootlets of cilia, and enzymes. Some of these proteins are also found in the soma membrane, but their distinct and critical functions are in the cilia. *Paramecium* has 4000 or more cilia per cell, giving it an advantage for biochemical studies over cells that have one primarily cilium per cell. Nonetheless, a challenge for studies of many ciliary proteins in *Paramecium* is their low abundance. We discuss here several strategies to overcome this challenge and other challenges such as working with very large channel proteins. We also include for completeness other techniques that are critical to the study of swimming behavior, such as genetic crosses, recording of swimming patterns, electrical recordings, expression of very large channel proteins, RNA Interference, among others.

1 Introduction

By using *Paramecium* we take advantage of more than a century of fundamental and transformative research (Plattner, 2002, 2022; Van Houten, 2019). In a recent special issue of Journal of Eukaryotic Microbiology, Plattner gives an overview on ciliates including *Paramecium* as a model organism and reinforces how this research has evolved from early microscopy to transmission genetics, electrophysiology, modern molecular biology, genome analysis, biochemistry, and the most powerful electron microscopy techniques today such as cryoelectron tomography (Bouhouche et al., 2022; Plattner, 2022; Valentine & Houten, 2021, 2022; Van Houten, 2019).

Paramecia swim forward by beating their cilia at rates controlled by the membrane potential and they turn in their swimming paths, likewise, through a process controlled by membrane potential (Eckert, 1972; Machemer, 1988a, 1988b). The calcium action potential is fundamental to this change of direction as reviewed in (Eckert, 1972; Kung, Chang, Satow, Van Houten, & Hansma, 1975; Valentine & Van Houten, 2022). During the action potential, triggered when the membrane potential exceeds the threshold for the action potential, the ciliary voltage gated Ca^{2+} channels (Ca_V) open and the resulting in-rush of Ca^{2+} depolarizes the cell and causes the ciliary axoneme to change power stroke reversing the beat. The cell quickly returns the membrane potential to rest with voltage-dependent (K_v) and Ca^{2+} dependent (K_{Ca}) K channels, but the cell continues to move in the backward direction until the Ca^{2+} levels have returned to rest.

Our interest has been in the proteins of the cilia that participate in these processes. For a long time the physiology of these proteins has been known through extraordinary electrophysiological studies (reviewed in (Valentine & Van Houten, 2022)). The voltage-gated Ca^{2+} (Ca_V) channels and voltage-dependent (K_v) and Ca^{2+} dependent (K_{Ca}) K channels are thought to be exclusively in the cilia (Brehm, Dunlap, & Eckert, 1978; Brehm & Eckert, 1978b; Dunlap, 1977; Eckert, 1972; Saimi & Kung, 1982). Our interest has been in identifying the proteins that are responsible for these ciliary conductances.

The most abundant protein class in *Paramecium* cilia are the peripheral glycosyl phosphatidyl inositol anchored (GPI) proteins. These proteins are so abundant that they hinder protein identification by LC MS/MS. Therefore, separating the membrane and other proteins of interest from the GPI fractions and further concentration is critical for successful analysis by LC MS/MS and other methods. While we have sometimes directly mined the annotated genome when hunting for proteins of interest, such as the homologs of the ciliopathy gene PKD2 (Valentine, Yano, & Van Houten, 2019), other genes cannot be approached this way. Because of three whole genome duplications in its past, *P. tetraurelia* has a large number of homologs and ohnologs in its genome, making it difficult to assign function to each, and making RNAi challenging because of off-target effects. Therefore, as an example, whittling down the potential K channels in the cilia by first identifying the ciliary proteins by LC MS/MS allows for the genetic analysis of the relevant proteins and genes from among almost 800 in the genome (Haynes, Ling, Saimi, & Kung, 2003). Once proteins are identified by LC MS/MS, especially if they are highly enriched or exclusive to the ciliary membrane, the process of identifying and then manipulating relevant genes is greatly simplified.

To begin, in Section 2 we discuss the isolation, concentration and study of ciliary membrane proteins, especially in preparation for LC MS/MS mass spectrometry. Differential protein solubilization is critical; we describe the use of detergents and phase separation (Section 2). To further separate ciliary membrane proteins, we take advantage of their differential partition into detergent resistant membranes (DRMs) (Section 3). DRMs can be separated by their density using techniques such as sucrose or Optiprep gradients. We take a diversion in Section 3 to describe the use of density gradients to study the ciliary striated rootlets, which are not membrane proteins but are critical for the proper patterning and beating of *Paramecium* cilia (Nabi, Yano, Valentine, Picariello, & Van Houten, 2019).

In Section 4, we continue with biochemistry and include molecular biology for the design of fluorescent epitope tagging of proteins and over-expression, techniques that complement the section on isolation and concentration of ciliary proteins. These techniques allow us to confirm whether proteins locate differentially in membranes, and use selective pathways for trafficking to their membrane destination (Valentine & Houten, 2021; Valentine et al., 2012). Microinjection is necessary for transformation of cells with expression constructs, and has other uses as shown by Haga (2022).

Continuing with molecular genetics, we will discuss the use of RNA Interference (RNAi) in Section 5 to probe for protein function. Section 6 covers techniques that follow on from the creation of cells expressing tagged proteins or the effects of RNAi: RT-PCR, Scanning electron microscopy, transmission electron microscopy.

Section 7 details an example of a genetic cross, still useful in the creation of cell lines with specific genotypes. For a modern expansion of forward genetics, we will refer to the work of others on combining mutagenesis with whole genome sequencing (Bhullar et al., 2018; Marker, Carradec, Tanty, Arnaiz, & Meyer, 2014). *P. tetraurelia* has long been studied with classical Mendelian crosses, and because this technique remains relevant as we describe it here.

Last, we will bring all of these discussions back together in Section 8 with a short primer on electrophysiology to study indirectly how ciliary channels control swimming behavior, and techniques to analyze behavior.

2 Triton X-114-phase separation for the analysis of membrane proteins by mass spectrometry

Introduction. The identification of *Paramecium* ciliary membrane proteins using mass spectrometry can be challenging because many proteins are in low abundance, such as calcium channels (Yano et al., 2013). However, this challenge for membrane integral proteins, transmembrane proteins and membrane-associated proteins modified (lipidated) by myristoylation, palmitoylation, prenylation or GPI-anchor can be partially overcome using differential solubilization in 1% Triton X114 at 4 °C, as we did in our immunoprecipitation (IP) protocol (Lodh, Yano, Valentine, & Van Houten, 2016; Yano, Wells, Lam, & Van Houten, 2021). The Triton X-114 solution separates into an aqueous phase and a detergent phase above 20 °C, with integral membrane proteins partitioning into the detergent phase and hydrophilic proteins into the aqueous phase (Bordier, 1981). Therefore, the phase separation of proteins using Triton X-114 is very useful for the concentration or purification of the ciliary membrane proteins, especially low abundance proteins that partition into the detergent phase (Fig. 1) (Yano et al., 2013). The large GPI-anchored proteins, including the large surface antigens, make up the majority of ciliary membrane proteins (Capdeville & Benwakrim, 1996), and are cleaved from the lipid moiety of their GPI-anchor by a treatment of cold Triton X-114. These cleaved proteins are then partitioned into the aqueous phase, not into detergent one (Paquette, Rakochy, Bush, & Van Houten, 2001; Yano et al., 2013). In contrast, small less abundant GPI-anchored proteins such as the GPI-anchored folate receptor (Valentine et al., 2012; Weeraratne, 2007) are considered resistant to cold Triton X-114 and separate into the detergent phase (Fig. 2) (Yano et al., 2013). Thus, the removal of the high GPI-anchored proteins into the aqueous phase allows us to detect and analyze with mass spectrometry the low abundance membrane proteins such as Ca_V channels and K channels from the detergent phase (Yano et al., 2013).

We describe here the protocols that we have used in publications and also provide more current methods from the Proteomics Facility of the Vermont INBRE (Vermont Biomedical Research Network) at the University of Vermont (https://vbrn.org/proteomics/).

(1) *Cell Culture and Isolation of Cilia* (modified from Adoutte et al., 1980)

1. Grow wild type cells or cells expressing tagged proteins to late log phase in the volumes of wheat culture required (Sasner & Van Houten, 1989). Do not allow the cells to starve, as they will start autogamy and lose any linearized expression plasmids.
2. Collect cells using a continuous flow clinical centrifuge at speed 5 ($\sim$500 × *g*). Further concentrate the cells using pear-shaped centrifuge tubes

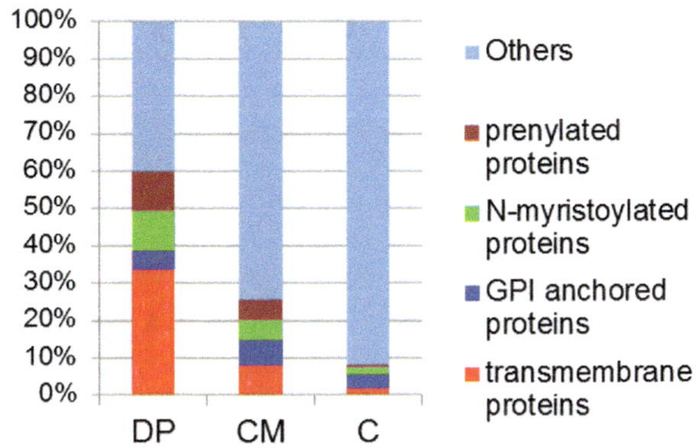

FIG. 1

Frequency of integral membrane proteins identified by mass spectrometry from detergent phase (DP), pure ciliary membrane (CM), and whole cilia (C) using Triton X-114 phase separation. Lipidated proteins, such myristoylated and prenylated proteins, are associated with the ciliary membrane and partitioned into the DP. The ratio of transmembrane proteins is increased in the DP of the Triton X-114 phase separation.

From Yano, J., Rajendran, A., Valentine, M. S., Saha, M., Ballif, B. A., & Van Houten, J. L. (2013). Proteomic analysis of the cilia membrane of Paramecium tetraurelia. Journal of Proteomics, 78, *113–122. doi: 10.1016/j.jprot.2012.09.040, with permissions.*

in an IEC NHS-II clinical centrifuge with pear-shaped vessels (see *Note 1*) using an oil testing centrifuge at speed ¾ ($\sim 350 \times g$) for 2 min.

3. Quickly remove cells using a Pasteur pipette and resuspend the cells in a beaker containing 100–200 mL of Dryl's solution (1 mM NaH_2PO_4, 1 mM Na_2HPO_4, 2 mM Na_3 citrate, 1.5 mM $CaCl_2$, pH 6.8) at 4 °C. Spin and collect the cells in the clinical centrifuge at $\sim 350 \times g$ for 2 min after removing the discharged trichocysts (see *Note 2*).
4. Remove the trichocyst layer and wash the cells in fresh 100–200 mL of Dryl's solution at 4 °C. Continue to centrifuge and remove trichocysts as described, until the fluffy layer is gone.
5. Suspend the washed cells with a 1:1 mixture of Dryl's solution and STEN Buffer (500 mM sucrose, 30 mM Tris-Cl, 2 mM EDTA, 6 mM NaCl, pH 7.5) at 4 °C for 10 min. Cells should be immobilized within 10 min.
6. Add enough cilia shock buffer (60 mM $CaCl_2$ and 180 mM KCl) so the final concentration is 10 mM Ca^{2+} and 30 mM K^+ to the cell suspension (1:6). Swirl the mixture gently and keep on ice for 5–10 min. It is important to monitor whether the cells are deciliated and lysed under a phase microscope. If lysed cells are observed during this process, cilia isolation should be canceled because intracellular components such as ER and mitochondria will contaminate the isolated cilia preparation.
7. When 80–90% cells are deciliated, centrifuge the suspension in the NHS-II clinical centrifuge at $1350 \times g$ for 2 min to pellet cell bodies using the

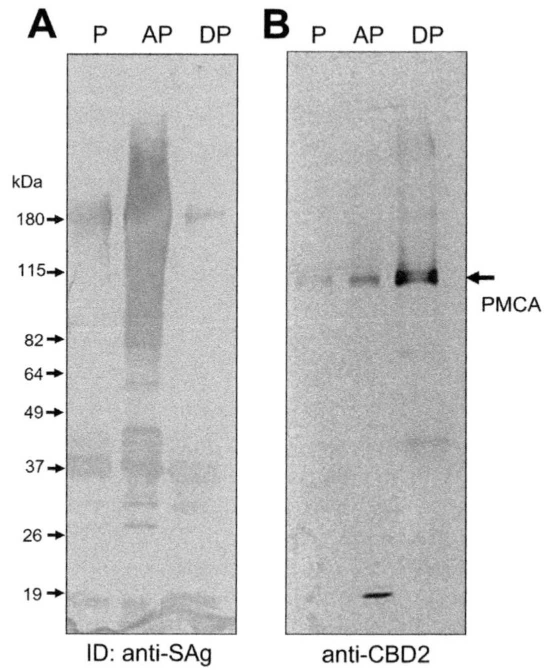

FIG. 2

Western blot analysis of ciliary membrane proteins partitioned into the aqueous phase (AP) and the detergent phase (DP) using Triton X-114 phase separation. The ciliary membrane is solubilized with 1% Triton X-114 at 4 °C for 1 h then centrifuged at 14,000 × *g* for 10 min at 4 °C. The supernatant is then used for Triton X-114 phase separation. Total protein from each sample is loaded on the gel and transblotted. Insoluble proteins (pellet, P) are also used. (A) GPI-anchored proteins are detected on the Western blot by anti-surface antigen antisera (anti-SAg). Most of GPI-anchored proteins are partitioned into the AP. (B) After stripping the blot (A), the plasma membrane calcium ATPases (PMCAs) were detected using anti-CBD2 against the calmodulin binding domain of PMCA2 (ptPMCA1a) and detected using ECL (arrow). Most of the PMCA detected is partitioned into the DP.

From Yano, J., Rajendran, A., Valentine, M. S., Saha, M., Ballif, B. A., & Van Houten, J. L. (2013). Proteomic analysis of the cilia membrane of Paramecium tetraurelia. Journal of Proteomics, *78, 113–122. doi: 10.1016/j.jprot.2012.09.040, with permissions.*

pear-shaped tubes. Pour the supernatant into new pear-shaped tubes and centrifuge under the same conditions. The supernatant here should be cloudy as it contains the cilia.

8. Suspend the deciliated cell pellet in 100 mL of HM buffer (20 mM Tris-HCl, 1 mM EDTA, 25 mM KCl, and 50 mM Sucrose, pH 7.5) at 4 °C to isolate the crude pellicle (see Section 3, *Preparing Crude Pellicle Containing Plasma Membrane and Alveolar Membrane, Without the Endoplasmic Reticulum (ER)*).

9. Centrifuge the collected supernatant at 28,000 × *g* for 20 min at 4 °C to pellet cilia. Suspend cilia to a suitable buffer for the next desired step, ciliary membrane preparation, immunoprecipitation (IP) or mass spectrometry, etc.

(2) *Preparing Ciliary Membrane* (modified from Adoutte et al., 1980)

1. Suspend cilia in a freshly prepared solution (1 mM Tris, 0.1 mM EDTA, pH 8.0) at 4 °C and vortex vigorously for 2 min to separate the ciliary membrane from the axoneme. Centrifuge the suspension at 48,000 × *g* for 30 min at 4 °C to recover all ciliary components.
2. Prepare sucrose step gradient: load 0.7 mL of 66% (*w*/w) sucrose in 10 mM Tris (pH 8.0) on the bottom of ultracentrifuge tube, and overlay 1.7 mL of 55% sucrose, 1.7 mL of 40% sucrose, and 0.7 mL of 20% sucrose in order from the bottom. Allow the sucrose step gradient to sit for about 1 h before using.
3. Before loading the cilia mixture, measure the protein concentration of each sample with a protein assay kit (e.g., Pierce Micro BCA Protein Assay (Thermo Fisher Scientific, Waltham, MA, USA)) following the protocol provided by the company. Adjust the suspension to a concentration to 5–10 mg/mL of protein before loading 0.2–0.3 mL of that suspension on top of the step gradients.
4. Place the centrifuge tubes in a Beckman SW 60 Ti swinging bucket rotor and centrifuge at 200,000 × *g* in a Beckman LB-70M ultracentrifuge for 90 min at 4 °C to separate the ciliary membrane from the axonemal components. The resulting band of material observed within the 45% sucrose is the ciliary membrane. Partially demembranated cilia is within the 55% sucrose, and the axonemes, trichocysts, and some bacteria are found at the interface between 55% and 66% sucrose.
5. Collect the band of ciliary membrane carefully, usually done using an insulin syringe to puncture the side of the tube and not mix the suspension with others. Resuspend the collected sample in 20 times the collected volume of 10 mM Tris, pH 8.0. Check whether the ciliary membrane is pure under phase contrast microscope.
6. Centrifuge at 48,000 × *g* for 30 min at 4 °C to concentrate the pure ciliary membrane. Suspend the ciliary membrane in Tris buffered saline (TBS) (20 mM Tris, 150 mM NaCl, pH 7.5) or membrane buffer (10 mM Tris buffer, pH 7.4, 50 mM KCl, 5 mM $MgCl_2$, 1 mM EGTA) for next steps: Triton X-114 phase separation, Western blot analysis, or mass spectrometry, etc.

(3) *Triton X-114 Phase Separation; Membrane Proteins Partitioned into Detergent Phase, but Large GPI-Anchored Proteins are Mostly in the Aqueous Phase* (from Yano et al., 2013)

1. Treat the ciliary membrane with 1 mL of 1% Triton X-114 in TBS or membrane buffer at 4 °C for 2 h to solubilize the ciliary membrane proteins.
2. Centrifuge the solubilized sample at 14,000 × *g* for 10 min at 4 °C to remove the insoluble materials. Suspend the insoluble materials in 50 μL of the membrane buffer at 4 °C for Western blot.

3. Overlay the supernatant on 100 μL of 6% sucrose in TBS or membrane buffer, and kept at 37 °C for 4 min. The supernatant becomes cloudy during the warming process.
4. Centrifuge the sample at 8000 × *g* for 2 min at room temperature to separate two phases. The aqueous phase (top layer) containing hydrophilic proteins is clear and the detergent phase (bottom layer) containing hydrophobic proteins is partitioned into 6% sucrose layer, which appears cloudy. The aqueous phase is collected to a new tube for the next analysis such as Western blot.
5. Repeat Triton X-114 phase separation as in previous step. Add 1 mL of TBS or membrane buffer to the detergent phase and then mix. Overlay the mixture on 100 μL of 6% sucrose and kept at 37 °C for 4 min. This step is necessary for removing the leftover from the aqueous phase in first round.
6. Centrifuge at 8000 × *g* for 2 min at room temperature. Discard the aqueous phase. Transfer the detergent phase to the new tube.
7. To precipitate the proteins, add four times the volume of cold acetone (−20 °C) to first aqueous and second detergent phase and then kept at 4 °C for 30 min. Centrifuge at 14,000 × *g* for 15 min at 4 °C to pellet the precipitated proteins.
8. Wash the pellet in cold 90% acetone three times. After briefly air-drying, dissolve the pelleted proteins in membrane buffer for next steps, such as protein concentration, Western blot analysis, or mass spectrometry (see *Note 3*).

(4) *Polyacrylamide Gel Electrophoresis (PAGE): Coomassie Blue or Silver Staining, and In-Gel Digestion*

a. *Coomassie blue staining* (Yano et al., 2013).

1. Add 6× SDS sample buffer and 3% βME at the final concentration to protein sample and then boil for 5 min.
2. Run proteins on a 4–18% gradient SDS-polyacrylamide gel using electrophoresis.
3. Rinse the gel with deionized water for 1 min. Gently shake to stain gel with 0.1% (*w*/*v*) Coomassie brilliant blue R-250, 30% (*v*/v) ethanol, and 5% (v/v) acetic acid for 15 min or with Coomassie stain recommended by LC-MS/MS facility.
4. Rinse gel with deionized water once.
5. Gently shake gel with the deionized water or a destain solution with 10% ethanol and 7.5% acetic acid until desired background is achieved.
6. Discard the destain solution and pore the deionized water. Keep gel at 4 °C until it is to be processed.

b. *Silver staining* (e.g., FOCUS™FASTsilver, G-Bioscience, Cat. # 786-240 or 786-240T) (from Lodh et al., 2016; Valentine et al., 2012)

1. Follow the protocol provided by the company for staining the gel. In this instance, after electrophoresis, fix the gel in a fixing solution (30% ethanol and 10% glacial acid) for 30 min to 3 h.

2. Wash the gel twice in 10% ethanol for 5–10 min each with gentle rocking.
3. Wash the gel three times in deionized water for 5–10 min each with gentle rocking.
4. For the staining solution, add 5 mL FOCUS™ Silver Stain to 45 mL deionized water and then add 65 μL of Sensitizer-I.
5. Soak the gel in the staining solution for 20–30 min with gentle rocking of the gel.
6. During the silver stain, prepare the developer. Add two full heaping spoons (~3–4 mg) of FOCUS™ Developer to 100 mL of deionized water. After dissolving the developer, add 65 μL of Sensitizer-I and 65 μL of Sensitizer-II.
7. Rinse the gel in deionized water for 15–20 s.
8. Add the developer to the gel and gently shake. When the band density reaches to the desirable level, immediately stop development by rinsing the gel in 2% acetic acid. Transfer the gel to new acetic acid and incubate for 10 min. You can store the gel in a deionized water or 2% acetic acid at 4 °C until processing.

(5) *In-Gel Trypsin Digestion* (from Lodh et al., 2016; Valentine et al., 2012; Yano et al., 2013, 2021) (see *Note 4*)

1. Excise gel region containing proteins of interest according to the molecular weight markers on a clean glass place.
2. Dice each gel slice into 1 mm^3 cubes and put the cubes into 1.5 mL Eppendorf tube.
3. Wash the gel cubes with 1 mL HPLC-grade water once. The gel slices can also sit in the 1 mL of water while the entire gel is processed.
4. For Coomassie Blue stained gel bands, incubate gel pieces with destain buffer [50 mM ammonium bicarbonate (NH_4HCO_3) and 50% acetonitrile (MeCN)] to remove the Coomassie Blue. Repeat with additional rinses if needed until the blue color almost completely disappears.
5. For silver stained gels, incubate gel pieces with 30 mM $K_3Fe(CN)_6$ and 100 mM $Na_2S_2O_3$ in Ultra-Pure water for 15 min. Discard the solution and add 250 μL of 100 mM NH_4HCO_3. Incubate until the band (brownish yellow color) disappears. You may perform cysteine reduction and alkylation after destaining silver-stained gel pieces. Reduce disulfide bonds with 1 mL of 100 mM DTT in NH_4HCO_3 for 1 h at 56 °C. After discarding the DTT solution, alkylate cysteines by adding 1 mL of 10 mM iodoactamide (IAA) in 100 mM NH_4HCO_3 for 40 min in a dark place at room temperature. After discarding the IAA solution, wash gel pieces with 100 mM NH_4HCO_3.
6. After removing the destaining solution, incubate the gel pieces with 200 μL of 100% MeCN for 10 min to dehydrate the gel pieces until they are shrunk and appear opaque, and then remove the solution. Dry the gel pieces in a speed vac for 15 min to completely remove the MeCN.

7. Perform the in-gel trypsin digestion with 6 ng/mL sequencing grade modified trypsin (Promega, Madison, WI, USA) in 5% MeCN and 40 mM NH_4HCO_3. Ensure the gel pieces are fully hydrated by the trypsin solution by incubating them for 45 min on ice. Add more trypsin solution until the gel pieces are fully covered. Incubate at 37 °C for no more than 18 h.
8. Spin down the gel pieces at 10,000 × *g* using a microcentrifuge for 5 min. Collect any liquid and transfer to a clean 1.5 mL Eppendorf tube. A gel loading tip can help to collect this liquid.
9. Next, extract the digested peptides by adding 150 μL of 5% Formic Acid (FA) and 50% MeCN and vortex. Spin down gel pieces at 16,000 × *g* for 10 min. Transfer the peptide extracted liquid to the same tube as the previous step.
10. Add enough 100% of MeCN to gel pieces so they are covered and vortex. Let sit for 5 min and then spin at 16,000 × *g* for 10 min. Transfer the peptide extracts to the same tube as in the previous steps. Dry the combined peptide extracts in a speed vac.
11. Once dry, resuspend the dried peptides in 10 μL of a solution of 2.5% MeCN and 2.5% FA.

(6) *Analyzing Peptides*

The following method was described in (Yano et al., 2021) for LTQ XL MS (Thermo Fisher Scientific) coupled to a conventional HPLC. However, methods that are developed for specific mass spectrometers (e.g., high resolution high mass accuracy instrumentation Q-Exactive coupled to a UPLC) should be used instead.

1. Depending on the concentration of the samples and instrument sensitivity, a portion of the peptide-suspension is loaded onto a 100 μm inner diameter capillary fused silica column packed with MAGIC C18 (5 μm particle size, 20 nm pore size, Michrom BioResources, Auburn, CA, USA) (chromatographic materials of smaller bead size (e.g., 1.8 μm) should be used for UPPLC).
2. Peptides are separated at a flow rate of 300 nL/min for 45 min by a gradient of 2.5–35% MeCN/0.1% FA, 35–100% MeCN/0.1% FA and 100% MeCN/0.1% FA, followed by an immediate return to 2.5% MeCN/0.1% FA and a hold of 2.5% MeCN/0.1% FA.
3. Separated peptides are introduced into the mass spectrometer via a nanospray ionization source and a laser pulled ~3 μm orifice at a spray voltage of 1.8 kV.
4. MS data are acquired in a data-dependent "top 10" acquisition mode, in which a survey scan from *m/z* 400–1600 is followed by 10 collision-induced dissociation (CID) tandem mass spectrometry (MS/MS) scans of the most abundant ions. Parameters for MS/MS scans of LTQ-XL are (i) isolation width = 2 m/z, (ii) normalized collision energy = 35%, (iii) activation

Q=0.250 and activation time=30ms, and (iv) Dynamic Exclusion enabled. Parameters for MS/MS scans of QE are acquired from m/z 350 to 1600 at 70,000 resolution (automatic gain control [AGC] target, $1e^6$; maximum ion time [max IT], 100ms; profile mode), and the lock mass function is activated (m/z, 371.1012; use lock masses, best; lock mass injection, full MS). Resolution for dd-MS^2 spectra is set to 17,500 (AGC target: $1e^5$) with a maximum ion injection time of 50ms. the normalized collision energy is 27eV.

5. Product ion spectra are searched with SEQUEST using Proteome Discoverer 2.4 (Thermo Fisher Scientific) against protein sequences of *P. tetraurelia* stock d4-2 or 51s downloaded from ParameciumDB (http://paramecium.i2bc.paris-saclay.fr/; Arnaiz, Meyer, & Sperling, 2020) in a concatenated target-decoy fashion. Suggested search parameters for the instrument type of LTQ and QE are (i) full enzyme activity; (ii) maximum missed cleavages=2; (iii) minimum peptide length=6; (iv) mass tolerance (LTQ) of 2Da for precursor ions and 0.8Da for fragmented ions or a mass tolerance (QE) of 10ppm for precursor and 0.02Da for MS/MS tolerance; (v) static modification on cysteine (+57.0215Da: carbamidomethylation), dynamic modifications on methionine (+15.9949Da: oxidation) and (vi) four maximum dynamic modifications allowed per peptide. The percolator node is included in the workflow to limit the false positive (FP) rates to less than 1% in the dataset. If two or more peptides match to a given protein, the protein is considered to be identified.

(7) *Data Analysis Using Scaffold 4.11* (Proteome Software, Portland, OR, USA)

The software Scaffold was used in the study described in Yano et al. (2021).

1. As multiple gel bands are cut out from the same lane or the same sample in a typical experiment, all the raw files derived from all the gel bands of each sample lane are searched against the database as one contiguous input file, resulting in one result file (.msf) for each sample (e.g., CONTROL, TEST). The digest from each band is analyzed individually by mass spectrometry. As these bands belong to the same sample, the raw data generated for individual bands are imported into the software as "fractions." The raw files are thus combined before analyzing, resulting in one .msf result file for each sample.
2. The .msf files are incorporated into Scaffold 4.11 (Proteome Software, Portland, OR, USA) as mudPIT with "prefiltered mode" and protein cluster analysis to compare the unique peptide counts between "CONTROL" and "TEST" with respect to specific protein isoforms. False discovery rate (FDR) at protein and peptide levels at 1% and "Minimum number of peptides"=1 are selected to achieve less than 1% protein and peptide decoy FDR in the filtered dataset. Peptide identifications are shown with the corresponding XCorr values. If two or more peptides match to a certain protein, the protein is considered to be identified.

2.1 Conclusion

Even though the *P. tetraurelia* genome was annotated (ParameciumDB (Arnaiz et al., 2020)), it was not feasible just to use these sequence data to identify which channel or pump gene coded for a protein that was responsible for a physiological function. After three genome-wide duplications, the *P. tetraurelia* genome had an abundance of related proteins, such as almost 800K channel sequences (Haynes et al., 2003). Sorting through these by using RNA interference (RNAi) was not feasible; it was essential to find which proteins were physically present in the ciliary membrane or other cell fractions.

Three research approaches came together to make progress on these proteins: we developed methods to concentrate the low abundance large ciliary proteins; we identified the membrane proteins by mass spectrometry; and we became proficient at epitope tagging and expressing the large channel (>250kd) and pump proteins. As a consequence, we identified which Ca_V channels were those exclusively in the cilia and participate in the action potential (Lodh et al., 2016). We also identified which calcium pumps were highly concentrated in the cilia and likely participate in the removal of Ca^{2+} after the action potential (Yano et al., 2021).

We combined our knowledge of proteins of the cilia with techniques like RNAi to address other issues of the ciliary proteins. For example, we found that mutants called PawnA or PawnB cannot reverse their ciliary beat because their Ca_V channels fail to reach the cilia (Lodh et al., 2016). The K_V and K_{Ca} channels for repolarizing the membrane potential require the BBSome for trafficking to the cilia (Valentine et al., 2012). Proteomics shows that a possible Mg exchanger XntA (Haynes, Kung, Saimi, & Preston, 2002) is localized in the ciliary membrane and associated with the PKD2 channel, which is permeable to Mg^{2+} (Valentine et al., 2019).

3 Analysis of detergent resistant membranes (DRMs) using sucrose and OptiPrep™ density gradient ultracentrifugation

Introduction. The cell body and ciliary membranes have distinct arrays of proteins that participate in signal transduction. As we discussed above, some proteins are exclusively in the cilia, making it possible to separate signaling into surface compartments. Likewise, the lipids of the ciliary and cell body membranes of *P. tetraurelia* differ and possibly contribute to the differential function of membrane proteins of the cell. For example, the ciliary membrane of *P. tetraurelia* is enriched in sterols and sphingolipids as compared to its somatic membrane (Kaneshiro, 1987).

Small membrane lipid microdomains called rafts are sphingolipid/cholesterol-enriched (Brown, 2006; Simons & Ikonen, 1997). They can organize biological membranes into liquid-ordered regions with specific protein compositions and functions (Levental, Levental, & Heberle, 2020; Lingwood & Simons, 2010). These regions are dynamic platforms for protein–protein interactions that are necessary for signal transduction (Simons & Toomre, 2000).

The whole cilia prepared from each treatment are treated with cold non-ionic detergent Triton X-100 to isolate DRMs. The DRMs are separated and fractionated using OptiPrep™ (Sigma-Aldrich, St Louis, Mo, USA) density gradient ultracentrifugation. Methyl beta-cyclodextrin (MβCD), which binds to sterols (including cholesterol and *Paramecium's* stigmasterol) and removes them from the membrane, can be used to manipulate membrane lipid composition (Mahammad & Parmryd, 2015). In this example, cells expressing FLAG-Ca_V1c (former name VGCC1c) are treated with MβCD or mock treated).

(1) *Detergent Resistant Membranes (DRMs) from Cilia Treated with 1% Triton X-100 at 4 °C for 1 h* (from Yano et al., 2021)

1. Suspend collected cilia of 5–10 mg in the solubilization buffer (1% Triton X-100 in 150 mM NaCl, 5 mM dithiothreitol (DTT), 5 mM EDTA and 25 mM Tris; pH 7.4) containing protease inhibitors (100 μM phenylmethlsulfonyl fluoride, 1 μg/mL Leupeptin (Research Products International, Mount Prospect, IL, USA) and 1 μg/mL Pepstatin (Research Products International)) and vortex for 1 min two times to separate ciliary membrane from axonemes, and then agitate for 1 h at 4 °C. For an alternative method, see *Note 5*.
2. Centrifuge the vortexed cilia at 48,400 × *g* at 4 °C for 30 min to collect the detergent-resistant membrane (DRMs). Suspend the resulting pellet containing DRMs in 500 μL solubilization buffer with protease inhibitors at 4 °C and vortex vigorously until completely suspended.
3. Mix the DRM suspension with 60% iodixanol (OptiPrep™) at a final concentration of 40% (*v*/v) (total 1500 μL).
4. Place 750 μL of 40% OptiPrep™ with DRMs on the bottom of an Ultra-clear ultracentrifuge tube (11 × 60 mm, 4.5 mL, Beckman Coulter Life Science, Indianapolis, IN, USA). Overlay aliquots of 650 μL each of 35, 30, 25, 20 and 0% (v/v) OptiPrep™ prepared in the solubilization buffer with protease inhibitors in order from the bottom.

(2) *Analysis of DRMs by OptiPrep™ Density Gradient Ultracentrifugation and Western blots* (from Yano et al., 2021)

a. *OptiPrep™ Density Gradient Ultracentrifugation*

1. Place the centrifuge tubes in a Beckman SW 60 Ti swinging bucket rotor and centrifuge at 160,000 × *g* at 4 °C for 4 h in Beckman LB-70M ultracentrifuge to fractionate the DRMs.
2. Harvest each of the eight fractions each with equal volumes of 500 μL from top to bottom from each tube carefully. Combine same number fractions from the centrifuge tubes and mix with enough solubilization buffer (containing protease inhibitors) so each fraction is one tenth of the final volume.
3. Centrifuged samples at 48,400 × *g* at 4 °C for 30 min to collect the DRMs' pellets.
4. Suspend each pellet in 50 μL membrane buffer.

b. *Western blot analysis*

1. Add 6× SDS sample buffer (375 mM Tris-HCl, pH 6.8, 12% SDS, 60% glycerol, and 0.06% bromophenol blue) and 3% βME at the final concentration to each protein sample and boil for 5 min before placing on ice.
2. Separate the proteins on 4–18% gradient polyacrylamide gel (PAGE).
3. Electro-transfer the separated proteins to a nitrocellulose membrane at 250 mA and 390 mA for 60 min each (or a membrane of choice).
4. Incubate the blotted membrane in blocking buffer [TBST buffer (16 mM Tris-HCl, 4 mM Tris, 137 mM NaCl, and 0.1% Tween-20, pH 7.5) with 5% non-fat milk, 2% teleost fish gelatin, and 3% normal goat serum] for 1 h. If different molecular weight portions of the gel need to be analyzed using different antibodies, cut the blocked membrane into the desired molecular mass sections (see *Note 6*).
5. Incubate the blocked membrane or section of the membrane with the desired primary antibody against the protein(s) of interest (i.e., ptPMCA2a/2b, FLAG-Ca_V1c, or α-tubulin) in the blocking buffer overnight at 4 °C.
6. Wash the membrane in TBST buffer, first wash for 15 min followed by two additional washes for 5 min each. Then, incubate with alkaline phosphatase (AP) or HRP-conjugated secondary antibody in TBST buffer for 1 h at room temperature.
7. Wash the membrane in TBST buffer, first wash for 15 min followed by two additional washes for 5 min each. Then, develop the blot in 10 mL of nitro-blue tetrazolium/5-bromo-4-chloro-3-indolyl phosphate (NBT/BCIP) until the band color reaches to the desirable level or proceed with ECL development as per the company protocols. After development, transfer the membrane to Ultra-Pure water to stop development and wash with Ultra-Pure water several times. Air-dry the blot and keep in a plastic sleeve in the dark.

(3) *Analysis of DRMs From Somatic Membrane Using Sucrose Continuous Gradient Ultracentrifugation*

Preparing Crude Pellicle Containing Plasma Membrane and Alveolar Membrane, Without the Endoplasmic Reticulum (ER) (from Wright & van Houten, 1990)

1. Rinse deciliated cells with cold HM buffer (20 mM Tris-HCl, 1 mM EDTA, 25 mM KCl, and 50 mM Sucrose, pH 7.5) at 4 °C twice by centrifuging as described previously in Section 2, *(1) Cell Culture and Isolating Cilia.*
2. Homogenize the deciliated-cell pellet in cold HMI buffer (HM buffer with protease inhibitors such as leupeptin, pepstatin, and phenylmethyl sulfonyl fluoride (PMSF)) using a Potter-Elvehjem homogenizer to disrupt cells by hand or connect the pestle to a drill. Cells should be homogenized until no intact cells are observed via a phase contrast microscope.
3. Centrifuge homogenates at 1240 × *g* for 5 min at 4 °C.

4. Discard the supernatant and wash the pellet in cold HMI buffer by vortexing the pellet in the buffer for 30 s and centrifuging at 1240 × *g* for 5 min at 4 °C. The supernatant should be discarded, and the pellet resuspended in fresh cold HMI buffer several times until the supernatant appears clear (see *Note 7*).
5. Suspend the final pellet in membrane buffer with protease inhibitors and check the protein concentration using an appropriate method. Adjust final protein to ~5 mg/mL protein.

(4) *Detergent Resistant Membranes (DRMs) From Crude Pellicle Treated With 1% Triton X-100 and Fractionation of DRMs by 5–40% Continuous Linear Density Gradient* (from Pan, 2008)

1. Add membrane buffer with 1% Triton X-100 and protease inhibitors to the isolated crude pellicle and vortex for 2 min twice. Incubate the suspension on a shaker for 1 h at 4 °C.
2. Centrifuge at 28,200 × *g* for 10 min at 4 °C to collect the DRMs.
3. Suspend DRMs (resulting pellet) with 500 μL of membrane buffer. Add 500 μL of TNE buffer (10 mM Tris, 100 mM NaCl, and 1 mM EDTA, pH 7.4) with 80% (*w*/*v*) sucrose to the DRMs suspension.
4. Load 500 μL of DRM suspension in 40% sucrose onto the bottom of an Ultra-clear ultracentrifuge tube and overlay with 4 mL of a 5–40% linear sucrose gradient with the TNE buffer made using a continuous gradient maker and a Buchler Duostltic pump.
5. Place the centrifuge tubes in a Beckman SW 60 Ti swinging-bucket rotor and centrifuge at 160,000 × *g* at 4 °C for 4 h in Beckman LB-70M ultracentrifuge to fractionate DRMs (see *Note 8*).
6. Harvest nine fractions each with equal volumes of 500 μL from top to bottom from each tube carefully. Combine the same number fractions and dilute with enough TNE buffer that the collected fractions are one-tenth of the final volume.
7. Centrifuge at 48,400 × *g* at 4 °C for 30 min to get the DRMs' pellet (see *Note 9*).
8. Suspend the pellets in membrane buffer with protease inhibitors at 4 °C. Be sure to also check the protein concentration and if needed, adjust the final concentration to ~5 mg/mL protein with membrane buffer.
9. Follow by Western blot analysis as already described.

(5) *Depletion of Sterols by Methyl-Beta-Cyclodextrin* (from Pan, 2008; Ray, 2008)

1. To deplete the sterols, treat cells with (test) and treat cells without (control) 4.6 mM methyl-β cyclodextrin (MβCD) in 2 mM NaCl chemokinesis buffer (1 mM calcium citrate, 1 mM Tris, 2 mM NaCl, pH 7.05 adjusted with 100 mM Tris Base) for 2 h at 28 °C. Cells should be checked for viability under this condition, as the cells should not die.
2. Collect cells immediately after treatment and suspend in Dryl's solution as described earlier to collect cilia (see Section 2, *(1) Cell Culture and Isolation of Cilia*) (see *Note 10*).

The combination with the analyses of DRMs and co-IP is useful for examining potential protein-protein interactions. As an example, plasma membrane Ca ATPase (ptPMCA2a and ptPMCA2b) are in the same lipid fraction and close enough to ciliary Ca_V channels to co-IP (Ca_V1a, Ca_V1b, and Ca_V1c) in DRMs and these pumps may regulate Ca_V channel activity by affecting local Ca^{2+} levels that can feed back on the channels (Yano et al., 2021).

The ciliary lipid composition may affect the ciliary Ca^{2+} or K^+ channel activity involved in Ca^{2+} action potential (Forte, Satow, Nelson, & Kung, 1981). We found that MβCD depleted 17% of sterols from ciliary membranes and about 40% of sterols from the cell body (i.e., the pellicle containing somatic membrane and alveolar membrane) (Pan, 2008; Ray, 2008). The reduction of sterols by MβCD causes a very small but significant decrease of backward swimming duration under depolarization by high K^+. The ciliary membrane proteins such as ptPMCA2a/2b and $Ca_V1a/1b/1c$ are located in the MβCD-resistant lipid domain of ciliary membrane. Ciliary ptPMCA1a is also located in the MβCD-resistant lipid domain, but somatic one in the MβCD-sensitive (Pan, 2008).

(6) *More on the use of density gradients: Using density gradients to isolate striated rootlets (kinetodesmal fibers) of the cilia*

Cilia arise from the surface of the cell, and just below that surface are important structural anchors that help to stabilize the cilia and organize them into longitudinal rows (reviewed in (Tassin, Lemullois, & Aubusson-Fleury, 2016)). Two of these appendages are composed of microtubules, the transverse microtubule and the post-ciliary rootlet, while the third is composed of non-tubulin materials called the striated rootlet (SR) (also called kinetodesmal fiber (KF) (Iftode, Adoutte, & Fleury, 1996; Nabi et al., 2019; Sperling, 1989)). The SR projects anteriorly from its basal body past numerous other basal bodies. The proteins that comprise these rootlets have a molecular mass of 30–36 kDa and appear striated, some with phosphoprotein properties (Sperling, 1989; Sperling, Keryer, Ruiz, & Beisson, 1991). These rootlets are composed of overlapping proteins called SF-assemblins that organize into five structural groups, and by using the techniques described below, we were able to isolate these rootlets and examine them using immunofluorescence (Fig. 3, immunostaining techniques described in Section 4) and using transmission electron microscopy (TEM). Additionally, RNA interference (RNAi) was used (described in Section 5) to tease apart the different paralogues and structural groups that make up this large family of SF-Assemblin proteins in *Paramecium* (Nabi et al., 2019).

Here, we followed the protocols published by Sperling (Sperling, 1989; Sperling et al., 1991) with some modifications, primarily the use of Optiprep™ instead of Percoll gradients.

(1) *Isolation of striated rootlets*

1. Harvest cells from mid-late log phase culture using a continuous flow centrifuge (IEC clinical centrifuge, 300 × *g*). Wash cells three times each in

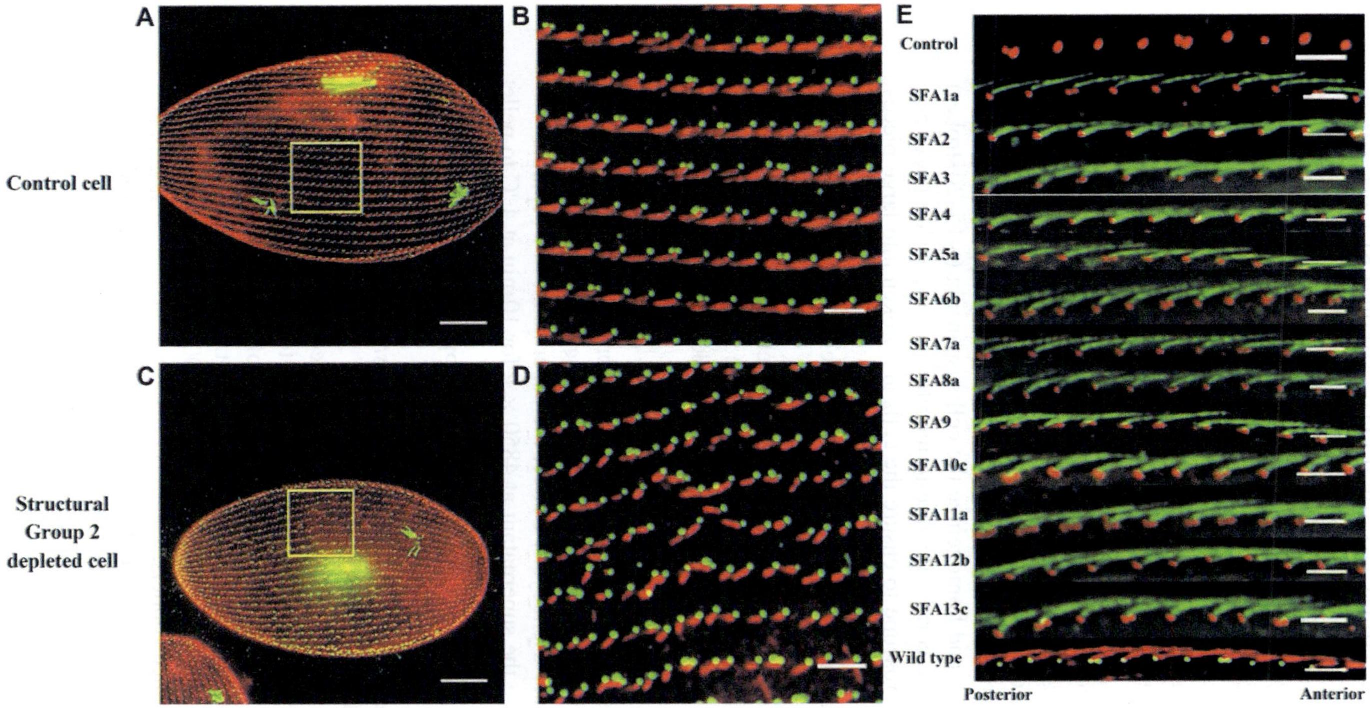

FIG. 3

Use of Immunofluorescence to monitor the location and orientation of SRs. Also a demonstration of the use of RNAi that is discussed in Section 5. The depletion of structural groups using RNAi leads to basal body row misalignment and abnormal SR appearance. (A) Control and (C) Structural Group 2 (also Paralog Group 7)-depleted cells. The yellow box show in each shows the enlarged area shown in (B & D) to highlight basal body rows. The basal bodies (green) are stained with ID5 antibody and the SRs (red) are stained with anti-SR antibody. (A & B) show straight rows of basal bodies and SRs rows that extend between the posterior and the anterior pole of the cell. (C & D) show severely misaligned basal body rows as well as abnormal SRs (shorter, not pointed toward the anterior pole) and disrupted SR rows. All images are of the dorsal surface of cells, but misalignments can occur anywhere on the surface, except the oral groove. (E) shows the SF-Assemblin (SFA) proteins in and distributed throughout the SR structures. Cells expressing only the FLAG epitope (control) or FLAG-SFA representative of each paralog group were immunostained with anti-basal body centrin (red) to highlight the basal bodies and anti-FLAG (green) to highlight the FLAG-SFA proteins. Panels show the side-view images of control, SFA1a, SFA2, SFA3, SFA4, SFA5a, SFA6b, SFA7a, SFA8a, SFA9, SFA10c, SFA11a, SFA12b, SFA13c cells expressing FLAG or FLAG-SFA. An additional control is included (bottom) showing a wild-type cell with green basal bodies (Anti-ID5) and red SRs (anti-SR). Scale bars are 15 μm (A, C) and 3 μm (B, D, E).

Images from Nabi, A., Yano, J., Valentine, M. S., Picariello, T., & Van Houten, J. L. (2019). SF-Assemblin genes in Paramecium: Phylogeny and phenotypes of RNAi silencing on the ciliary-striated rootlets and surface organization. Cilia, *8(1), 2. doi: 10.1186/s13630-019-0062-y, with permission.*

100 mL TEK buffer (20 mM Tris, 5 mM EGTA, 100 mM KCl, pH 7.0) by centrifuging (500 × *g*) using pear-shaped tubes.

2. To prepare cell cortex without membrane, resuspend the washed cells in PHEM-1% Triton X100 buffer (60 mM PIPES, 25 mM HEPES, 10 mM EGTA, 2 mM $MgCl_2$, pH 6.9 with 1% Triton X100) in a 10:1 volume ratio (buffer: cell pellet). Incubate mixture on ice for 10 min. Add protease inhibitors leupeptin (final concentration 5 μg/mL), pepstatin (final concentration 10 μg/mL) and PMSF (final concentration 1 mM/mL) to all buffers.
3. After incubation, collect the cells by centrifugation in a clinical centrifuge for 5 min at 650 × *g* and resuspend the cells in 2 volumes of PHEM buffer without Triton-X100.
4. Place the cell suspension in a 55 mL glass-Teflon homogenizer (Wheaton, Millville, NJ, USA) and homogenize the cell suspension (~15–20 strokes) to generate the crude cortex. Be sure to monitor the suspension under a phase contrast microscope to ensure the cells are broken into large pieces of cortex.
5. After homogenization, collect the crude cortex pellet centrifugation in a clinical centrifuge (1000 × *g*) for 5 min.
6. Resuspend the crude cortex pellet in PHEM buffer with protease inhibitors containing 100 mM $MgCl_2$ (2 mL per 1 mL of the original cell pellet) and homogenize very extensively, at least 70–80 strokes, at 4 °C to detach the ciliary rootlets.
7. Dilute the resulting homogenate in PHEM buffer ((1:2 volume ratio (homogenized suspension: PHEM buffer)) with protease inhibitors.
8. Centrifuge the homogenate suspension at 1940 × *g* for 7 min at 4 °C to pellet the macronuclei and trichocysts. Collect the supernatant in a separate tube and centrifuge the supernatant at 4850 × *g* for 10 min at 4 °C to pellet the crude striated rootlets.
9. Resuspend the crude striated rootlets in 600 μL of PHEM buffer and follow to OptiPrep™ step density gradient.

(2) *Fractionation of striated rootlets using OptiPrep™ step density gradient ultracentrifugation* (from Nabi et al., 2019)

1. Create a step density gradient of Optiprep™ mixed with PHEM buffer and protease inhibitors by layering, starting at the bottom, 500 μL each of 40%, 35%, 30%, 25%, 20%, 15%, and finally 500 μL of the crude rootlets suspension on the top of the step density gradient.
2. Centrifuge in a Beckman ultracentrifuge at 272,000 × *g* for 3 h at 4 °C to fractionate the striated rootlets.
3. Carefully transfer each layer (500 μL) to a new centrifuge tube. Add 10-fold of membrane buffer (10 mM Tris-Cl, 10 mM Tris-base, 50 mM KCl, 5 mM $MgCl_2$, 1 mM EGTA, pH 7.4) with protease inhibitors to each tube and mix well.

4. Centrifuge the suspension at 49,500 × *g* for 30 min at 4 °C to remove the OptiPrep™. Resuspend the resulting pellet in 100 μL PHEM buffer with protease inhibitors. The 30% Optiprep™-PHEM layer in the gradient should have a whitish diffuse band that indicates the presence of SRs. This should be confirmed by western blot and should be used for further analysis.
5. See Section 2, *(4) Polyacrylamide Gel Electrophoresis (PAGE): Coomassie Blue or Silver Staining, and In-Gel Digestion* for next steps.

(3) *Mass Spectrometry Analysis Using Q-Exactive Mass Spectrometer Coupled to an EASY-nLC system* (Nabi et al., 2019)

1. Dissolve the peptides obtained from in-gel trypsin digestion in 7 μL of 0.1% FA and 2.5% MeCN. Load 2 μL of peptide suspension onto a fused silica capillary (12 cm × 100 μm inner diameter) packed with Halo C18 (2.7 μm particle size, 90 nm pore size, Michrom Bioresources).
2. Peptides are separated through the capillary at a flow rate of 300 nL/min. The separated peptides are introduced into the mass spectrometer via a nanospray ionization source at a spray voltage of 2.2 kV.
3. Mass spectrometry data are acquired in a data-dependent top-10 mode and the lock mass function is activated (*m*/*z*, 371.1012). Full scans should be acquired from m/z 350 to 1600 at 70,000 resolutions (automatic gain control (AGC) target, 1e6; maximum ion time (max IT), 100 ms; profile mode). Resolution for dd-MS2 spectra should be set to 17,500 (AGC target: $1e^5$) with a maximum ion injection time of 50 ms. The normalized collision energy is 27 eV. A gradient of 0 to 40% acetonitrile (0.1% FA) over 55 min should be applied.
4. Search the resulting spectra against the protein sequences of *P. tetraurelia* wild type d4-2 or 51s downloaded from ParameciumDB (http://paramecium.i2bc.paris-saclay.fr/; (Arnaiz et al., 2020)) by Proteome Discoverer (PD) 1.4. The search parameters permit a 10 ppm precursor MS tolerance and a 0.02 Da MS/MS tolerance. Carboxymethylation of cysteines is set up as fixed modifications and oxidation of methionine (M) are allowed as variable modifications. Up to three missed tryptic cleavages of peptides are considered with the false-discovery rate set to 1% at the peptide level.

4 Epitope tagging, expression of proteins, and immunofluorescence

The addition of epitope or fluorescent tags on proteins has been incredibly useful to localize and immunoprecipitate proteins of interest. The creation and expression of these fluorescent tags requires some planning, and similar to other organisms, specific expression plasmids. *Paramecium* sp. have a slightly different codon usage than other eukaryotes (Hauser, Haynes, Kung, Plattner, & Kissmehl, 2000), requiring these specific plasmids and targeted modification of fluorescent proteins. The pPXV plasmid (Hauser et al., 2000) has been the go-to for protein expression and epitope-tagging in *Paramecium* sp. This plasmid has been modified to also contain epitope tags, including a 3× FLAG tag or a 3× myc tag at either the N- or C-terminus of the expressed protein (Picariello, Valentine, Yano, & Van Houten, 2014; Valentine, 2015; Valentine et al., 2012).

(1) *Plasmid Creation*

1. Design primers to amplify the gene of interest. These primers must include the start (ATG) codon and the stop (TGA) and should add additional sequence to generate restriction enzyme cut sites for the insertion of the gene using conventional methods, or Isothermal Assembly. The available cut sites for each plasmid are shown below (Fig. 4).
2. The desired gene should be amplified from genomic DNA using a proof-reading Taq polymerase (such as New England Biolabs Q5) and the suggested annealing temperatures. For example, a T_m calculator (https://tmcalculator.neb.com/#!/main) can be used if using NEB's Q5 *Taq* polymerase.
3. Visualize results using a 0.8% agarose gel stained with Ethidium Bromide. If the resulting band is the desired size, remaining PCR reaction can be used for subcloning into a blunt end vector or if needed, the amplified DNA fragment can be gel purified before subcloning into desired blunt-end vector for screening.
4. Screen resulting colonies of bacteria (~10) using colony PCR and the primers used to generate the insert. One or two positive clones should be selected to be grown in 6mL Luria broth (LB) cultures with the needed antibiotics for selective pressure at 37 °C overnight.
5. If desired, generate a glycerol stock for long-term storage at −80 °C (850 μL of culture and 150 μL of glycerol). The remaining culture should be used for plasmid extraction as per kit instructions. Assess final concentration and purity using a NanoDrop or other DNA assessment tool.
6. Remove the desired gene from the blunt-end plasmid using the desired restriction enzymes. At the same time, open the desired cloning plasmid, 1 μg in 20 μL (50 ng/μL), with the same enzymes (see *Note 11*).
7. Run digested sample on a 0.8% agarose gel stained with ethidium bromide. Extracted band should gel purified as per kit instruction and the resulting eluate concentration determined using a NanoDrop or other DNA assessment tool.
8. Ligate the insert into the opened (linear) expression plasmid using a 3:1 insert:plasmid ratio, taking into account the length or size of the plasmid and the insert. After ligation, clone into appropriate competent *E. coli* bacterial line, such as DH5α, and spread on LB plates with 100 ng/mL ampicillin. Grow plates overnight at 37 °C.

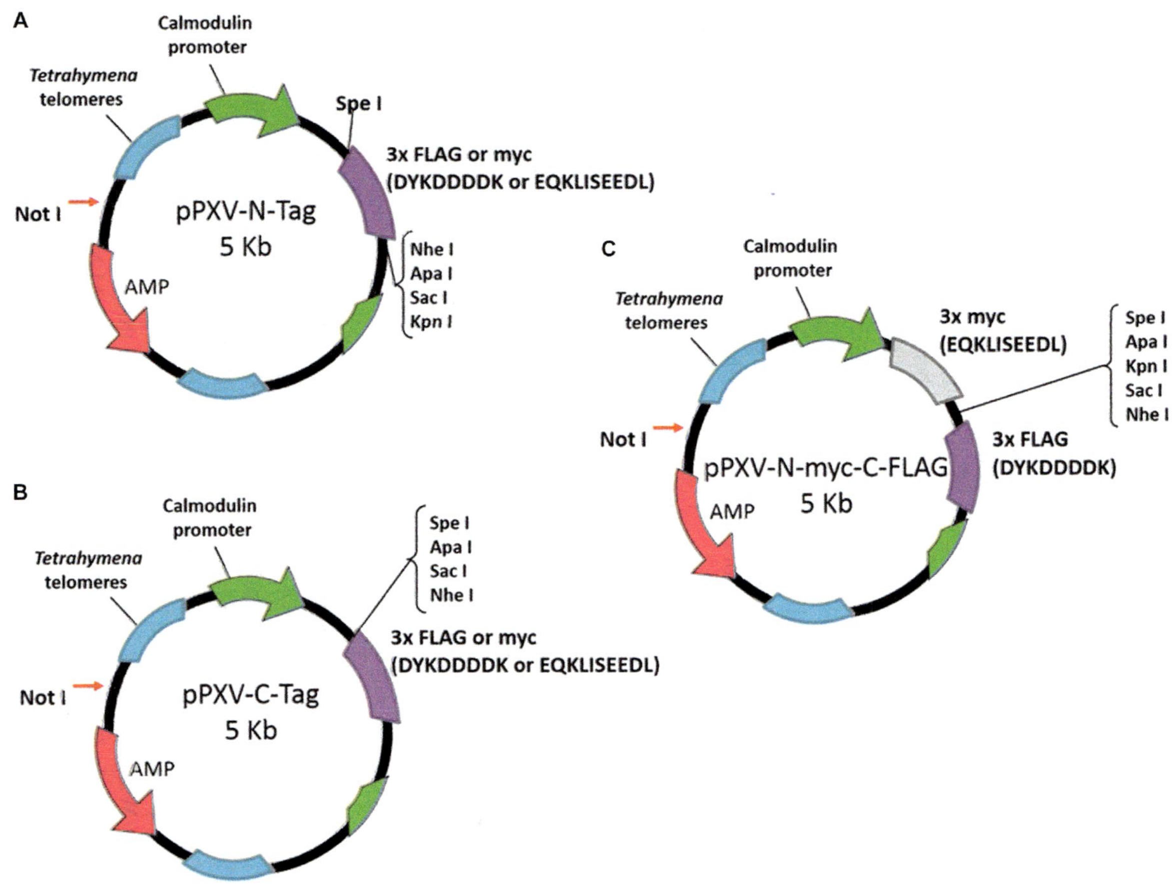

FIG. 4

Plasmid maps for 3× FLAG and 3× myc tags. (A) The multiple cloning site for the pPXV-N tag (either 3× FLAG or 3× myc) is shown while the restriction enzyme cut sites for the C-terminal 3× tag (FLAG or myc) (B) has slightly different sites, where the *Nhe*I site has been switched out for *Spe*I. (C) A double tag is also available, where 3× myc can be placed at the N-terminus and 3× FLAG at the C terminus. If only the N-terminal 3× myc tag is desired, the reverse primer for the gene amplification should be designed such that it includes at least one stop codon.

Adapted from Valentine, M., Rajendran, A., Yano, J., Weeraratne, S., Beisson, J., Cohen, J., Koll, F., & Van Houten, J. (2012). Paramecium BBS genes are key to presence of channels in Cilia. Cilia, 1*(1), 16. doi: 10.1186/2046-2530-1-16, with permissions.*

9. Screen resulting clones (~8) using PCR or a cracking gel. One or two positive clones should be used to again start 6 mL overnight shaking cultures in LB with 100 ng/mL ampicillin, grown at 37 °C while shaking. The following day, extract the plasmid as per kit instructions.
10. Plasmid must be linearized with *Not*I (Fig. 4) before microinjection or electroporation into the cells. To ensure plasmid is exceptionally clean, plasmid should be dried after NotI digestion and resuspended in UltraPure water to a concentration of 5–10 µg/mL (Valentine et al., 2012).
11. A total of 5–9 pL should be microinjected into the macronucleus of multiple cells suspended under mineral oil. After injection, cells were individually moved to depressions containing 500–1000 µL of inoculated culture fluid in a humidification chamber for 24–48 h.
12. Injected cells should be maintained as individual clones and screened though organic DNA extraction and PCR using a primer that will also amplify the tag as part of the gene.
13. Maintain clonal lines at 15 °C to slow cell growth. Transfer 4–9 cells to fresh culture fluid every 4–5 days. *Re*-screen clones for the linear plasmid as described in previous step every week (see *Note 12*).

(2) *Immunostaining*

Once the clones expressing the tagged protein are identified, larger cultures can be grown to isolate the cilia from the expressing cells or to harvest the cell membrane. Immunostaining can be used to look for the expressed protein in whole cells or particular intracellular structures can be isolated, as described for examining the striated rootlets (Nabi et al., 2019). Using cellular fractionation, different structures can be isolated and the proteins expressing these tags can then be identified or visualized. The same basic techniques are used to visualize structures like the SRs and basal bodies using antibodies specifically against their proteins, as seen with the SF-Assemblin genes of the Striated Rootlets (see Fig. 3, and Sections 3 and 7).

1. When using cells, a 100 mL culture is all that is needed and cells should be washed twice in 100 mL Dryl's solution (1 mM Na_2HPO_4, 1 mM NaH_2PO_4, 1.5 mM $CaCl_2$, 2 mM Na-citrate, pH 6.8).
2. Permeabilize collected cells in 1 mL PHEM buffer with 0.1% Triton-X100 for 60 min while remaining still at room temperature.
3. Fix cells or embedded material for 1 h in 1 mL freshly made 4% paraformaldehyde in PHEM while rocking at room temperature.
4. Wash the fixed cells or embedded material three times in 1 mL blocking buffer (2 mM NaH_2PO_4, 8 mM Na_2HPO_4, 150 mM NaCl, 1% Tween20, 1% BSA, 10 mM EGTA, 2 mM $MgCl_2$, pH 7.4).
5. Primary antibodies should then be incubated with the cells or the embedded material in the appropriate dilution while rocking gently at room temperature for 1 h. (See Nabi et al., 2019; Picariello et al., 2014; Valentine et al., 2012 for suggestions.)

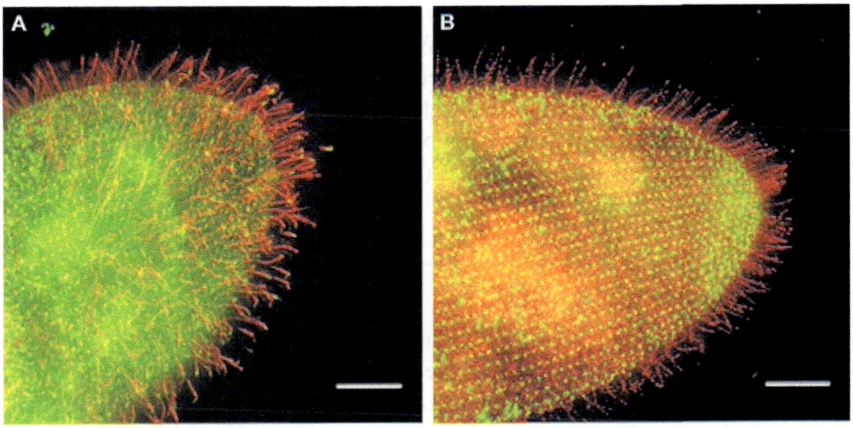

FIG. 5

Expression of epitope tagged membrane proteins in *Paramecium.* (A) a cell fed the empty vector for RNAi (L4440) and expressing the tagged potassium channel, FLAG-SK1a. The green staining shows the GPI anchored folate binding protein (Anti-FBP) and the red staining shows the FLAG-SK1a channel in the cilia. (B) *Paramecium* cell fed the empty vector for RNAi while expressing the FLAG-tagged PKD2 channel, PKD2-FLAG. The PKD2-FLAG can be seen at the cell surface and in the cilia in red (Anti-FLAG) and the basal bodies are highlighted in green using Anti-*Tetrahymena* centrin. Scale bars represent 15 μm.

From Valentine, M., Rajendran, A., Yano, J., Weeraratne, S., Beisson, J., Cohen, J., Koll, F., & Van Houten, J. (2012). Paramecium BBS genes are key to presence of channels in Cilia. Cilia, 1*(1), 16. doi: 10.1186/2046-2530-1-16.*

6. After 1 h, collect cells via brief centrifugation and wash three times with 1 mL PBS-T (2 mM NaH_2PO_4, 8 mM Na_2HPO_4, 150 mM NaCl, 1% Tween20, pH 7.4) each wash.
7. Incubate cells or embedded material in 1:10,000 dilutions of desired secondary antibodies while rocking at room temperature for 1 h while covered in aluminum foil to protect from light.
8. Wash final cells 5 times with 1 mL of the PBS-T solution, resuspend the final bit of cells in 1–2 drops of VectaShield® mounting medium for imaging using available fluorescence microscope.

Immunostaining has allowed us to visualize the location of proteins within the cell, whether the proteins are in the cell membrane or if they are in the cilia. See Fig. 5 for an example expressed proteins locating in the cell body membrane vs ciliary membrane. In addition, if fluorescent proteins, such as GFP do not fold correctly and fail to fluoresce, the use of immunostaining can illuminate the location of the tagged proteins (Hauser et al., 2000). In some cases, especially with intracellular proteins, as in the case with Rabs, the GFP tag fluoresces and can be used for live-cell imaging (Bright, Gout, & Lynch, 2017). When immunostaining is required, it is very beneficial to illuminate the proteins within different membranes and can be used in isolated structures as well, as shown by the investigation into the SF-assemblin genes that are important for the structure and interaction of the striated rootlets in *Paramecium* (Nabi et al., 2019).

5 RNA interference (RNAi)

The use of fluorescence has allowed us, and others, to find and identify the locations of different expressed proteins within the cell, in addition to using those tags for the immunoprecipitation and co-immunoprecipitation of proteins. In most cases, there were no adverse effects from the overexpression of genes with an N- or C-terminus epitope tag. At the same time, the reduction of genes in *Paramecium* has long been used to investigate the depletion of gene products and the affects the depletion has on the cells. RNAi has allowed for the temporary knock-down of many different genes and has allowed many researchers to identify roles and mechanisms for proteins, that in the case of knockouts, is lethal (Picariello et al., 2014).

Designing an RNAi construct begins with searching the annotated *Paramecium* genome (https://paramecium.i2bc.paris-saclay.fr/) for potential homologous proteins (Arnaiz et al., 2020). Those proteins that are identified can then be linked back to the genes of interest. Because of multiple genome duplications in *Paramecium* sp. (Aury et al., 2006), there are often paralogs that need to be considered when searching for genes. These gene paralogues can maintain the same function or could be divergent and important for the evolution of gene function (Gout, Kahn, Duret, & Consortium, 2010). Often, sequences used for RNAi construct design will target closely related paralogs but must be taken into consideration when choosing a sequence for RNAi design.

1. Identify potential DNA sequence for target and whether any paralogs exist within the genome.
2. Chose a portion of the sequence that can be used for RNAi design. Typically, 400–800 bp in length is sufficient, introns can be included.
3. Check chosen sequence for restriction enzyme cut sites, off-target possibilities (RNAi Off-Target tool: https://paramecium.i2bc.paris-saclay.fr/cgi/tool/rnai_off_target), and should not encompass the 3′ end of the gene (that region will be needed for the design of primers that will amplify the endogenous mRNA if desired). (See *Note 13*.)
4. Use genomic DNA to amplify the construct region (Picariello et al., 2014). Analyze PCR samples using gel electrophoresis. If bands are clean, product can be subcloned directly into the plasmid. If not, band should be excised, and gel extracted as per kit instructions (see *Note 14*).

5. Subclone into a stable bacterial cell line, such as DH5α, on plates with selective pressure. PCR can be used to screen resulting colonies for positive clones.
6. Grow at least one positive clone overnight in a shaking liquid culture (LB with antibiotics) at 37 °C for plasmid extraction. The following morning, extract the plasmid, a minimum of 1 μg should be used for digestion with the appropriate restriction enzymes as per the instructions.
7. After digestion, separate the sample(s) on an agarose gel and cut the construct pieces out of the gel and clean them using a DNA Gel extraction kit.
8. As previously described (plasmid construction), the gel extracted DNA should be ligated into the linearized plasmid using a 3:1 ratio of the insert to the plasmid and again, cloned into a stable line of bacteria such as DH5α.
9. Screen resulting colonies (~7–10) using PCR to identify positives. Pick one or two positives to be grown in overnight shaking cultures for plasmid extraction (see *Note 15*).
10. Resulting plasmid and L4440 plasmid should then be used to transform competent Ht115 bacteria and spread on LB plates with 100 μg/mL ampicillin and 12.5 μg/mL tetracycline (see *Note 16*).
11. Generate liquid overnight shaking cultures of the bacteria to make small glycerol stocks (store at −80 °C, 25 μL glycerol mixed with 225 μL bacteria) and to grow bacterial cultures to feed to the paramecia.
12. Use 1 mL of the overnight bacterial cultures to start 50 mL LB cultures with 100 μg/mL ampicillin and 12.5 μg/mL tetracycline. Shake cultures for approximately 2 h at 37 °C until the OD_{600} reaches 0.3–0.4.
13. Induce cultures with 125 μL 0.2 M IPTG (final concentration of 0.125 mg/mL) and allow the cultures to shake for an additional 3–5 h at 37 °C (see *Note 17*).
14. Collect bacteria through centrifugation at 3500 × *g* for 10 min and resuspend in 100 mL sterile wheat culture medium (2.5 g/L wheat grass boiled and filtered, buffered with 3.75 g mM Na_2HPO_4 and 3 mg/L stigmasterol) enriched with 100 μg/mL ampicillin and 0.125 mg/mL IPTG.
15. Add 100–1000 purged paramecia (see *Notes 18* and *19*) to each culture and incubate at room temperature for 48–72 h before observing/collecting cells. Every 24 h, an additional 0.125 mg/mL IPTG and 100 μg/mL ampicillin should be added. The number of cells, length of incubation with the RNAi bacteria, and strains of *Paramecium* may have an impact on results (see *Note 20*).

The use of these RNAi depleted cells for immunostaining has allowed for a better understanding of the location of proteins within the cell, both visually and through the use of Western blots and other methods. In the case of the SRs, immunostaining combined with epitope tagging and RNAi was instrumental in deciphering how these different proteins interact and where they were located within the cell and within the rootlets (Fig. 3) (Nabi et al., 2019).

6 Semi-quantitative reverse transcriptase PCR, SEM, and TEM

Once RNAi cells are produced, there are numerous tests that can be done on the depleted cells to observe the cells for changes in their phenotype, cilia, intracellular structures, and more. One such method to assess the cells for the efficacy of RNAi is to extract the total RNA from test and control cells in order to examine the mRNA levels for the targeted gene. It is vital that an internal control is used to examine the produced cDNA levels using a housekeeping gene, one that is ubiquitously expressed and would remain unaffected by the RNAi. Common genes include G-beta (Jacobs, 2007) or calmodulin (Nabi et al., 2019; Picariello et al., 2014; Valentine et al., 2012; Valentine, 2015). Cells can also be examined using immunostaining to probe for changes in cellular structures such as the cilia (Picariello et al., 2014; Valentine et al., 2012) or the anchors for those cilia (Nabi et al., 2019; Picariello et al., 2014). Extensive studies of basal body structure, development, and function in *Paramecium* have been done where the use of RNAi has been instrumental (Reviewed in Bouhouche et al., 2022; Tassin et al., 2016). To ensure the targeted gene product has been depleted, extracting RNA from the control and depleted cells after RNAi and using a semi-quantitative approach supports the validity of the experiments. This approach equalizes the amount of RNA used to generate cDNA and the resulting cDNA is tested using a ubiquitously expressed gene to support the depleted or absent mRNA transcript.

(1) *Semi-Quantitative Reverse Transcriptase (RT) PCR*

1. Collect cells by filtering through a KimWipe™ into a pear-shaped tube. Spin at 650 × *g* and collect cells using a Pasteur pipette.
2. Wash cells in 100 mL Dryl's solution and spin again as just described. Final pellet of cells should be collected in a 1.5 mL microcentrifuge tube and reduce cells to 100 μL.
3. Extract total RNA from the test and control cells according to kit protocols. Final RNA concentrations should be assessed using a Nanodrop or other concentration measuring device (see *Note 21*).
4. To make cDNA, equal amounts of RNA should be used for all samples. Either all samples should receive the same amount of RNA (depending on the reverse transcriptase being used) or if the volume is restrictive, then all samples should have the same amount (in ng or μg) as the restrictive sample.
5. Produce cDNA as per the kit protocols using an OligodT or PolydT primer (5′-CGGCTCGAGTTTTTTTTTTTTTTTTTTTT 3′). Store cDNA at −20 °C and use within 2 weeks. Serial dilutions of the cDNA should be used in PCRs to observe a reducing intensity of the produced cDNA for the intended transcript.
6. If calmodulin is to be used to test the control and test cDNA samples for similar concentrations, using cDNA that is UD (undiluted), diluted 1:10, and

1:100 are recommended. The calmodulin primers (forward: 5′-CTGAAGCT GAACTTCAAG-3′ and reverse, 5′-CAGAATGATGGTTTCTAAATGA-3′) should be used with an initial denaturation step of 5 min at 95 °C followed by 24 to 27 cycles of 95 °C for 1 min, 42 °C for 1 min, and 68 °C for 30 s. Lastly, a final extension of 68 °C for 10 min.

7. Assuming cDNA of test and control samples appears approximately equal, cDNA can then be assessed using gene-specific primers (see *Note 22*).

(2) *Scanning Electron Microscopy (SEM)*

The use of scanning electron microscopy (SEM) has provided the ability for close-up analysis of cells and their cilia (Picariello et al., 2014; Valentine et al., 2012). The use of RNAi depleted cells is simple and can show changes in the cell surface morphology as well as changes in the distribution of cilia across the cell surface or in the shape or length of the cilia themselves.

1. 200 mL of cells should be prepared by washing in Dryl's solution as already described, twice. Move cells to a 1.5 mL microcentrifuge tube with minimal fluid.
2. Treat cells with 1 mL 1% osmium tetroxide in 10 mM sodium cacodylate while remaining still on the bench top for 1 min.
3. Spin cells at $4000 \times g$ briefly (~20–30 s), remove the supernatant, and resuspend cells quickly in 1 mL 2% glutaraldehyde in 10 mM sodium cacodylate buffer.
4. Allow cells to sit at room temperature for 10 min with occasional inversion of the tube. Spin cells at $4000 \times g$ and remove supernatant. Resuspend cells in 1 mL 2% glutaraldehyde in 10 mM sodium cacodylate and incubate cells at room temperature for 1 h with occasional inversion.
5. Collect cells by centrifugation and place drops of cells on 13-mm round glass cover slips that were coated with 0.1% poly-L-lysine (high molecular weight) overnight and rinsed in PBS (137 mM NaCl, 2.7 mM KCl, 10.4 mM Na_2HPO_4, 1.7 mM KH_2PO_4, pH 7.4).
6. Allow cells to settle on the coverslips for 15 min. After, gently rinse cells that have not attached to the coverslip with PBS. Process coverslips and dry at critical point.
7. Glue coverslips to an aluminum chuck using graphite cement, allow to dry, and sputter coat before imaging.

(3) *Transmission Electron Microscopy (TEM)*

The use of TEM can help to observe changes in the subcellular structure, the structures of the basal bodies, or alterations in the cilia. TEM also provided contrast to samples after OptiPrep™ gradients to observe changes in the striations of the striated rootlets of *Paramecium* (Nabi et al., 2019). The preparations of the samples are similar, but to provide negative staining to collected structures (i.e., striated rootlets), the samples may need high-pressure freezing followed by freeze-substitution (Nabi et al., 2019; Pearson, Giddings, & Winey, 2009). By using TEM on cells that have been treated with RNAi, important clues as to changes in the smallest of cell structures of the cells or the composition of those structures can be closely investigated.

1. Wash cells in Dryl's solution as already described. Resuspend the washed cells in 1 mL 1% glutaraldehyde in 50 mM sodium cacodylate, pH 7.2.
2. Rock cells gently while on ice for 30 min. Wash pelleted cells three times, each wash for 10 min while rocking on ice with 1 mL of the same solution.
3. Collect cells and resuspend in 1 mL 1% glutaraldehyde in 50 mM sodium cacodylate with 1% osmium tetroxide, pH 7.2.
4. Rock cells gently while on ice for 30 min and wash the pelleted cells three times, just as described above.
5. Pre-embed the cells in 2% agarose gel made in 50 mM odium cacodylate in a 1.5 mL microcentrifuge tube. Once set, the embedded cells can be sliced into 1 mm × 1 mm blocks.
6. Place blocks in 50% ethanol and rock on a nutator for 30 min. Remove the ethanol and replace with an additional 1 mL of 50% ethanol.
7. Wash blocks with increasing concentrations of ethanol, so that blocks are washed in 1 mL of 70%, 90%, and twice in 100% ethanol. Two more additional washes with propylene oxide should also be completed.
8. Leave cells overnight on a nutator in 1 mL 1:1 propylene oxide to Spurr's solution.
9. Wash samples for 6 h in 1 mL fresh Spurr's before placing the blocks in flat embedding molds with fresh Spurr's. Incubate samples at 60 °C for 48 h.
10. Cut sections to desired thickness (or 90 mm thickness) and place on copper 200-mesh grids. Contrast using droplets of 2% uranyl acetate in 50% ethanol for 6 min, followed by lead citrate solution (120 mM sodium citrate, 2.66% lead nitrate, and 0.65% sodium hydroxide in water) for 4 min. Once dry, grids can be imaged.

7 Genetic crosses

A genetic cross is a useful tool to generate cells with desired genotype and phenotype. The following procedure is explained using a well-known mutant *pwA* for easy understanding. A common cell line to include in these crosses are the non-discharge (nd) mutants that provide an excellent easily testable phenotype: these cells cannot discharge their trichocysts, a calcium-dependent defense mechanism used by the cells (Buonanno, Harumoto, & Ortenzi, 2013; Harumoto & Miyake, 1991). The ability of cells to discharge their trichocysts can easily be tested by placing the cells in picric acid and observing them under a phase contrast or powerful light microscope (Fig. 6).

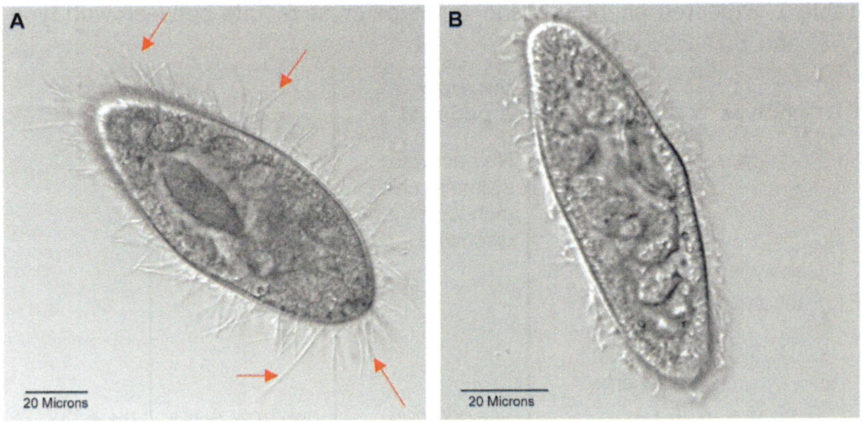

FIG. 6

Trichocyst discharge in a wild type cell (A) and the non-discharge mutant, nd6 (B). Cells were exposed to picric acid and imaged using a multiphoton microscope, scale bars are 20μm. The trichocysts are seen all around the edge of the wild type cell (red arrows). The nd6 cells only shows cilia around the edge as these cells are unable to discharge the trichocysts.

7.1 Characterization of the genotype by observing the phenotype

To begin the crosses, determine the suitable phenotypes that can be tested to confirm the genotype of the progeny cells during the process of genetic crosses. In this example, the cross is being performed between wild type and *pwA-nd6* cells. The swimming behavior of the daughter cells is tested in an 8 mM $BaCl_2$ in base buffer solution (1 mM calcium citrate, 1 mM Tris, 8 mM $BaCl_2$, pH 7.05 adjusted with 100 mM Tris Base). The trichocyst discharging ability of the cells is tested using picric acid. The expected phenotypes and genotypes of F1 and F2 generations are shown in Table 1.

The desired cell line from this cross is *pwA*/*pwA nd6+*/*nd6+* homozygote. The *pwA*/*pwA nd6+*/*nd6+* homozygotes are expected to discharge trichocysts with no backward swimming. Multiple crosses can be performed simultaneously to obtain the desired outcome. The cells showing no backward swimming but are able to discharge trichocysts are collected and grown individually in depression slides with fresh culture fluid. The cell lines should be maintained individually for future downstream usages.

(1) *Genetic Cross Between Wild Type and pwA-nd6* (from Lodh, 2012)

Wild type and *pwA-nd6* cells were crossed to obtain *pwA* cells without *nd6* mutation. Please refer to Fig. 7 to see an illustration of this process.

1. At first, transfer single cells from both strains to depression slides with fresh culture medium.

Table 1 Expected phenotype and genotype cross results between wild type (WT) and *pwA-nd6 cells.*

P1: Genotype	**WT (+/+ +/+)**	***pw A/pw A nd6/nd6***		
P1: Phenotype	Backward swimming trichocyst discharge	No backward swimming No trichocyst discharge		
F1: Genotype	+/*pw A* +/*nd6*			
F1: Phenotype	Backward swimming trichocyst discharge			
F2: Genotype	+/+ +/+	+/+ *nd6/nd6*	*pw A/pw A* +/+	*pw A/pw A nd6/nd6*
F2: Phenotype	Backward swimming trichocyst discharge	Backward swimming No trichocyst discharge	No backward swimming trichocyst discharge	No backward swimming No trichocyst discharge
Expected ratio	1	1	1	1

2. Let the cells grow in the depressions and check for autogamy. See below for procedure to check autogamy (see *Note 23*).
3. Once the cells are in autogamy, transfer single cells again into separate depression slides with fresh culture media.
4. The day before the actual cross, at around 9:00 am, add one volume of *pwA-nd6* cell culture to one volume of fresh culture medium. Do the same with the wild type cells by noon (see *Note 24*).
5. The next day, add 200 μL of wild type and *pwA-nd6* cells to the left and right depressions of a new depression slide, respectively, and add 100 μL of each strain to the middle depression.
6. Place the slide in a moisture chamber (see *Note 25*).
7. Check the depressions for mating pairs (see *Note 26*).
8. Six tight pairs from the middle depression should be transferred to six middle depressions of six new slides with 200 μL of fresh culture fluid.
9. The next day, isolate the exconjugants to the side depressions of each slide.
10. Allow the cells to grow for another day.
11. The next day, check the swimming behavior of the cells (F1). The two clones from one mating pair should swim backward and discharge their trichocysts (Table 1).
12. For the next cross, choose three pairs with good growth and strong phenotypes.
13. Perform three single cell isolations for each of the three clones by moving the cells to new depression slides.

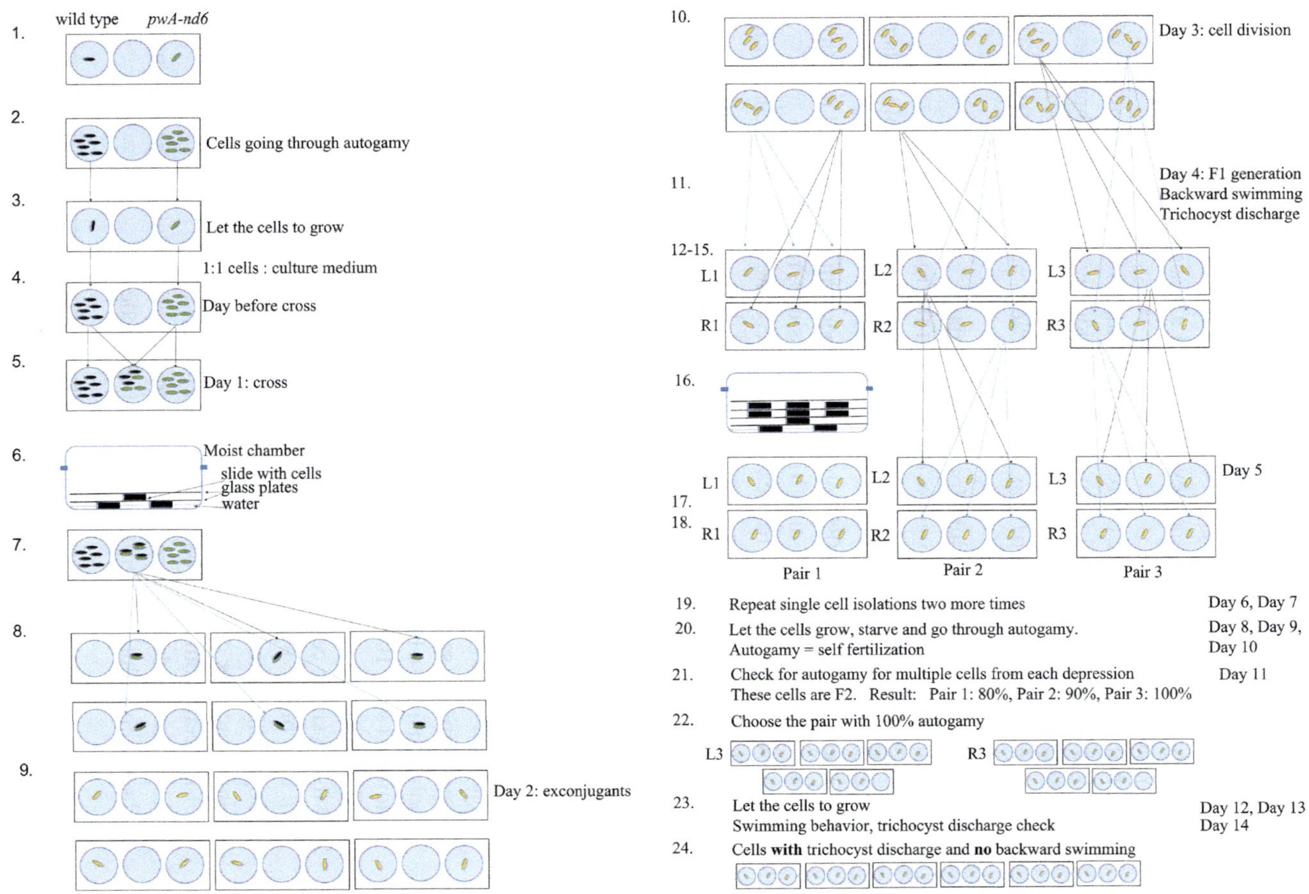

FIG. 7

Crosses of wild type (+/+ +/+) and pwA-nd6 (pxA/pwA nd6+/nd6+) cells using depression slides. Steps correspond to the protocol in Section 7, *(1) Genetic Cross Between Wild Type and pwA-nd6.*

14. Transfer three single cells from each left depression to three depressions of a slide with fresh culture medium and label as "left."
15. Repeat the same with the right depressions and label as "right."
16. Maintain the slides in the moist chamber. Keep placing a new glass plate over the previous sets of depression slides and place the new slides on it. Remember to cover the new slides with a glass plate to prevent evaporation.
17. The next day, from three depressions of a slide, choose one depression with the best cell growth and from that depression, again perform three single cell isolations.
18. Repeat this previous step (step 16) for each "left" and "right" slides for each three pairs.
19. Repeat the same procedure for the next 2 days.
20. Then, allow the cells to grow for 3 days in the moist chamber.
21. Check the cells for autogamy on the fourth day. Sample multiple cells per depression.
22. Choose the pair with 100% autogamy and perform 28 single cell isolations, 14 each from the "left" and "right" depressions, respectively.
23. Allow the cells to grow for next 2 days. Check the swimming behavior and trichocyst discharge (see *Note 27*) on third day.
24. Transfer the cells *with* trichocyst discharge and *no* backward swimming to fresh culture fluid separately.

(2) *Checking swimming behavior using 8 mM $BaCl_2$ solution*

This $BaCl_2$ solution induces all-or-none action potentials that cause repeated bouts of backward swimming in wild type cells. Pawn mutants cannot reverse their ciliary beat and swim backward in this solution (Kung et al., 1975).

1. Place an aliquot of cells in a depression slide and using a hand-pulled micropipette (see *Note 28*), transfer cells to resting buffer, 4 mM KCl solution in base buffer (1 mM calcium citrate, 1 mM Tris, 4 mM KCl, pH 7.05 adjusted with 100 mM Tris Base) to test backward swimming (see *Note 29*). Allow cells to incubate for 10–30 min.
2. Move a single cell at a time using a hand-pulled micropipette to 8 mM $BaCl_2$ in base buffer while observing the cell through a dissecting microscope. Begin timing the cell as soon as it begins swimming backward and stop the timer once the cell returns to forward swimming. Jerking back and forth of the cell should also be recorded and noted.
3. Repeat as needed with cells, at minimum 10 cells should be tested. Change the testing solution every three cells or as needed.

8 Electrophysiology to measure membrane potential (V_m)

The beating of the cilia on the surface of paramecia is a highly coordinated event that is governed by the membrane potential of the cell. When *Paramecium* is attracted to a stimuli, the cell becomes hyperpolarized and the cilia beat with force toward the posterior of the cell, propelling it forward with few turns (Brehm & Eckert, 1978a, 1978b; Machemer, 1988a, 1988b, 1989). If *Paramecium* encounters an obstacle or a repellant, the membrane becomes depolarized. If the depolarization is sufficient to open the voltage gated Ca channels of the cilia, the ciliary beat will change the power stroke toward the anterior of the cell causing the cell to swim backward. This change in the ciliary beat is caused by the graded opening of calcium channels that are exclusively in the cilia (discussed earlier) (Dunlap, 1977; Machemer & Ogura, 1979). These changes in membrane potential can be measured using electrophysiology. This technique is helpful in the analysis of mutant cell lines, such as the Pawn mutants (pwA and pwB) (Kung et al., 1975; Oertel, Schein, & Kung, 1977) or the XntA1 mutant that is impermeable to Mg^{2+} ions (Haynes et al., 2002; Preston, 1990; Preston & Kung, 1994a, 1994b; Valentine et al., 2019), as well as cells treated with RNAi to deplete channels (Valentine et al., 2019). Additionally, both ciliated and deciliated cells can be used for electrophysiology and to tease apart the contributions of ciliary channels to the membrane potential of the cells (Dunlap, 1977; Machemer & Ogura, 1979; Valentine et al., 2019). By measuring the changes in the membrane potential, changes in the activity or presence of membrane channels can be inferred (reviewed in (Valentine & Van Houten, 2022).

In our work, electrophysiology was used to better understand the contributions of a probable Mg^{2+} channel PKD2, a transient receptor potential channel (TRP), to the membrane potential of cells. Using both ciliated and deciliated cells in our study, we were able to tease apart the contributions of the channel in the cilia to versus the whole cell. In addition, we have used these techniques to better understand the contributions of other proteins, such as the glutamate receptor (Romanovitch, 2012) and the eccentric (XntA1) protein.

(1) *Recording from cells*

1. A glass electrode puller (Narishige) or a puller of choice should be used to pull ~10 electrodes using 3 in. 1.0 mm 1/0.58 OD/ID standard glass capillary tubes with filament (see *Note 30*) that have been wiped with 95% ethanol before pulling.
2. *Paramecium* should be grown to early log phase so cells are large and not starved.
3. Place 10–20 cells in a depression slide with 5 mM KCl in basic buffer (5 mM KCl, 1 mM $Ca(OH_2)$, 1 mM citric acid, 1.3 mM Tris base, adjusted to pH 7 with Tris base) for at least 10 min before use (see *Note 29*).
4. A pulled electrode should be backfilled with 500 mM KCl along with the holder.
5. An agar bridge (see *Note 31*) should be placed between the recording chamber that contains an Ag/AgCl wire and the reference chamber and dammed using Vaseline to prevent leakage between the two chambers.
6. Properly position the recording electrode in the field of view using the micromanipulator. An inverted microscope with a 40× magnification and dual stands are needed. Position tubing in both sides of the chamber so

fluid can be simultaneously added and removed from the recording chamber at a flow rate of approximately 1 mL/min using a Polystaltic pump.

7. Fill the recording chamber with 5 mM KCl in base buffer and test the resistance of the electrode. The resistance should be between 40 and 100 MΩ. If it is not, then a new electrode backfilled with 500 mM KCl should be used.
8. Once the resistance of the electrode is obtained, the fluid should be removed from the chamber. Have a Pasteur pipette filled with the control solution (5 mM KCl in base buffer) filled and ready to flood the chamber.
9. Using a hand-pulled micropipette (see *Note 28*), a single *Paramecium* should be removed from the depression slide where cells have been acclimating to 5 mM KCl in base buffer with as little liquid as possible. If needed, deposit the cell on the side of the depression slide and repeat picking up the cell with the micropipette with a minimal amount of solution.
10. Place the *Paramecium* in the center of the chamber where the recording electrode enters with as little liquid as possible so the cell becomes still. Quickly, but carefully, impale the cell in its center with the recording electrode and rapidly, but gently, flood the chamber with 5 mM KCl in buffer.
11. Ensure the cell is not blistering and is living, then turn on the Polystaltic pump to move liquid through the chamber and begin recording.
12. To change solutions, briefly turn the pump off and move the intake tube to the next solution. Cells should be in a solution for 4 min before the membrane potential is noted and cells should return to their resting solution for 4 min before changing the solution again. If cells do not return to approximately the same resting potential in the resting solution, they should be discarded. Likewise, if the cell blisters or begins dividing, the cell should be discarded.
13. Once recording is complete, the electrode should be lifted out of the bath and then placed back into the bath to test the drift in potential. New electrodes should be used for every cell.
14. The change in membrane potential (ΔV_m) can be calculated by subtracting the membrane potential in the control solution (the steady state potential) from the membrane potential in the test solution. If a cell is hyperpolarized, this ΔV_m calculation will be negative while if the cell is depolarized in the solution, the calculation will be positive.

(2) *Using Deciliated Cells for Recording*

1. To use deciliated cells for recording, the deciliation must be fresh for each recording as cilia regenerate over time.
2. To deciliated, place ~50 cells in a 5% ethanol (*v*/v) in 5 mM KCl in base buffer solution. Using a Pasteur pipette, triturate (pipette up and down) the cells 20 times with as few air bubbles as possible. Immediately move cells that are unable to swim to fresh 5 mM KCl in base buffer in a new depression.

3. Proceed as described above to record from cells. Because these cells are deciliated, impaling with the electrode is very easy. However, it is critical to watch the cells for blistering. If the cell begins to blister or leak, it should be discarded and begin again.
4. After recording from deciliated cells, 30 mM KCl in base buffer should be run through the chamber to ensure no action potentials are produced by the cell. If any action potentials are produced at any point during recording, it would indicate the presence of cilia, and the cell and the recordings from that cell should be ignored.

9 Conclusions

We would like to point out an older but very useful set of protocols for *Paramecium* research that has helpful directions for culturing and cloning cells as well as microinjection, immunocytochemistry and RNAi (Beisson et al., 2010). However, this chapter includes more current techniques and goes beyond previous reviews of techniques.

Throughout this chapter, we have described multiple approaches to isolate proteins from the cilia or the cell body from *Paramecium* along with multiple techniques for studying these proteins. We have presented methods for demembranating the cilia, separating the abundant GPI-anchored proteins from other less abundant membrane proteins, such as the channels, using Triton X-114 phase separation. The use of RNAi and epitope tagging have been instrumental for deciphering protein localization and potential interacting partners. The use of proteomics to identify the channels and proteins present in these different membranes has been central to our research and understanding of ciliary composition. Lastly, the use of genetic crosses provides a method for creating cells with a desired phenotype and genotype while electrophysiology allows for the study of proteins in a whole cell, as well as the activity of those channels in a cell that has been deciliated, teasing apart the contributions of the channels in the cilia vs the whole cell. This collection of protocols provides an abundance of approaches for identifying, isolating, and understanding ciliary proteins in *Paramecium*.

Notes

Note 1: Pear-shaped tubes allow for the fast and easy collection of the cells using 9″ Pasteur pipettes. The cells will quickly begin to swim upward, and by collecting in the long stem of a pear-shaped tubes, it is more difficult for the cells to swim away.

Note 2: *Paramecium* discharge their trichocycts in cold Dryl's solution. You can see two distinct layers after the centrifugation unless using a non-discharge (nd) line of cells. The fluffy trichocyst layer (top) can easily be collected and separated from the cell pellet (bottom).

Note 3: Before processing to the next step after phase separation or from collecting ciliary membranes, it is very important to measure the protein concentration of each sample using an acceptable protein assay method.
Note 4: Please note, we followed the protocols of in-gel trypsin digestion and mass spectrometry provided by the VBRN Proteomic Facility at University of Vermont. It is best to contact the facility of choice that will analyze your samples via LC-MS/MS and ask for their protocols to be provided. The reagent volumes are small and should be adjusted if digesting larger regions of a gel.
Note 5: As an alternative method, collected cilia are suspended in 1% Triton X-100 in membrane buffer with protease inhibitors and vortexed for 2 min, and agitated at 4 °C for 1 h. Centrifuge the vortexed cilia at 48,400 × *g* at 4 °C for 30 min to collect the detergent-resistant membrane (DRMs) and resuspend in 500 μL of membrane buffer and vortexed vigorously. Similarly, the DRM suspension is mixed with OptiPrep™ to the final concentration of 40%. Six hundred μL of DRM in 40% OptiPrep™ is placed on the bottom of centrifuge tube and 600 μL each of 35, 30, 25, and 20% OptiPrep™ prepared in the solubilization buffer is overlaid in order from the bottom. After ultracentrifuging at 160,000 × *g* at 4 °C for 4 h in Beckman LB-70M ultracentrifuge to fractionate DRMs, the nine fractions each with equal volume, are harvested (Ray, 2008).
Note 6: If you plan to use enhanced chemiluminescence to develop the blot, use TBST with 5% non-fat milk as a blocking buffer and follow the protocol provided by a company.
Note 7: The cytoplasmic factors such as the ER and mitochondria that adhere to the pellicle are removed during these processes. The crude pellicle predominantly consist of somatic plasma membrane and the alveolar membrane (Bilinski, Plattner, & Tiggemann, 1981).
Note 8: Use 5% sucrose in the TNE buffer to balance the centrifuge tubes. Keep swinging-bucket rotor at 4 °C (cold room) before using.
Note 9: As an alternative method, dialyze the nine collected fractions against Ultra-Pure water at 4 °C for 24 h. Then, precipitate proteins with 100% acetone by adding four times the volume of each fraction in acetone. Intubate the fractions at −20 °C overnight. Centrifuge at 15,200 × *g* for 15 min at 4 °C. Wash the protein pellets with 90% acetone at 4 °C and centrifuge again. After discarding the supernatant carefully, air-dry protein pellets.
Note 10: The amount of sterol depletion from the isolated cilia or cell membranes can be examined using a sterol assay kit (e.g., Amplex Red Kit, Thermo Fisher Scientific) following the protocol provided by the company.
Note 11: If desired, the linearized plasmid can be dephosphorylated using calf intestinal alkaline phosphatase (CIAP) by adding 1 U and incubating at 37 °C for 5–10 min. Because of the restriction enzyme buffers already present, no additional buffer should be added. Importantly, treated plasmid should be treated to heat inactivate the restriction enzymes after adding the required amount of 0.5 M Na_2EDTA to chelate magnesium.

Note 12: Cells must be maintained in a well-fed state. If cells are brought to the level of starvation, they will begin the process of autogamy where the macronucleus (and therefore the injected tagged gene) will be destroyed to generate a new macronucleus.
Note 13: The TOPO pCR2.1® plasmid (ThermoFisher Scientific: https://www.thermofisher.com/order/catalog/product/K450002) works well for subcloning and the restriction enzymes *Spe*I, *Sac*I, or *Kpn*I can be used in combination with either *Xho*I, *Xba*I, or *Apa*I to remove the construct. The same pair of enzymes should then be used to open the L4440 plasmid (Addgene, https://www.addgene.org/1654/). The orientation of the construct in L4440 does not matter.
Note 14: ZymoClean Gel DNA Recovery is one suggested Gel extraction kit available (https://www.zymoresearch.com/products/zymoclean-gel-dna-recovery-kit).
Note 15: A glycerol stock of the overnight shaking culture should be made (850 μL of the bacterial culture and 150 μL of glycerol) and stored at −80 °C for future use.
Note 16: To make competent Ht115 bacteria, start an overnight shaking culture in LB-TET (tetracycline, 12.5 μg/L) from a single isolated colony on a plate. The following day, start a 25 mL LB culture with 12.5 μg/mL tetracycline using 500 μL of the overnight culture. Shake cultures at 37 °C for ~1.5 h (until OD_{600} reaches 0.3–0.4). Collect bacteria via centrifugation at 2500 × *g* for 10 min. Resuspend the pellet in 12.5 mL ice cold 50 mM $CaCl_2$. After incubating on ice for 30 min, the cells should be spun again for 10 min at 2500 × *g* and the final pellet resuspended in 2.5 mL ice cold 50 mM $CaCl_2$. Cells are only stable for ~2 weeks at 4 °C.
Note 17: Induction times are dependent on the construct and for feeding, trying different induction times, typically of 3, 4, and 5 h is best to see which produces the most potent bacteria for feeding and producing an observable phenotype.
Note 18: To purge paramecia, a starved culture of cells that have recently gone through autogamy should be collected by brief centrifugation and placed in Dryl's solution for at least 1 h before being added to the RNAi cultures. This 1 h allows all food vacuoles to be cleared from the cells.
Note 19: The number of cells added to the cultures varies for each RNAi construct and may need to be tested. Too many cells will consume all of the bacteria while two few may not produce sufficient numbers of paramecia for study.
Note 20: Our work with RNAi has been highly reproducible among different lab members when we attend to Note 20 and other details of preparation of the paramecia. We have been able to reproduce results from the Cohen lab for trichocyst discharge. However, there is an instance of another lab not reproducing our work on depleting MKS3. The use of different *P. tetraurelia* strains and the production of culture media may have an impact, but as of yet, we have not narrowed down the reason for the different results of RNAi (Picariello et al., 2014).

Note 21: Treatment of the RNA to remove genomic contamination may be required. If treatment is needed, RNA samples should be equalized (see Step 4 below) before treatment. DNase must be deactivated before moving forward.
Note 22: primers designed to amplify the mRNA levels of the targeted gene should be designed to amplify sequence that is outside of the construct region. Primers designed for the 3′ end of the gene, closest to the Poly-A tail, are ideal and the amplicon size should not be larger than 500 bp.
Note 23: To check the cells for autogamy, the macronuclei of the cells needs to be observed. A small sample of cells in their culture fluid can be moved to a slide where 50–100 μL of DAPI stain solution is applied. After incubating the cells for 10 min, observe the cells using a fluorescence microscope. If the macronucleus is fragmented, then the cells have entered autogamy. If the cells are not in autogamy, a single macronucleus will be observed.
Note 24: The time difference is to compensate for the differences in the growing rates of the cell types. Cells with similar growth rates can be transferred to fresh culture medium at the same time.
Note 25: To create a moist chamber, pour 2 in. of water into a rectangular plastic storage container with a lid. Place two or three depression slides in the bottom of the container and a glass slab on top of the depression slides to create a raised platform. Place the depression slides with cells on the glass plate and cover with another glass plate. Make sure that the water in the bottom of the container doesn't enter the depression slides with the cells. Cover the container with the lid and keep at room temperature and add more water to the bottom of the bin as needed. This prevents the culture media from evaporating.
Note 26: Check the original slide (step 5) for cells joined at the anterior end. There should be none in these depressions. If pairs were found in these side depressions, then the whole depression slide should be discarded.
Note 27: To check cells for trichocyst discharge, transfer concentrated cells with a small amount of culture fluid to a depression slide with a shallow well that can be observed using a light microscope. Place one drop of saturated picric acid onto the cells. The +/+ and +/nd6 cells should discharge their trichocysts (Fig. 7).
Note 28: To make hand-pulled micropipettes, heat the wide part of a glass Pasteur pipette over a Bunsen burner flame. Rotate the pipette in the flame approximately 2/3 of the way up the pipette until the glass becomes soft and malleable. At the same time as removing the Pasteur pipette from the flame, pull both ends of the pipette briefly. Cool briefly, and then return the thinned pipette to the flame near the wide part and heat briefly. As soon as the glass is hot, quickly remove the pipette and pull to generate a fine pipette that is capable of collecting one cell at a time.
Note 29: All buffers should contain the same base of 1 mM $Ca(OH_2)$, 1 mM citric acid, 1.3 mM Tris base. The salt, e.g., acetate, folate, barium, potassium, etc., are then included. All buffers should be made from 100 mM liquid stocks, and not by adding salts to an already made buffer to keep the ions in the solutions controlled. To further control salt concentrations, all solutions should have their

pH adjusted with tris base or citric acid (if needed) so additional ions are not added to the solution (see below). Also, ions must be balanced between control and test solutions. Excellent attractants for *Paramecium* include acetate, folate, glutamate, ammonium, biotin, and lactate. All solutions should be stored at 4 °C and brought to room temperature before use. Test solutions regularly for contamination. Bacterial metabolites are strong attractants and can affect swimming behavior.

The final pH of these testing solutions is brought to 7 with Tris Base, making the final Tris concentration slightly higher than 1.3 mM. A common mistake is to adjust the buffer pH with NaOH or some other base or acid. By doing so, additional salts would be added to the solution which changes the ionic composition of the buffer. This difference in salt can greatly affect the behavior and membrane potential of the cells and makes interpreting the results very difficult.

Note 30: The capillary tubes recommended are from World Precision Instruments, cat. #1B100F-3. After cleaning with ethanol, allow tubes to dry and use tweezers to place the tubes in the puller and to remove from puller to prevent grease and oil from hands contaminating the electrode. Store pulled electrodes in a petri dish with a thin strip of putty or modeling clay until use.

Note 31: Agar bridges are created using the narrow end of a glass Pasteur pipette that has been heated until it makes a bend, becoming U-shaped. A glass cutting pencil should be used to scar the glass, making a bridge that is ~3–4 mm on one side of the bend and ~5–6 mm on the other. To fill these bridges, using a small beaker, bring 3% agar in 10 mL of 3 M KCl to a gentle boil to dissolve the agar. Place sterile bridges (~10–20) into the solution and stir using a glass Pasteur pipette. Cool to room temperature and once the agar hardens, keep a layer of 3 M KCl over the surface to prevent drying. Cover the beaker with Parafilm and store at 4 °C.

Bridges should be used for only a single day of recording but can be cleaned and reused. To clean them, wipe away any Vaseline after use and store in a beaker. Cover used bridges in distilled water and boil briefly. Next, place the bridges in a beaker with 95% ethanol briefly, and remove bridges to a Kimwipe to dry. Repeat the boiling processes as needed. Store wrapped and protected from dust until use.

Acknowledgment

Vermont Biomedical Research Network (VBRN) Publication Citation for VBRN Funded Research from 1/1/2012 to 05/31/2021

Research reported in this publication was supported by an Institutional Development Award (IDeA) from the National Institute of General Medical Sciences of the National Institutes of Health under grant number P20GM103449. Its contents are solely the responsibility of the authors and do not necessarily represent the official views of NIGMS or NIH.

References

Adoutte, A., Ramanathan, R., Lewis, R. M., Dute, R. R., Ling, K. Y., Kung, C., et al. (1980). Biochemical studies of the excitable membrane of Paramecium tetraurelia. III. Proteins of cilia and ciliary membranes. *The Journal of Cell Biology*, *84*(3), 717–738. https://doi.org/10.1083/jcb.84.3.717.

Arnaiz, O., Meyer, E., & Sperling, L. (2020). ParameciumDB 2019: Integrating genomic data across the genus for functional and evolutionary biology. *Nucleic Acids Research*, *48*(D1), D599–D605. https://doi.org/10.1093/nar/gkz948.

Aury, J.-M., Jaillon, O., Duret, L., Noel, B., Jubin, C., Porcel, B. M., et al. (2006). Global trends of whole-genome duplications revealed by the ciliate Paramecium tetraurelia. *Nature*, *444*(7116), 7116. https://doi.org/10.1038/nature05230.

Beisson, J., Bétermier, M., Bré, M.-H., Cohen, J., Duharcourt, S., Duret, L., et al. (2010). Paramecium tetraurelia: The renaissance of an early unicellular model. *Cold Spring Harbor Protocols*, *2*, 1–32.

Bhullar, S., Denby Wilkes, C., Arnaiz, O., Nowacki, M., Sperling, L., & Meyer, E. (2018). A mating-type mutagenesis screen identifies a zinc-finger protein required for specific DNA excision events in Paramecium. *Nucleic Acids Research*, *46*(18), 9550–9562. https://doi.org/10.1093/nar/gky772.

Bilinski, M., Plattner, H., & Tiggemann, R. (1981). Isolation of surface membranes from normal and exocytotic mutant strains of Paramecium tetraurelia. Ultrastructural and biochemical characterization. *European Journal of Cell Biology*, *24*(1), 108–115.

Bordier, C. (1981). Phase separation of integral membrane proteins in Triton X-114 solution. *The Journal of Biological Chemistry*, *256*(4), 1604–1607.

Bouhouche, K., Valentine, M. S., Le Borgne, P., Lemullois, M., Yano, J., Lodh, S., et al. (2022). Paramecium, a model to study ciliary beating and ciliogenesis: Insights from cutting-edge approaches. *Frontiers in Cell and Development Biology*, *10*. https://www.frontiersin.org/article/10.3389/fcell.2022.847908.

Brehm, P., Dunlap, K., & Eckert, R. (1978). Calcium-dependent repolarization in Paramecium. *The Journal of Physiology*, *274*, 639–654.

Brehm, P., & Eckert, R. (1978a). An electrophysiological study of the regulation of ciliary beating frequency in Paramecium. *The Journal of Physiology*, *283*(1), 557–568. https://doi.org/10.1113/jphysiol.1978.sp012519.

Brehm, P., & Eckert, R. (1978b). Calcium entry leads to inactivation of calcium channel in paramecium. *Science*, *202*(4373), 1203–1206. https://doi.org/10.1126/science.103199.

Bright, L. J., Gout, J.-F., & Lynch, M. (2017). Early stages of functional diversification in the Rab GTPase gene family revealed by genomic and localization studies in Paramecium species. *Molecular Biology of the Cell*, *28*(8), 1101–1110. https://doi.org/10.1091/mbc.e16-06-0361.

Brown, D. A. (2006). Lipid rafts, detergent-resistant membranes, and raft targeting signals. *Physiology (Bethesda, Md.)*, *21*, 430–439. https://doi.org/10.1152/physiol.00032.2006.

Buonanno, F., Harumoto, T., & Ortenzi, C. (2013). The defensive function of trichocysts in Paramecium tetraurelia against metazoan predators compared with the chemical defense of two species of toxin-containing ciliates. *Zoological Science*, *30*(2), 255–261. https://doi.org/10.2108/zsj.30.255.

Capdeville, Y., & Benwakrim, A. (1996). The major ciliary membrane proteins in Paramecium primaurelia are all glycosylphosphatidylinositol-anchored proteins. *European Journal of Cell Biology*, *70*(4), 339–346.

Dunlap, K. (1977). Localization of calcium channels in Paramecium caudatum. *The Journal of Physiology*, *271*(1), 119–133.

Eckert, R. (1972). Bioelectric control of ciliary activity. *Science*, *176*(4034), 473–481. https://doi.org/10.1126/science.176.4034.473.

Forte, M., Satow, Y., Nelson, D., & Kung, C. (1981). Mutational alteration of membrane phospholipid composition and voltage-sensitive ion channel function in paramecium. *Proceedings of the National Academy of Sciences of the United States of America*, *78*(11), 7195–7199.

Gout, J.-F., Kahn, D., Duret, L., & Consortium, P. P.-G. (2010). The relationship among gene expression, the evolution of gene dosage, and the rate of protein evolution. *PLoS Genetics*, *6*(5), e1000944. https://doi.org/10.1371/journal.pgen.1000944.

Haga, N. (2022). Micromanipulation in Paramecium: From non-mendelian inheritance to the outlook for versatile micromachines. *Journal of Eukaryotic Microbiology*, *69*, e12909. https://doi.org/10.1111/jeu.12909. n/a(n/a).

Harumoto, T., & Miyake, A. (1991). Defensive function of trichocysts in Paramecium. *Journal of Experimental Biology*, *260*(1), 84–92.

Hauser, K., Haynes, W. J., Kung, C., Plattner, H., & Kissmehl, R. (2000). Expression of the green fluorescent protein in Paramecium tetraurelia. *European Journal of Cell Biology*, *79*(2), 144–149. https://doi.org/10.1078/S0171-9335(04)70016-3.

Haynes, W. J., Kung, C., Saimi, Y., & Preston, R. R. (2002). An exchanger-like protein underlies the large Mg2+ current in Paramecium. *Proceedings of the National Academy of Sciences*, *99*(24), 15717–15722. https://doi.org/10.1073/pnas.242603999.

Haynes, W. J., Ling, K.-Y., Saimi, Y., & Kung, C. (2003). PAK paradox: Paramecium appears to have more K+-channel genes than humans. *Eukaryotic Cell*, *2*(4), 737–745. https://doi.org/10.1128/EC.2.4.737-745.2003.

Iftode, F., Adoutte, A., & Fleury, A. (1996). The surface pattern of Paramecium tetraurelia in interphase: An electron microscopic study of basal body variability, connections with associated ribbons and their epiplasmic environment. *European Journal of Protistology*, *32*, 46–57. https://doi.org/10.1016/S0932-4739(96)80076-9.

Jacobs, C. (2007). *NMDA receptor associated protein in Paramecium and its inveolvement in glutamate chemoresponse*. Masters University of Vermont. https://library.uvm.edu/collections/theses?search_type=item&bid=1602922.

Kaneshiro, E. S. (1987). Lipids of Paramecium. *Journal of Lipid Research*, *28*(11), 1241–1258.

Kung, C., Chang, S.-Y., Satow, Y., Van Houten, J., & Hansma, H. (1975). Genetic dissection of behavior in paramecium. *Science*, *188*(4191), 898–904. https://doi.org/10.1126/science.188.4191.898.

Levental, I., Levental, K. R., & Heberle, F. A. (2020). Lipid rafts: Controversies resolved, mysteries remain. *Trends in Cell Biology*, *30*(5), 341–353. https://doi.org/10.1016/j.tcb.2020.01.009.

Lingwood, D., & Simons, K. (2010). Lipid rafts as a membrane-organizing principle. *Science (New York, N.Y.)*, *327*(5961), 46–50. https://doi.org/10.1126/science.1174621.

Lodh, S. (2012). *Characterization of PwA and PwB proteins in Paramecium [PhD]*. University of Vermont.

Lodh, S., Yano, J., Valentine, M. S., & Van Houten, J. L. (2016). Voltage-gated calcium channels of Paramecium cilia. *The Journal of Experimental Biology*, *219*(19), 3028–3038. https://doi.org/10.1242/jeb.141234.

Machemer, H. (1988a). Electrophysiology. In *Paramecium* (1st ed., pp. 186–215). Springer-Verlag.

Machemer, H. (1988b). Motor control of cilia. In *Paramecium* (1st ed., pp. 216–235). Springer-Verlag.

Machemer, H. (1989). Cellular behaviour modulated by ions: Electrophysiological implications. *The Journal of Protozoology*, *36*(5), 463–487. https://doi.org/10.1111/j.1550-7408.1989.tb01082.x.

Machemer, H., & Ogura, A. (1979). Ionic conductances of membranes in ciliated and deciliated Paramecium. *The Journal of Physiology*, *296*(1), 46–60.

Mahammad, S., & Parmryd, I. (2015). Cholesterol depletion using methyl-β-cyclodextrin. *Methods in Molecular Biology (Clifton, N.J.)*, *1232*, 91–102. https://doi.org/10.1007/978-1-4939-1752-5_8.

Marker, S., Carradec, Q., Tanty, V., Arnaiz, O., & Meyer, E. (2014). A forward genetic screen reveals essential and non-essential RNAi factors in Paramecium tetraurelia. *Nucleic Acids Research*, *42*(11), 7268–7280. https://doi.org/10.1093/nar/gku223.

Nabi, A., Yano, J., Valentine, M. S., Picariello, T., & Van Houten, J. L. (2019). SF-Assemblin genes in Paramecium: Phylogeny and phenotypes of RNAi silencing on the ciliary-striated rootlets and surface organization. *Cilia*, *8*(1), 2. https://doi.org/10.1186/s13630-019-0062-y.

Oertel, D., Schein, S. J., & Kung, C. (1977). Separation of membrane currents using a Paramecium mutant. *Nature*, *268*(5616), 120–124. https://doi.org/10.1038/268120a0.

Pan, Y. (2008). *The role of plasma membrane calcium ATPase and its association with lipid rafts in chemoattraction in Paramecium [Masters]*. University of Vermont.

Paquette, C. A., Rakochy, V., Bush, A., & Van Houten, J. L. (2001). Glycophosphatidylinositol-anchored proteins in Paramecium tetraurelia: Possible role in chemoresponse. *The Journal of Experimental Biology*, *204*(Pt. 16), 2899–2910. https://doi.org/10.1242/jeb.204.16.2899.

Pearson, C. G., Giddings, T. H., & Winey, M. (2009). Basal body components exhibit differential protein dynamics during nascent basal body assembly. *Molecular Biology of the Cell*, *20*(3), 904–914. https://doi.org/10.1091/mbc.e08-08-0835.

Picariello, T., Valentine, M. S., Yano, J., & Van Houten, J. (2014). Reduction of meckelin leads to general loss of cilia, ciliary microtubule misalignment and distorted cell surface organization. *Cilia*, *3*(1), 2. https://doi.org/10.1186/2046-2530-3-2.

Plattner, H. (2002). My favorite cell—Paramecium. *BioEssays*, *24*, 649–658.

Plattner, H. (2022). Ciliate research: From myth to trendsetting science. *The Journal of Eukaryotic Microbiology*, *69*, 1–21.

Preston, R. R. (1990). A magnesium current in Paramecium. *Science (New York, N.Y.)*, *250*(4978), 285–288. https://doi.org/10.1126/science.2218533.

Preston, R. R., & Kung, C. (1994a). Inhibition of Mg2+ current by single-gene mutation in Paramecium. *The Journal of Membrane Biology*, *139*(3), 203–212. https://doi.org/10.1007/BF00232624.

Preston, R. R., & Kung, C. (1994b). Isolation and characterization of paramecium mutants defective in their response to magnesium. *Genetics*, *137*(3), 759–769.

Ray, K. (2008). *Characterization of Paramecium tetraurelia ciliary membrane plasma membrane calcium pumps and lipid rafts [Masters]*. University of Vermont.

Romanovitch, M. (2012). *The L-glutamate receptor in paramecium [Masters]*. University of Vermont.

Saimi, Y., & Kung, C. (1982). Are ions involved in the gating of calcium channels? *Science*, *218*(4568), 153–156. https://doi.org/10.1126/science.6289432.

Sasner, J. M., & Van Houten, J. (1989). Evidence for a Paramecium folate chemoreceptor. *Chemical Senses | Oxford Academic*, *14*(4), 587–595. https://doi.org/10.1093/chemse/14.4.587.

Simons, K., & Ikonen, E. (1997). Functional rafts in cell membranes. *Nature*, *387*, 569–572.

Simons, K., & Toomre, D. (2000). Lipid rafts and signal transduction. *Nature Reviews Molecular Cell Biology*, *1*(1), 1. https://doi.org/10.1038/35036052.

Sperling, L. (1989). Isolation and partial characterization of ciliary rootlets from Paramecium tetraurelia. In U. Aebi, & J. Engel (Eds.), *Cytoskeletal and extracellular proteins* (pp. 316–318). Springer. https://doi.org/10.1007/978-3-642-73925-5_57.

Sperling, L., Keryer, G., Ruiz, F., & Beisson, J. (1991). Cortical morphogenesis in Paramecium: A transcellular wave of protein phosphorylation involved in ciliary rootlet disassembly. *Developmental Biology*, *148*(1), 205–218. https://doi.org/10.1016/0012-1606(91)90330-6.

Tassin, A.-M., Lemullois, M., & Aubusson-Fleury, A. (2016). Paramecium tetraurelia basal body structure. *Cilia*, *5*(1), 6. https://doi.org/10.1186/s13630-016-0026-4.

Valentine, M. S. (2015). *Polycystin-2 (PKD2), eccentric (XNTA), and meckelin (MKS3) in the ciliated model organism Paramecium tetraurelia*. University of Vermont.

Valentine, M., & Houten, J. V. (2021). Using Paramecium as a model for ciliopathies. *Genes*, *12*(10), 1493. https://doi.org/10.3390/genes12101493.

Valentine, M., Rajendran, A., Yano, J., Weeraratne, S., Beisson, J., Cohen, J., et al. (2012). Paramecium BBS genes are key to presence of channels in Cilia. *Cilia*, *1*(1), 16. https://doi.org/10.1186/2046-2530-1-16.

Valentine, M. S., & Van Houten, J. (2022). Ion channels of cilia: Paramecium as a model. *The Journal of Eukaryotic Microbiology*, *69*, e12884. https://doi.org/10.1111/jeu.12884.

Valentine, M. S., Yano, J., & Van Houten, J. (2019). A novel role for polycystin-2 (Pkd2) in P. tetraurelia as a probable Mg2+ channel necessary for Mg2+-induced behavior. *Genes*, *10*(6), 455. https://doi.org/10.3390/genes10060455.

Van Houten, J. (2019). *Paramecium* biology. In W. Tworzydlo, S. M. Bilinski, & Nature (Eds.), *Vol. 68. Evo-Devo: Non-model species in cell and developmental biology, results and problems in differentiarion* (pp. 291–318). http://doi.org/10.1007/978-030-23459-1_13.

Weeraratne, S. D. (2007). *GPI receptors in folate chemosensortransduction in Paramecium tetraurelia*. PhD University of Vermont. https://library.uvm.edu/collections/theses?search_type=item&bid=1605510.

Wright, M. V., & van Houten, J. L. (1990). Characterization of a putative Ca2+-transporting Ca2+-ATPase in the pellicles of Paramecium tetraurelia. *Biochimica et Biophysica Acta (BBA)—Biomembranes*, *1029*(2), 241–251. https://doi.org/10.1016/0005-2736(90)90160-P.

Yano, J., Rajendran, A., Valentine, M. S., Saha, M., Ballif, B. A., & Van Houten, J. L. (2013). Proteomic analysis of the cilia membrane of Paramecium tetraurelia. *Journal of Proteomics*, *78*, 113–122. https://doi.org/10.1016/j.jprot.2012.09.040.

Yano, J., Wells, R., Lam, Y.-W., & Van Houten, J. L. (2021). Ciliary Ca2+ pumps regulate intraciliary Ca2+ from the action potential and may co-localize with ciliary voltage-gated Ca2+ channels. *Journal of Experimental Biology*, *224*(9), jeb232074. https://doi.org/10.1242/jeb.232074.

CHAPTER

Using mammary organoids to study cilia

11

Aurore M.M. Dupuy, Philippe P. Juin, and Vincent J. Guen*

Nantes Université, Inserm, CNRS, Université d'Angers, CRCI[2]NA, Nantes, France

**Corresponding author: e-mail address: vincent.guen@univ-nantes.fr*

Chapter outline

Abstract

Cilia are hair-like projections that assemble at the surface of cells in various tissues of multicellular organisms through a complex cell biological process called ciliogenesis. Cilia can assemble as single structures per cell (i.e. non-motile primary cilia), which act as cell signaling centers that dictate cell fate, or can be assembled in distinct cell types as many copies per cell (i.e. motile cilia) that beat to move fluids at the cell surface. The mechanisms that orchestrate formation and function of cilia, which are dysregulated in pathological settings such as ciliopathies, remain incompletely understood. Stem cell-derived organoids represent valuable models to study the mechanisms of ciliogenesis, ciliary signaling, and ciliary beating that collectively promote tissue development and homeostasis. Here, we present a comprehensive protocol for the growth of mammary organoids derived from mouse mammary stem cells and for immunofluorescence staining of primary cilia in these three-dimensional structures.

Methods in Cell Biology, Volume 175, ISSN 0091-679X, https://doi.org/10.1016/bs.mcb.2022.09.010
Copyright © 2023 Elsevier Inc. All rights reserved.

1 Introduction

Cilia are microtubule-based organelles that protrude from the cell surface in various tissues of multicellular organisms (Bloodgood, 2009; Carvalho-Santos, Azimzadeh, Pereira-Leal, & Bettencourt-Dias, 2011; Satir, Pedersen, & Christensen, 2010). Cilia can have distinct ultrastructures and functions (Carvalho-Santos et al., 2011; Ishikawa & Marshall, 2011), and are classified based on their motility (Bloodgood, 2009). Motile cilia are generally made of 9+2 axonemal microtubules and are assembled in many copies in the same cell, with a few exceptions that include nodal cilia and insect chordotonal organ cilia (Bloodgood, 2009). The main function of motile cilia is to move fluids at the surface of tissues (Spassky & Meunier, 2017). For example, motile cilia in multiciliated differentiated cells of the adult airway epithelium move mucus at the surface of the lung (Spassky & Meunier, 2017). Additionally, multiciliated ependymal cells of the brain ventricles move cerebrospinal fluids for proper brain functioning (Spassky & Meunier, 2017). Motile cilia in multiciliated ovarian epithelial cells enable the transport of gametes in the oviduct (Spassky & Meunier, 2017).

Non-motile primary cilia represent the second class of cilia that are assembled in distinct tissues of multicellular organisms (Satir et al., 2010). They are generally made of 9 + 0 axonemal microtubules and are assembled as a single structure per cell (Guen & Prigent, 2020). The main function of the primary cilium is to act as a cell signaling center in stem and/or differentiated cells, depending on the tissue and the developmental stage (Goetz & Anderson, 2010; Guen et al., 2017; Satir et al., 2010; Wilson et al., 2021). Indeed, the primary cilium is thought to regulate proliferation, differentiation, and migration of cells to coordinate embryonic and postnatal development, as well as adult tissue homeostasis (Anvarian, Mykytyn, Mukhopadhyay, Pedersen, & Christensen, 2019; Guen, Chavarria, et al., 2017; Satir et al., 2010; Wilson et al., 2021).

Ciliogenesis is the dynamic process of assembling cilia (Ishikawa & Marshall, 2011). The molecular and cellular events that induce multiciliogenesis differ from those that induce primary ciliogenesis. Multiciliogenesis occurs in fully differentiated cells and begins with massive biogenesis of centrioles that serve as the basal bodies of cilia (Yan, Zhao, & Zhu, 2016). In multiciliogenesis, centriole biogenesis occurs mainly through the formation of a structure called the deuterosome (Spassky & Meunier, 2017; Yan et al., 2016), which assembles at the daughter centriole in a precursor cell that commits to the multiciliated cell fate. The daughter centriole serves as the template for deuterosome formation. The deuterosome then acts as the platform for centriole biogenesis (Spassky & Meunier, 2017; Yan et al., 2016). In the first step of this process, pro-centrioles are generated in deuterosomes through sequential rounds of deuterosome and pro-centriole biogenesis in the same cell (Hazan et al., 2021; Spassky & Meunier, 2017; Yan et al., 2016). In the second step, pro-centrioles undergo growth and maturation. In the third and final step, centrioles disengage from their platform and migrate to the plasma membrane where they anchor and then orchestrate motile cilium assembly (Hazan et al., 2021;

Spassky & Meunier, 2017; Yan et al., 2016). While the vast majority of centrioles are generated in a deuterosome-dependent pathway, centriole biogenesis can also occur in a deuterosome-independent pathway (Spassky & Meunier, 2017; Yan et al., 2016). In this setting, centrioles are generated by both the mother and the daughter centriole of the precursor cell (Spassky & Meunier, 2017). These two pathways involve distinct molecules that initiate centriole biogenesis, but they share downstream effectors (Hazan et al., 2021; Spassky & Meunier, 2017).

Primary ciliogenesis is the distinct process of assembling the primary cilium (Guen & Prigent, 2020). It can occur in both stem cells and differentiated cells (Satir et al., 2010). In proliferating cells, primary ciliogenesis is dynamically regulated. Primary cilia are assembled in interphase and disassembled in mitosis (Ford et al., 2018). Primary ciliogenesis is a multi-step process that is coordinated by the centrosome, which is composed of a mother and a daughter centriole surrounded by pericentriolar material (Guen & Prigent, 2020). During interphase, the mother centriole differentiates to form the basal body through maturation of subdistal and distal appendages (Guen & Prigent, 2020). Following intense cytoskeletal reorganization, the centrosome migrates toward the plasma membrane, where it anchors through its distal end and nucleates microtubules for axoneme formation and elongation (Guen et al., 2016; Guen & Prigent, 2020). A vesicle can associate with the distal end of the basal body to initiate axoneme formation prior to plasma membrane fusion in many but not all cell types (Guen & Prigent, 2020). Active vesicular trafficking at the basal body supplies essential components for primary cilium assembly that enter into the cilium by passing through a transition zone (Guen & Prigent, 2020).

At the cell surface, motile cilia and primary cilia elongation requires constant equilibrium between anterograde and retrograde intraflagellar transport (IFT), a molecular motor-driven process that tightly controls axonemal assembly and disassembly. IFT promotes cilium assembly and maintains cilium length by controlling the distribution and removal of the ciliary building blocks throughout the cilium (Ishikawa & Marshall, 2011).

Ciliogenesis defects or dysfunctional cilia can lead to severe multisystemic developmental disorders collectively named ciliopathies that impact most human organ systems, most notably including the brain, kidney, and skeleton (Guen et al., 2016, 2017; Reiter & Leroux, 2017). These defects are thought to result from improper movement of fluids at the surface of tissues as a consequence of dysfunctional motile cilia, or from improper activation or repression of key ciliary signaling pathways, such as Hedgehog, at the appropriate developmental stage within the appropriate cell lineages (Reiter & Leroux, 2017). More recently, improper regulation of ciliogenesis and ciliary signaling was demonstrated to contribute to the development and progression of distinct tumor types (Eguether & Hahne, 2018; Guen & Prigent, 2020).

Despite the tremendous progress over the last two decades that have led to a better understanding of the biology of cilia in development and disease, we still have much to learn about the mechanisms of assembly and function of these organelles in multicellular organisms. The reasons why primary cilia are assembled in some tissues and not in others or why primary cilia are assembled only at specific developmental stages in some tissues remain elusive. Moreover, the biology of primary cilia in embryonic stem cells and in adult stem cells that control tissue development, regeneration, and repair remains to be studied more largely. To address these questions, establishment of models such as organoids that recapitulate *ex vivo* the physiology and complex architecture of organs, as well as methods for their analysis, are needed.

Recently, the biology of cilia has been studied in airway, endometrial, kidney, retinal, cerebral, fallopian tube, lung, liver and mammary organoids (An, Kuo, & Tang, 2022; Boutaud et al., 2022; Chen, Kaya, Dong, & Swaroop, 2016; Cruz et al., 2022; Duclos, Prigent, Le Borgne, & Guen, 2021; Forbes et al., 2018; Ford et al., 2018; Gabriel et al., 2016; Guen, Chavarria, et al., 2017; Haider et al., 2019; Kessler et al., 2015; Sachs et al., 2019; Schembs et al., 2022; Tadokoro et al., 2014; van der Vaart et al., 2021; Wilson et al., 2021; Zhang et al., 2019; Zheng et al., 2022; Zhou et al., 2018). Airway, endometrial, fallopian tube, lung, liver and mammary organoids were all derived from adult stem cells of humans or mice (Ford et al., 2018; Haider et al., 2019; Kessler et al., 2015; Sachs et al., 2019; Tadokoro et al., 2014; van der Vaart et al., 2021; Wilson et al., 2021; Zheng et al., 2022; Zhou et al., 2018). Kidney, retinal and cerebral organoids were derived from human induced-pluripotent stem cells (iPSCs) or human embryonic stem cells (hES) (An et al., 2022; Boutaud et al., 2022; Chen et al., 2016; Cruz et al., 2022; Forbes et al., 2018; Gabriel et al., 2016; Schembs et al., 2022; Zhang et al., 2019). In these studies, stem cells were isolated from healthy humans and mice or from individuals affected by ciliopathies. Notably, airway organoids derived from stem cells of primary ciliary dyskinesia patients were composed of multiciliated cells that displayed a ciliary beating defect similar to that observed *in vivo* in human patients (van der Vaart et al., 2021). Cerebral organoids derived from stem cells that carried mutations in ciliogenesis genes were composed of abnormally differentiated cells due to defects in the emergence and maintenance of neural stem cells, which was previously reported *in vivo* (Gabriel et al., 2016; Schembs et al., 2022). Additionally, genetic perturbation of ciliogenesis and of ciliary signaling in mammary stem cells resulted in mammary organoid formation defects, in agreement with the mammogenesis defects observed *in vivo* after such perturbations (Guen, Chavarria, et al., 2017; Wilson et al., 2021). Altogether, these studies have illustrated the importance of organoids for understanding the biology of cilia in physiology and disease. Importantly, some of these studies also emphasize the value of organoids for the development of therapeutic approaches for ciliopathies (Forbes et al., 2018; van der Vaart et al., 2021; Zheng et al., 2022). Notably, genetic correction of patient-specific mutations in airway or kidney organoids were proven to be possible in these mini-organs *ex vivo*, but the ability to transplant these structures *in vivo* for therapeutic purposes remain to be tested (Forbes et al., 2018; van der Vaart et al., 2021). Airway organoids were also recently used for the identification of drugs that promote motile cilia beating (Zheng et al., 2022). Such pharmacological compounds may be useful for the treatment of primary ciliary dyskinesia.

Altogether, these studies highlight the value of organoids for basic research on cilia and for the development of therapeutic strategies for cilia-related diseases. Importantly, the analysis of cilia in organoids relies on the development of cutting-edge imaging techniques to visualize motile and non-motile cilia in distinct cell types at the subcellular scale. We present here a protocol for immunofluorescence staining of primary cilia in mammary stem cell-derived organoids.

2 Buffers

1. Ammonium Chloride: 0.8 g of ammonium chloride are dissolved in 100 mL of a solution of 0.1 mM EDTA. The pH is adjusted to 7.4 with NaOH.
2. Dissociation buffer: 10 mL of fetal bovine serum (FBS) are added to 500 mL of Hank's Balanced Salt Solution with 10 mM Hepes (HBSS, Gibco).
3. Fixative solution: 125 μL of 16% paraformaldehyde (PFA, Electron Microscopy Sciences) are diluted in 375 μL of phosphate-buffered saline (PBS) to generate a 4% PFA solution.
4. Permeabilization buffer: 1.5 μL of Triton X100 are diluted in 500 μL of PBS, to produce a 0.3% Triton X100 solution.
5. Blocking buffer: 1 μL of Tween-20 and 50 μL of normal goat serum are added in 950 μL PBS, to generate a 5% goat serum-0.1% Tween 20 solution.

3 Mammary stem cell isolation

1. Mammary glands from adult female mice (8–16 weeks old) are collected and stored in DMEM/F-12 (Note 1-2, Fig. 1A).
2. Glands are transferred with a 10 mL pipette with the extremity cut in a 10 cm Petri dish.
3. Glands are chopped in small pieces with a razor blade.
4. Glands are incubated in collagenase/hyaluronidase (STEMCELL Technologies) solution diluted 1:10 in DMEM/F-12 medium (Gibco) for 5 h at 37 °C on a shaker at 200 rpm (Note 3).
5. Dissociated glands are vortexed vigorously twice for 30 s and pelleted by centrifuging them at 450 × g for 5 min.
6. Tissue fragments are resuspended in a solution of ammonium chloride diluted 4:1 in dissociation buffer and pelleted (Note 4).
7. Tissue fragments are resuspended with a solution of Trypsin-EDTA and mixed by pipetting them up and down for 2 min (0.25%, STEMCELL Technologies). Dissociation buffer is added in trypsin solution before centrifugation (Note 5).
8. The cell aggregates are resuspended in a solution of dispase (5 U/mL, STEMCELL Technologies) containing DNAse I (0.1 mg/mL) and mixed by pipetting them up and down for 1 min.

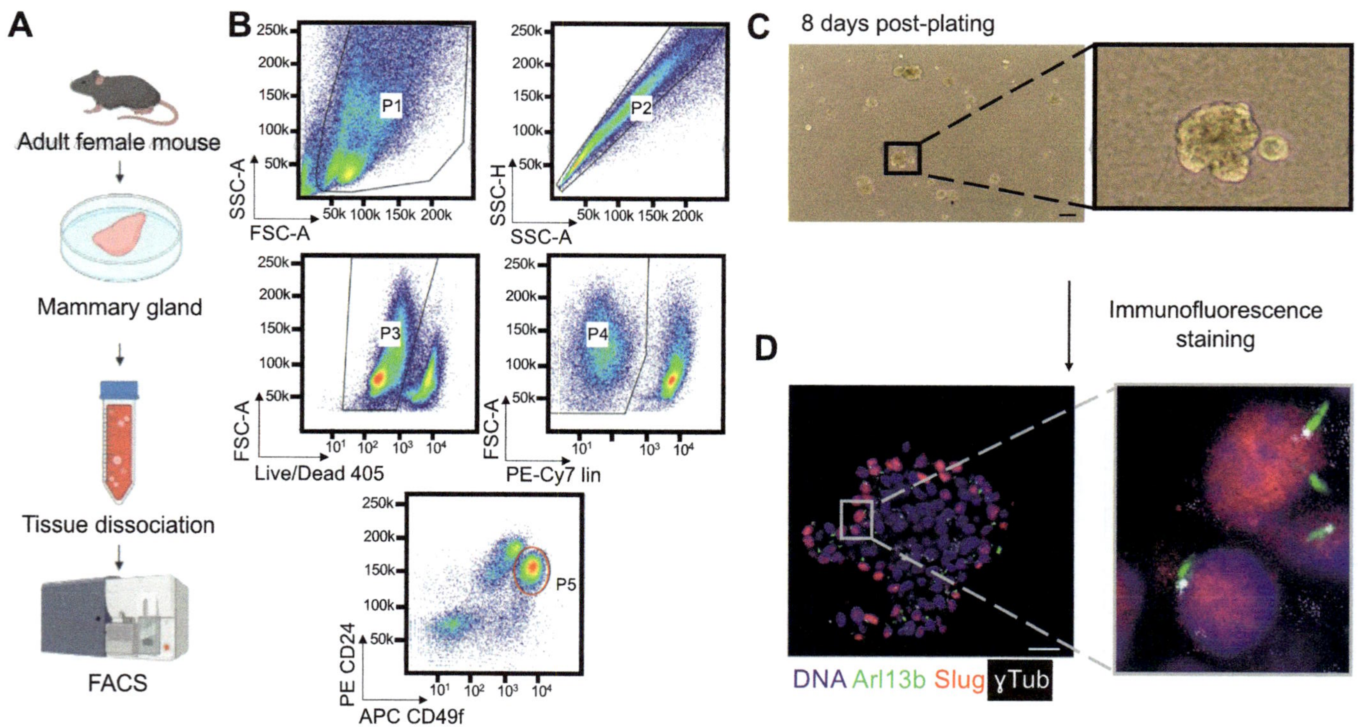

FIG. 1

Mammary-stem cell derived organoids are composed of cells that assemble primary cilia. (A) Schematic representation of the mammary stem cell isolation procedure by FACS (fluorescence activated cell sorting). (B) Gating strategy for the isolation of lin^{-}/$CD49f^{high}$/$CD24^{med}$ cells that correspond to mammary-stem cell-enriched basal cells (circled in red in the bottom plot). Lin: lineage, SSC-A: Side scatter area, FSC-A: Forward scatter area (C) Representative brightfield images of growing mammary organoids 8 days post-plating of mammary stem cells. Scale bar: 100μm, Inset: 4× (D) Mammary organoids were stained for the indicated proteins. DNA was labeled with Hoechst. A representative image is shown. Organoids were analyzed using a light sheet microscope and the image was processed using Fiji. Primary cilia are clearly apparent on a subset of Slug-expressing cells (12.97%±0.7769%), scale bar 20 μm, inset: 10×.

9. The dissociated cells are diluted in dissociation buffer, filtered through a 40 μm cell strainer, and pelleted (Note 6).
10. Cells are resuspended in dissociation buffer and counted (Note 7).
11. Dyes are added to the cell suspension (Note 8).
12. Cells are incubated 20 min on ice in the dark.
13. Cells are washed once with dissociation buffer (Note 9).
14. Cells are resuspended in dissociation buffer at a concentration of 15 million cells per mL for FACS sorting of the $lin^{-}/CD49f^{high}/CD24^{med}$ cells, which correspond to mammary stem cells (MaSC)-enriched basal cells (Fig. 1B).

4 Mammary organoid culture and long-term maintenance

1. MaSC-enriched basal cells are resuspended in epicult medium (STEMCELL Technologies) supplemented with 5% FBS, 10 ng/mL Human Recombinant EGF, 10 ng/mL Human Recombinant bFGF, 4 μg/mL heparin, 5 μM Y-27632, penicillin/streptomycin, 5% matrigel (Corning) and seeded in ultra-low attachment plate (Note 10).
2. 500 μL of fresh medium is added every 2–3 days.
3. Organoids generally reach 100 μm in size 7–10 days post-plating (Fig. 1C). Organoids can be analyzed or dissociated for long-term culture.
4. For long-term culture, organoids are transferred from the well to a low binding polymer tube (1.5 mL, Eppendorf), with a 1 mL tip, for which the extremity is cut.
5. The tube is filled with PBS.
6. Organoids are pelleted by centrifuging them at 2000 rpm for 30 s.
7. The pelleted organoids are resuspended in 500 μL of TrypLE Express (Gibco) by pipetting them up and down and incubated for 10 min at 37 °C.
8. Dissociated cells are diluted in 500 μL of dissociation buffer, mixed by pipetting them up and down, and pelleted by centrifuging them at 1000 rpm for 3 min.
9. Cells are resuspended in 500 μL of fresh medium and seeded in ultra-low attachment plate.
10. 500 μL of fresh medium is added every 2–3 days.
11. Organoids reach once again 100 μm in size 7–10 days post-plating.

5 Immunofluorescence staining of primary cilia

1. Organoids are pipetted up and down in the culture well with a 1 mL tip, for which the extremity is cut, and are transferred from the well to a low binding polymer tube (1.5 mL, Eppendorf).
2. Organoids are pelleted by low-speed centrifugation at 2000 rpm for 30 s and washed with 500 μL of PBS before another centrifugation step (Note 11).
3. Pelleted organoids are resuspended in 500 μL of fixative solution and incubated at room temperature for 30 min before centrifugation.

4. Fixed organoids are washed with 500 μL of PBS for 5 min and pelleted by centrifugation.
5. Organoids are permeabilized by resuspending them in 500 μL of PBS-Triton X100 0.3% and incubating them for 30 min at room temperature. Organoids are pelleted by centrifugation and washed with 500 μL of PBS for 5 min and pelleted again by centrifugation.
6. Non-specific antibody binding sites are blocked in organoids by resuspending them with 500 μL of blocking buffer (PBS, 5% goat serum, 0.1% Tween 20) and incubating them 1 h and 30 min at room temperature. Organoids are pelleted by centrifugation and washed with 500 μL of PBS for 5 min and pelleted again by centrifugation.
7. Organoids are labeled with 200 μL of blocking buffer with diluted primary antibodies (anti-Arl13b NeuroMab 73-287 1:200, anti-γTub Sigma-Aldrich T6557 1:1000, anti-Slug Cell Signaling Technology 9585 1:500, Note 12) and incubated overnight at 4 °C with mild shaking (60 rpm on a horizontal shaker). The tubes are placed with a 45 degree angle with the horizontal plan of the shaker.
8. Organoids are pelleted by centrifugation and washed in PBS for 5 min. The PBS wash is repeated twice (Note 13).
9. Organoids are labeled with 200 μL of blocking buffer containing diluted secondary antibodies (anti-mouse IgG2A 488 ThermoFisher Scientific A21131 1:1000, anti-rabbit 546 ThermoFisher Scientific A11035 1:1000, anti-mouse IgG1 647 ThermoFisher Scientific A21240 1:1000) and Hoechst 33342 (ThermoFisher Scientific 1:5000).
10. Labeled organoids are pelleted by centrifugation and washed in PBS for 5 min. The PBS wash is repeated twice.
11. Organoids can be mounted for imaging studies of cilia through confocal or light sheet microscopy (Fig. 1D).

6 Concluding remarks

The protocol presented here enables growth and immunofluorescence staining of MaSC-derived organoids for three-dimensional imaging studies of primary cilia in fixed three-dimensional structures at the single cell level and subcellular scale. It offers the possibility to study the biology of the cilium in MaSCs during *ex vivo* organogenesis. The protocol is presumably applicable to the staining of distinct types of organoids. Importantly, mammary organoids are 100–200 μm in size. Fixation, permeabilization and labeling steps of the immunofluorescence staining protocol must be adapted for the staining of organoids of different size and morphology. Triton X-100 is used here to permeabilize organoids in the presented protocol. However, distinct permeabilization agents, such as methanol or saponin, may be required depending on the type organoids used and on the desired type of staining.

The protocol discussed here does not include optical clearing of organoids. However, optical clearing is presumably compatible with the presented protocol. Optical clearing of organoids may be required for the analysis of organoids that are larger that 100 μm. Labeled organoids can be used for light sheet or confocal microscopy.

Organoid preparation for these two distinct types of microscopy will differ significantly, as described elsewhere (Duclos et al., 2021; van Ineveld, Ariese, Wehrens, Dekkers, & Rios, 2020). Light sheet microscopy enables faster image acquisition and limits photobleaching of samples in comparison to confocal microscopy. It provides the ability to analyze many organoids in a short period of time and allows for robust statistical analysis when quantitative measurements are performed during image processing. Although confocal microscopy is more time-consuming than light sheet microscopy, it enables three-dimensional analysis of organoids and cilia with better resolution. Thus, while light sheet enables standard quantification of ciliogenesis, more precise characterization of ciliogenesis defects may require analysis through confocal microscopy. The three-dimensional analysis of cilia in a large number of organoids will generate large data files that require time-consuming post-acquisition image processing and advanced computers, which is a limitation that should be considered prior to acquisition.

In conclusion, the emergence of distinct organoid models offers great promise for the development of our knowledge on mechanisms of ciliogenesis, ciliary signaling, and ciliary beating in space and time, and in distinct tissue contexts. These models will help reveal the mechanisms by which cilia contribute to tissue development, tissue homeostasis, and adult tissue regeneration in distinct physiological and pathological settings. Emerging studies have already demonstrated the value of organoids for cilia research. These studies have also highlighted the requirement for innovative methodologies to both genetically manipulate stem cell-derived organoids and analyze the phenotypes of these three-dimensional structures for groundbreaking research on the role of cilia in health and disease.

7 Notes

1. Mice are housed and handled in accordance with protocols approved by an Animal Care Committee.
2. Inguinal, abdominal and thoracic mammary glands from 2 or more mice are used for the experiment.
3. 10 mL of dissociation solution per mouse is used.
4. 7.5 mL of ammonium chloride solution per mouse is used
5. 10 mL of Trypsin-EDTA solution and 3 mL of dissociation buffer per mouse is used. The pellet is loose and caution is used when removing the supernatant.
6. 10 mL of dissociation buffer per mouse.
7. 2.5 mL of dissociation buffer is used per mouse.
8. 5 μL of APC anti-CD49f (Biolegend #313616), 1 μL PE anti CD24 (Biolegend #101808), 1 μL of cell viability dye (ThermoFisher Scientific, LIVE/DEAD 405), 1 μL PE/Cy7 anti-CD31 (Biolegend #102417), 1 μL PE/Cy7 anti-CD45 (Biolegend #103113), 1 μL PE/Cy7 anti-TER119 (Biolegend #116221) per million of cells. A complete list of antibodies used for FACS is presented in Table 1.
9. 2.5 mL of dissociation buffer per mouse.

Table 1 List of antibodies used for FACS purification of MaSCs and for organoid staining.

Name	Source	Company	Catalog number	Application	Dilution
Anti-Arl13b	Mouse	NeuroMab	73–287	Organoid staining	1:200
Anti-γTubulin	Mouse	Sigma Aldrich	T6557	Organoid staining	1:1000
Anti-Slug	Rabbit	Cell Signaling Technology	9585	Organoid staining	1:500
Anti-mouse IgG2A 488	Goat	ThermoFisher Scientific	A21131	Organoid staining	1:1000
Anti-rabbit 546	Goat	ThermoFisher Scientific	A11035	Organoid staining	1:1000
Anti-mouse IgG1 647	Goat	ThermoFisher Scientific	A21240	Organoid staining	1:1000
Hoechst 33342	/	ThermoFisher Scientific	B2261	Organoid staining	1:5000
APC anti-CD49f	Rat IgG2a	Biolegend	313616	FACS	5 μL per million cells
PE/Cy7 anti-TER-119	Rat IgG2b	Biolegend	116221	FACS	1 μL per million cells
PE/Cy7 anti-CD31	Rat IgG2a	Biolegend	102417	FACS	1 μL per million cells
PE/Cy7 anti-CD45	Rat IgG2b	Biolegend	103113	FACS	1 μL per million cells
PE anti-CD24	Rat IgG2b	Biolegend	101808	FACS	1 μL per million cells
LIVE/DEAD 405	/	ThermoFisher Scientific	L34965	FACS	1 μL per million cells

10. Matrigel is added to cold medium just before adding the cells. 15,000 cells are plated in 500 μL of organoid medium per well in 24-well ultra-low attachment plate.
11. Centrifugation is performed at the same speed for the entire experiment. Medium is added on the organoids without touching them to limit the attachment of the structures to the tip and progressive loss of the organoids during the experiment.
12. A list of antibodies used for immunofluorescence staining is presented in Table 1.
13. Organoids may stick to the side of the tube in the last step. Caution is used during aspiration of the supernatant to limit the loss of organoids.

Acknowledgments

We thank Molly M. Wilson for critical reading of the manuscript. This work was supported by the SIRIC Imaging and Longitudinal Investigations to Ameliorate Decision-making (ILIAD, INCa DGOS INSERM-12558) and the INSERM.

References

An, H. L., Kuo, H. C., & Tang, T. K. (2022). Modeling human primary microcephaly with hiPSC-derived brain organoids carrying CPAP-E1235V disease-associated mutant protein. *Frontiers in Cell and Development Biology*, *10*, 830432. https://doi.org/10.3389/fcell.2022.830432.

Anvarian, Z., Mykytyn, K., Mukhopadhyay, S., Pedersen, L. B., & Christensen, S. T. (2019). Cellular signalling by primary cilia in development, organ function and disease. *Nature Reviews. Nephrology*, *15*(4), 199–219. https://doi.org/10.1038/s41581-019-0116-9.

Bloodgood, R. A. (2009). From central to rudimentary to primary: The history of an underappreciated organelle whose time has come. The primary cilium. *Methods in Cell Biology*, *94*, 3–52. https://doi.org/10.1016/S0091-679X(08)94001-2.

Boutaud, L., Michael, M., Banal, C., Calderon, D., Farcy, S., Pernelle, J., et al. (2022). 2D and 3D Human induced pluripotent stem cell-based models to dissect primary cilium involvement during neocortical development. *Journal of Visualized Experiments*, *181*. https://doi.org/10.3791/62667.

Carvalho-Santos, Z., Azimzadeh, J., Pereira-Leal, J. B., & Bettencourt-Dias, M. (2011). Evolution: Tracing the origins of centrioles, cilia, and flagella. *The Journal of Cell Biology*, *194*(2), 165–175. https://doi.org/10.1083/jcb.201011152.

Chen, H. Y., Kaya, K. D., Dong, L., & Swaroop, A. (2016). Three-dimensional retinal organoids from mouse pluripotent stem cells mimic in vivo development with enhanced stratification and rod photoreceptor differentiation. *Molecular Vision*, *22*, 1077–1094. Retrieved from https://www.ncbi.nlm.nih.gov/pubmed/27667917.

Cruz, N. M., Reddy, R., McFaline-Figueroa, J. L., Tran, C., Fu, H., & Freedman, B. S. (2022). Modelling ciliopathy phenotypes in human tissues derived from pluripotent stem cells with genetically ablated cilia. *Nature Biomedical Engineering*, *6*(4), 463–475. https://doi.org/10.1038/s41551-022-00880-8.

Duclos, M., Prigent, C., Le Borgne, R., & Guen, V. J. (2021). Three-dimensional imaging of organoids to study primary ciliogenesis during ex vivo organogenesis. *Journal of Visualized Experiments*, *171*. https://doi.org/10.3791/62365.

Eguether, T., & Hahne, M. (2018). Mixed signals from the cell's antennae: Primary cilia in cancer. *EMBO Reports*, *19*(11). https://doi.org/10.15252/embr.201846589.

Forbes, T. A., Howden, S. E., Lawlor, K., Phipson, B., Maksimovic, J., Hale, L., et al. (2018). Patient-iPSC-derived kidney organoids show functional validation of a ciliopathic renal phenotype and reveal underlying pathogenetic mechanisms. *American Journal of Human Genetics*, *102*(5), 816–831. https://doi.org/10.1016/j.ajhg.2018.03.014.

Ford, M. J., Yeyati, P. L., Mali, G. R., Keighren, M. A., Waddell, S. H., Mjoseng, H. K., et al. (2018). A cell/cilia cycle biosensor for single-cell kinetics reveals persistence of cilia after G1/S transition is a general property in cells and mice. *Developmental Cell*, *47*(4), 509–523 e505. https://doi.org/10.1016/j.devcel.2018.10.027.

Gabriel, E., Wason, A., Ramani, A., Gooi, L. M., Keller, P., Pozniakovsky, A., et al. (2016). CPAP promotes timely cilium disassembly to maintain neural progenitor pool. *The EMBO Journal*, *35*(8), 803–819. https://doi.org/10.15252/embj.201593679.

Goetz, S. C., & Anderson, K. V. (2010). The primary cilium: A signalling centre during vertebrate development. *Nature Reviews. Genetics*, *11*(5), 331–344. https://doi.org/10.1038/nrg2774.

Guen, V. J., Chavarria, T. E., Kroger, C., Ye, X., Weinberg, R. A., & Lees, J. A. (2017). EMT programs promote basal mammary stem cell and tumor-initiating cell stemness by inducing primary ciliogenesis and Hedgehog signaling. *Proceedings of the National Academy of Sciences of the United States of America*. https://doi.org/10.1073/pnas.1711534114.

Guen, V. J., Edvardson, S., Fraenkel, N. D., Fattal-Valevski, A., Jalas, C., Anteby, I., et al. (2017). A homozygous deleterious CDK10 mutation in a patient with agenesis of corpus callosum, retinopathy, and deafness. *American Journal of Medical Genetics*. https://doi.org/10.1002/ajmg.a.38506.

Guen, V. J., Gamble, C., Perez, D. E., Bourassa, S., Zappel, H., Gartner, J., et al. (2016). STAR syndrome-associated CDK10/Cyclin M regulates actin network architecture and ciliogenesis. *Cell Cycle*, *15*(5), 678–688. https://doi.org/10.1080/15384101.2016.1147632.

Guen, V. J., & Prigent, C. (2020). Targeting primary ciliogenesis with small-molecule inhibitors. *Cell Chemical Biology*, *27*(10), 1224–1228. https://doi.org/10.1016/j.chembiol.2020.07.018.

Haider, S., Gamperl, M., Burkard, T. R., Kunihs, V., Kaindl, U., Junttila, S., et al. (2019). Estrogen signaling drives ciliogenesis in human endometrial organoids. *Endocrinology*, *160*(10), 2282–2297. https://doi.org/10.1210/en.2019-00314.

Hazan, R., Mori, M., Danielian, P. S., Guen, V. J., Rubin, S. M., Cardoso, W. V., et al. (2021). E2F4's cytoplasmic role in multiciliogenesis is mediated via an N-terminal domain that binds two components of the centriole replication machinery, Deup1 and SAS6. *Molecular Biology of the Cell*, *32*(20), ar1. https://doi.org/10.1091/mbc.E21-01-0039.

Ishikawa, H., & Marshall, W. F. (2011). Ciliogenesis: Building the cell's antenna. *Nature Reviews. Molecular Cell Biology*, *12*(4), 222–234. https://doi.org/10.1038/nrm3085.

Kessler, M., Hoffmann, K., Brinkmann, V., Thieck, O., Jackisch, S., Toelle, B., et al. (2015). The notch and Wnt pathways regulate stemness and differentiation in human fallopian tube organoids. *Nature Communications*, *6*, 8989. https://doi.org/10.1038/ncomms9989.

Reiter, J. F., & Leroux, M. R. (2017). Genes and molecular pathways underpinning ciliopathies. *Nature Reviews. Molecular Cell Biology*, *18*(9), 533–547. https://doi.org/10.1038/nrm.2017.60.

Sachs, N., Papaspyropoulos, A., Zomer-van Ommen, D. D., Heo, I., Bottinger, L., Klay, D., et al. (2019). Long-term expanding human airway organoids for disease modeling. *The EMBO Journal*, *38*(4). https://doi.org/10.15252/embj.2018100300.

Satir, P., Pedersen, L. B., & Christensen, S. T. (2010). The primary cilium at a glance. *Journal of Cell Science*, *123*(Pt. 4), 499–503. https://doi.org/10.1242/jcs.050377.

Schembs, L., Willems, A., Hasenpusch-Theil, K., Cooper, J. D., Whiting, K., Burr, K., et al. (2022). The ciliary gene INPP5E confers dorsal telencephalic identity to human cortical organoids by negatively regulating Sonic hedgehog signaling. *Cell Reports*, *39*(7), 110811. https://doi.org/10.1016/j.celrep.2022.110811.

Spassky, N., & Meunier, A. (2017). The development and functions of multiciliated epithelia. *Nature Reviews. Molecular Cell Biology*, *18*(7), 423–436. https://doi.org/10.1038/nrm.2017.21.

Tadokoro, T., Wang, Y., Barak, L. S., Bai, Y., Randell, S. H., & Hogan, B. L. (2014). IL-6/STAT3 promotes regeneration of airway ciliated cells from basal stem cells. *Proceedings of the National Academy of Sciences of the United States of America*, *111*(35), E3641–E3649. https://doi.org/10.1073/pnas.1409781111.

van der Vaart, J., Bottinger, L., Geurts, M. H., van de Wetering, W. J., Knoops, K., Sachs, N., et al. (2021). Modelling of primary ciliary dyskinesia using patient-derived airway organoids. *EMBO Reports*, *22*(12), e52058. https://doi.org/10.15252/embr.202052058.

van Ineveld, R. L., Ariese, H. C. R., Wehrens, E. J., Dekkers, J. F., & Rios, A. C. (2020). Single-cell resolution three-dimensional imaging of intact organoids. *Journal of Visualized Experiments*, *160*. https://doi.org/10.3791/60709.

Wilson, M. M., Callens, C., Le Gallo, M., Mironov, S., Ding, Q., Salamagnon, A., et al. (2021). An EMT-primary cilium-GLIS2 signaling axis regulates mammogenesis and claudin-low breast tumorigenesis. *Science Advances*, *7*(44), eabf6063. https://doi.org/10.1126/sciadv.abf6063.

Yan, X., Zhao, H., & Zhu, X. (2016). Production of basal bodies in bulk for dense multicilia formation. *F1000Res*, *5*. https://doi.org/10.12688/f1000research.8469.1.

Zhang, W., Yang, S. L., Yang, M., Herrlinger, S., Shao, Q., Collar, J. L., et al. (2019). Modeling microcephaly with cerebral organoids reveals a WDR62-CEP170-KIF2A pathway promoting cilium disassembly in neural progenitors. *Nature Communications*, *10*(1), 2612. https://doi.org/10.1038/s41467-019-10497-2.

Zheng, R., Yang, W., Wen, Y., Xie, L., Shi, F., Lu, D., et al. (2022). Dnah9 mutant mice and organoid models recapitulate the clinical features of patients with PCD and provide an excellent platform for drug screening. *Cell Death & Disease*, *13*(6), 559. https://doi.org/10.1038/s41419-022-05010-5.

Zhou, J., Li, C., Sachs, N., Chiu, M. C., Wong, B. H., Chu, H., et al. (2018). Differentiated human airway organoids to assess infectivity of emerging influenza virus. *Proceedings of the National Academy of Sciences of the United States of America*, *115*(26), 6822–6827. https://doi.org/10.1073/pnas.1806308115.

CHAPTER

12

Using in vivo cerebellar electroporation to study neuronal cell proliferation and differentiation in a Joubert syndrome mouse model

Chia-Hsiang Chang†, Ting-Yu Chen, and Tang K. Tang*

Institute of Biomedical Sciences, Academia Sinica, Taipei, Taiwan

**Corresponding author: e-mail address: tktang@ibms.sinica.edu.tw*

Chapter outline

†Present address: Department of Ophthalmology, UCSF, San Francisco, CA, United States.

Methods in Cell Biology, Volume 175, ISSN 0091-679X, https://doi.org/10.1016/bs.mcb.2022.07.021
Copyright © 2023 Elsevier Inc. All rights reserved.

Abstract

Joubert syndrome (JS) is an autosomal recessive ciliopathy that mainly affects the morphogenesis of the cerebellum and brain stem. To date, mutations in at least 39 genes have been identified in JS; all these gene-encoding proteins are involved in the biogenesis of the primary cilium and centrioles. Recent studies using the mouse model carrying deleted or mutated JS-related genes exhibited cerebellar hypoplasia with a reduction in neurogenesis; however, investigating specific neuronal behaviors during their development in vivo remains challenging. Here, we describe an in vivo cerebellar electroporation technique that can be used to deliver plasmids carrying GFP and/or shRNAs into the major cerebellar cell type, granule neurons, from their progenitor state to their maturation in a spatiotemporal-specific manner. By combining this method with cerebellar immunostaining and EdU incorporation, these approaches enable the investigation of the cell-autonomous effect of JS-related genes in granule neuron progenitors, including the pathogenesis of ectopic neurons and the defects in neuronal differentiation. This approach provides information toward understanding the multifaceted roles of JS-related genes during cerebellar development in vivo.

1 Introduction

The primary cilium is a microtubule-based organelle protruding from the basal body, originally derived from the mother centriole, serving as a cellular antenna for receiving extracellular stimuli. The granule neuron progenitors (GNPs), which are the main progenitor cells residing in the external granular layer (EGL) of the developing cerebellum, possess primary cilia. During cerebellum development, GNPs retract their primary cilia upon their differentiation and migration toward the internal granular layer (IGL), the destination of mature granule neurons (Chang et al., 2019). Because the transduction of sonic hedgehog (Shh) signaling relies on the proper function of primary cilia (Goetz & Anderson, 2010) and GNPs require Shh signaling for their proliferation (Kenney & Rowitch, 2000), the primary cilium-mediated transduction of Shh signaling is critical for proper cerebellar development. For example, impeding the function of Kif3a, a kinesin-II motor protein required for ciliogenesis, depletes the pool of cerebellar progenitors (Spassky et al., 2008). In line with this, defects in ciliogenesis in the developing cerebellum lead to Joubert syndrome (JS), an autosomal recessive ciliopathy that mainly affects the morphogenesis of the cerebellum and brain stem (Parisi, Doherty, Chance, & Glass, 2007).

The classical features of JS are congenital cerebellar ataxia with the molar tooth sign in radiology (Romani, Micalizzi, & Valente, 2013). At least 39 genes have been identified as causal genes in JS patients, all of which encode proteins in centriole–cilium complexes (OMIM:213300). In mouse models with deleting or mutating JS-related genes, the impairment of the global proliferation capacity of GNPs leads to cerebellar hypoplasia with decreased cerebellar foliation

(Bashford & Subramanian, 2019; Damerla et al., 2015; Wu et al., 2014). Reduced neurogenesis is not the sole feature of JS in several neuropathological examinations; additional phenotypes, such as cerebellar heterotopia, and defects in neuronal differentiation have also been reported (Friede & Boltshauser, 1978; Juric-Sekhar, Adkins, Doherty, & Hevner, 2012; ten Donkelaar, Hoevenaars, & Wesseling, 2000; van Dorp, Palan, Kwee, Barth, & van der Harten, 1991; Yachnis & Rorke, 1999). Therefore, a new approach for studying the pathogenesis of JS is crucial for understanding the diverse pathological aspects observed in JS patients.

Here, we describe an electroporation-based approach for studying the effects of JS-related genes in GNPs, which shows spatiotemporal precision during cerebellar development. Combining this technique with 5-ethynyl-2′-deoxyuridine (EdU) labeling enables the investigation of the cell-autonomous effects of genes of interest in the proliferation and differentiation of GNPs. Thus, using these advanced technologies, we can investigate the key events regulating neurogenesis and identify the underlining mechanism of the presence of ectopic neurons in JS-associated neurodevelopmental disorders (Chang et al., 2021).

2 In vivo cerebellar electroporation

The protocol described here is a modification of a previously published method (Chang et al., 2019; Chang et al., 2021; Umeshima, Hirano, & Kengaku, 2007).

2.1 Materials

The materials used included:

- Ice;
- Plasmids of interests (purified by QIAGEN EndoFree plasmids kits, see Note 1);
- Fast green FCF dye (Sigma);
- Sterilized surgical instruments (Scissors, Forceps, Mosquito Hemostats, 3-O Nylon surgical sutures);
- 26-gauge needle;
- Kite-R (KITE Manual Micromanipulator, World Precision Instrument);
- M10 Magnetic Stand (World Precision Instrument);
- Hook-type electrode;
- Bifurcate electric wire;
- NanoFil 10 μL syringe (World Precision Instrument);
- 33-gauge needles (NF33BL-2, World Precision Instrument);
- Platinum Tweezertrode (5 mm diameter, BTX);
- Heating pads;
- Electroporator (CUY21EDIT, BEX) (see the materials in Fig. 1A).

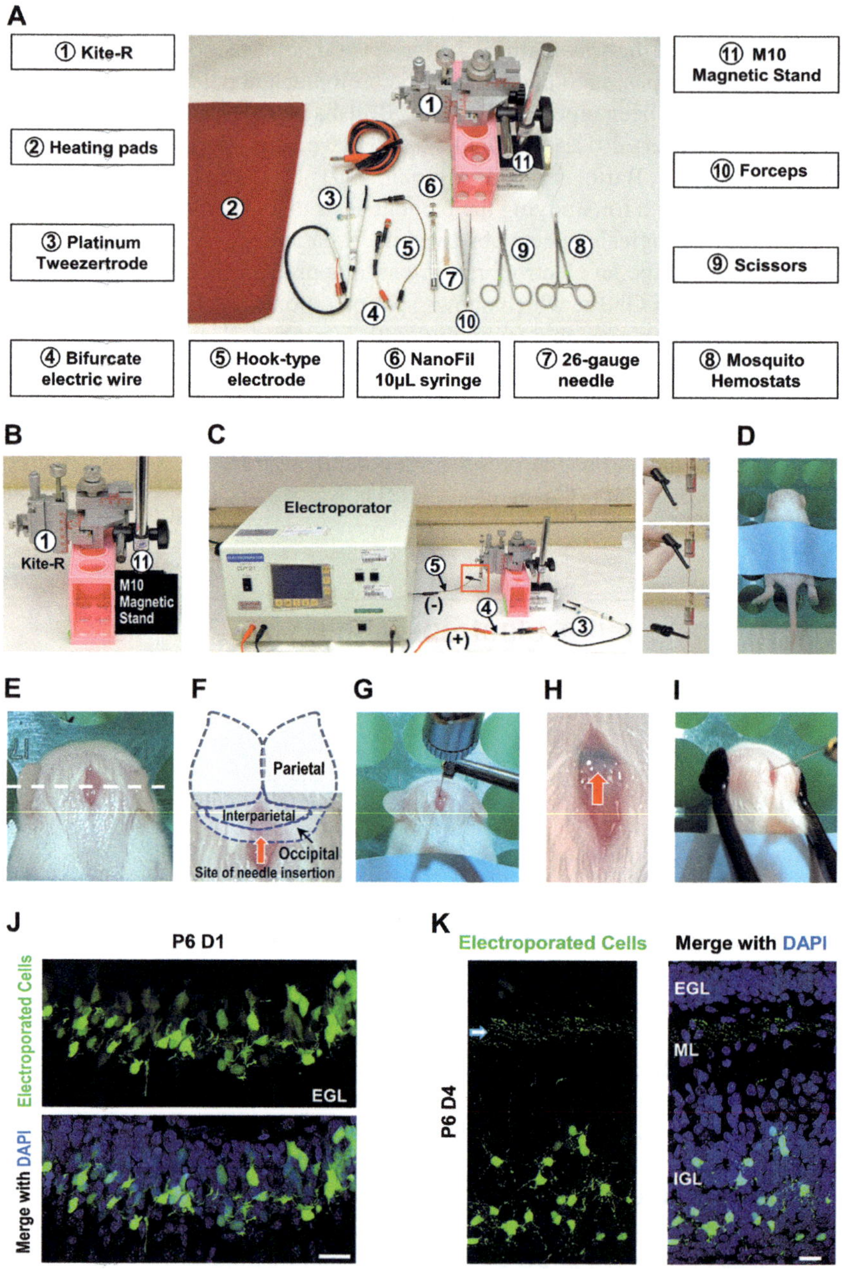

FIG. 1

See figure legend on opposite page.

2.2 Before you begin

The pre-process is as follows:

a. Mix the plasmids of interest with the Fast green FCF dye (see Note 2);
b. Set up Kite-R on a M10 Magnetic Stand (Fig. 1B);
c. Pull the plasmid/Fast green dye mixture into a NanoFil 10 μL syringe and insert a 33-gauge needle into the syringe;
d. Set up the NanoFil 10 μL syringe on the Kite-R;
e. Connect the Platinum Tweezertrode to the bifurcate electric wire and link to the positive pole (+) of the electroporator, while connecting the hook to the 33-gauge needle and linking the hook-type electrode to the negative pole (−) of the electroporator (Fig. 1C).

2.3 Procedure

The procedure is as follows:

a. Anesthetize mouse pups from postnatal day 6 (P6) or 7 (P7) (see Note 3) on ice for 2–4 min (see Note 4);

FIG. 1—Cont'd The instruments used for in vivo cerebellar electroporation and representative images after cerebellar electroporation. (A) Experimental instruments used for in vivo cerebellar electroporation. All instruments are labeled as indicated. (B) Instrument photos of the Kite-R and M10 magnetic stand set-up. (C) The connection of the platinum Tweezertrode, bifurcate electric wire, and hook-type electrode to the NanoFil syringe, and to the positive pole (+) and negative pole (−) of the electroporator. The red rectangle in (C) is magnified at right that shows the sequential order that connects the hook-type electrode to the needle of the NanoFil syringe. (D) A representative image depicting the location on the anesthetized mouse pup used for the in vivo cerebellar electroporation. Before cerebellar electroporation, the experimental incision position is determined (E, F). A representative image shows the incision site at the midline of two ears in an anesthetized mouse pup (E). The incision site is marked by a red arrow at the occipital bone (F). (G, H) After the incision site is determined, a 33-gauge needle is inserted into the incision site (G) and the DNA solution containing Fast green FCF dye (as a marker) is injected into the primary fissure of cerebellum. With a successful injection, a green line (marked by a red arrow) appears underneath the skull (H). (I) The set-up of tweezer-type electrodes for electroporation. (J) The appearance of cells at day 1 (D1) after electroporation. Postnatal day 6 (P6) cerebella are electroporated with plasmids encoding GFP (*green*). Cerebellar sections are counterstained with DAPI (*blue*). Scale bar: 20 μm. (K) The appearance of granule neurons at day 4 (D4) after electroporation. Postnatal day 6 (P6) cerebella are electroporated with plasmids encoding GFP (*green*) for 4 days. Cerebellar sections are counterstained with DAPI (*blue*). An arrow indicates the presence of parallel fibers of granule neurons between the EGL and ML. Scale bar: 20 μm. *EGL*, external granular layer; *ML*, molecular layer; *IGL*, internal granular layer.

b. After sedation (no painful response by pinching the tail and toes), ensure the head is perpendicular to the body axis (Fig. 1D);
c. Wipe the surgical area with 70% ethanol;
d. Make a 0.5 cm incision in the midline between the ears (Fig. 1E);
e. Dissect the underlying muscles to expose the skull above the cerebellum (see Note 5);
f. Identify the border between the interparietal and occipital bones, and make a hole with a 26-gauge needle at the occipital bone, around 1–2 mm adjacent to the midpoint of the identified border (Fig. 1F);
g. Slowly insert the 33-gauge needle into the hole made on the skull, but do not insert more than 5 mm from the top of the skull (see Note 6 and Fig. 1G);
h. Inject 0.2 μL of the DNA solution each time into the cerebellum and observe the distribution pattern of the DNA solution. Whenever a green line underneath the junction of the interparietal bone and occipital bone appears, inject more DNA solution (Fig. 1H). The total amount of injected DNA is approximately 0.8 μL per cerebellum. The DNA solution used in this procedure is usually injected into the primary fissure (between lobules IV/V and VIa) of the cerebellum;
i. Apply an electric current of 70 V in 50 ms intervals with a 150 ms duration in 6 pulses (see Note 7, Fig. 1I);
j. Carefully remove the needle from the skull and transfer the mouse pup to the heating pad;
k. Insert surgical ties in the skin to heal the wound, and quickly wake up the mouse pup by gently pinching its toes or tail;
l. Wait for more than 3 h after all the surgical procedures are complete, then place the mouse pups on their original nest bedding and transfer them back to their mother.

With a successful cerebellar electroporation in vivo, you will see the expression of your genes of interest in GNPs at the external granular layer (EGL) at day 1 (24 h after electroporation, Fig. 1J). These GNPs will differentiate and initiate their radial migration through the fibers of Bergmann glial cells in the following days until they reach the internal granular layer (IGL), which is their final destination (Fig. 1K).

2.4 Notes

Note 1: The concentration of plasmids should be above 3 μg/μL.
Note 2: We usually inject 0.8 μL of plasmid/fast green mixture per cerebellum.
Note 3: As the proliferation peak of GNPs is around P4–P8 (Roussel & Hatten, 2011), we usually use P6 mouse pups for our study.
Note 4: The duration of anesthesia on ice should be adjusted according to the age of the pups and the survival rate after the whole procedure. Prolonging anesthesia on ice generally decreases the survival rate of pups after surgery.
Note 5: Do not over-dissect as this will cause bleeding after surgery, resulting in cannibalism when the mouse pups return to their mother.

Note 6: The actual depth of needle insertion varies between animals. We try to inject 0.2 μL of plasmid solution each time and adjust the depth of needle insertion. The depth of needle insertion can be determined whenever a green line is observed underneath the border between the interparietal and occipital bone (Fig. 1H), indicating that the plasmid solution is well-distributed in the primary fissure of the cerebellum.

Note 7: Bubbles surrounding the hole of needle insertion may appear when a successful electric current is produced.

3 Immunostaining markers for studying neuronal differentiation

Once a successful in vivo cerebellar electroporation is achieved, the next step is defining the proliferation and differentiation status in electroporated cells. The GNPs undergo massive proliferation in response to Shh signaling in the EGL. The centriolar satellites disperse in GNPs when they undergo differentiation, resulting in the retraction of their primary cilium in the inner surface of the EGL (Chang et al., 2019; Chizhikov et al., 2007; Di Pietro et al., 2017). Therefore, the EGL can be subdivided into two layers on the basis of their gross morphology: the outer EGL (oEGL), carrying proliferating GNPs with intact primary cilium and clustered centriolar satellites, and the inner EGL (iEGL), carrying differentiating granule neurons with a shortened primary cilium and dispersed centriolar satellites (Chang et al., 2019). In addition to their morphological changes, several classical immunostaining markers can be used to delineate the states of proliferation, differentiation, and maturation in cerebellar granule neurons, including Ki67, NeuroD1, Tuj1, NeuN, and $p27^{Kip1}$. The detail immunostaining method can be found in Chang et al. (2021).

a. Ki67 is a pan-cell-cycle marker present in both interphase and mitotic cells (Sun & Kaufman, 2018). The presence of Ki67 in the oEGL is a reliable marker of proliferating GNPs (Nakashima, Umeshima, & Kengaku, 2015). In wild-type GNPs, with the electroporation labeling with GFP, approximately 92.97% ± 1.17% of GFP^+ cells were $Ki67^+$ at day 1 (24 h) after in vivo cerebellar electroporation, whereas only 52.01% ± 2.17% of GFP^+ cells were $Ki67^+$ at day 2, indicating that half of the electroporated GNPs left the cell cycle and underwent differentiation (Chang et al., 2021). On day 3, as most of the cells have already differentiated, few (less than 10%) electroporated cells are positive for Ki67 (Chang et al., 2019). An example of Ki67 immunostaining in electroporated cells is shown in Fig. 2A.

b. NeuroD1 is a basic helix-loop-helix (bHLH) transcription factor that mediates the differentiation of cerebellar granule neurons (Miyata, Maeda, & Lee, 1999). Conditional knockout of NeuroD1 in cerebellar GNPs reduces the genesis of granule neurons (Pan, Jahan, Lee, & Fritzsch, 2009). The expression pattern of NeuroD1 has been well-characterized in the iEGL by in situ hybridization

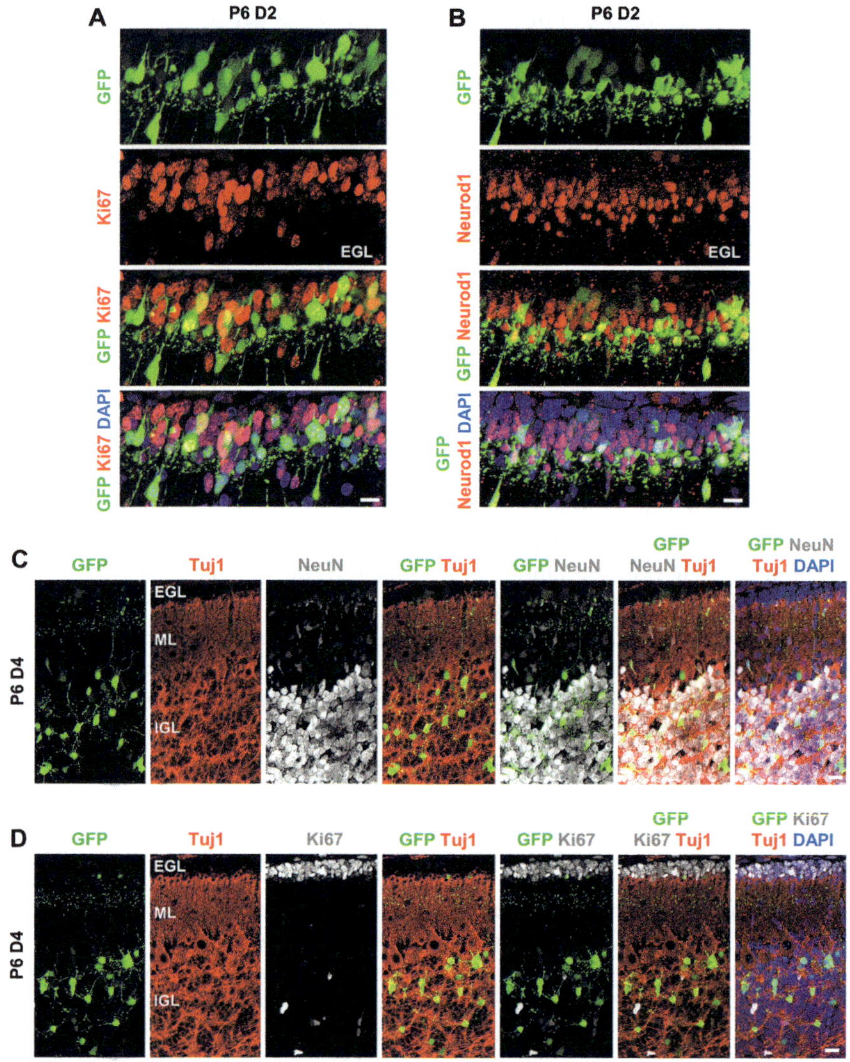

FIG. 2

Representative images of immunostaining markers used for examining neuronal cell proliferation and differentiation after cerebellar electroporation. (A, B) Representative images of Ki67 (A) and Neurod1 (B) immunostaining after cerebellar electroporation in vivo. Postnatal day 6 (P6) cerebella are electroporated with plasmids encoding GFP (*green*) for 2 days (D2), and then the sections are immunostained with Ki67 (*red* in A) or Neurod1 (*red* in B) and counterstained with DAPI (*blue*). Scale bar: 10 μm. (C, D) Representative images of immunostaining by combining Tuj1 with NeuN (C) or Ki67 (D) after cerebellar electroporation in vivo. Postnatal day 6 (P6) cerebella are electroporated with plasmids encoding GFP (*green*) for 4 days (D4), and then the sections are immunostained with Tuj1 (*red*) along with NeuN (*white* in C) or Ki67 (*white* in D) and counterstained with DAPI (*blue*). Scale bar: 20 μm. *EGL*, external granular layer; *ML*, molecular layer; *IGL*, internal granular layer.

(Miyata et al., 1999) and immunostaining (Ayrault et al., 2010), serving as an excellent marker to label newly differentiated neurons in the iEGL. In our previous study, we found that most wild-type neurons (92.88% ± 1.10%) expressed NeuroD1 48 h after in vivo cerebellar electroporation (Chang et al., 2021). An example of NeuroD1 immunostaining in electroporated cells is shown in Fig. 2B.

c. Tuj1 and NeuN: Tuj1, a neuronal-specific class III β tubulin, is expressed in newly differentiated postmitotic neurons (Menezes & Luskin, 1994). Thus, immunostaining with Tuj1 antibody can be used to label differentiated neurons and their fibers in the lower border of the iEGL, the molecular layer (ML), and the IGL of the developing cerebellum (Ayrault et al., 2010; Chang et al., 2019). The neuronal nuclear antigen, NeuN, is mostly expressed in differentiated, mature granule neurons in the IGL (Gusel'nikova & Korzhevskiy, 2015; Weyer & Schilling, 2003). Considering these two traits together, immunostaining by combining Tuj1 with either Ki67 or NeuN can reveal the identity of GNPs (labeled as $Ki67^+$ $Tuj1^-$ $NeuN^-$), granule neurons undergoing differentiation (labeled as $Ki67^-$ $Tuj1^+$ $NeuN^-$), and mature granule neurons (labeled as $Ki67^-$ $Tuj1^+$ $NeuN^+$) after a successful in vivo cerebellar electroporation (Chang et al., 2021). For example, 4 days after in vivo cerebellar electroporation, almost all wild-type neurons (93.00% ± 1.52%) were $NeuN^+$ $Tuj1^+$, while rare electroporated cells displayed as $NeuN^-$ $Tuj1^-$ (4.12% ± 2.16%) or $NeuN^-$ $Tuj1^+$ (2.88 ± 0.85%). Consistent with this result, most electroporated cells (91.08% ± 0.60%) in the wild-type background were $Ki67^-$ $Tuj1^+$ at the same time. Examples of electroporated cells coimmunostaining with Tuj1/NeuN and Tuj1/Ki67 are shown in Fig. 2C and D, respectively.

d. $p27^{Kip1}$, functioning as a universal cyclin-dependent kinase inhibitor (Lloyd et al., 1999), is highly expressed in the granule neuron lineage. The expression of $p27^{Kip1}$ can be identified in all post-mitotic granule neurons in the iEGL, ML, and IGL of the developing cerebellum (Ayrault et al., 2010; Miyazawa et al., 2000). Therefore, using $p27^{Kip1}$ as an immunostaining marker may provide similar information for differentiating granule neurons showing as $Ki67^-$.

After performing cerebellar electroporation in vivo, we usually apply Ki67 and NeuroD1 immunostaining to day 1 and day 2 cerebellar sections to reveal the proliferation and differentiation status in those electroporated cells. We typically take advantage of coimmunostaining Ki67/Tuj1 or NeuN/Tuj1 with day 3 and day 4 cerebellar slices for identifying their maturation stages in the electroporated cells. These methods can provide substantial information for studying neuronal differentiation in JS-related genes.

4 Edu incorporation and cell-cycle analysis

In addition to neuronal proliferation and differentiation, JS-related genes participate in the cell-cycle progression or the cell-cycle exit in cerebellar GNPs. Here, we describe a protocol using 5-ethynyl-2′-deoxyuridine (EdU) incorporation for studying the cell-cycle progression or the cell-cycle exit in GNPs after a successful cerebellar electroporation in vivo.

4.1 Materials

The required materials include:

- 26-gauge needle;
- Click-iT EdU Cell Proliferation Kit for Imaging, Alexa Fluor 555 dye;
- Ice;
- Weighing scale;
- Forceps;
- Heating pads;
- Butterfly scissors;
- 4% paraformaldehyde (PFA) and PBS.

4.2 Before you begin

Before you start, the following steps are required:

a. Dissolve EdU in DMSO to create a 5 mg/mL stock solution;
b. Weigh the mouse pups;
c. Prepare the EdU working solution by diluting the EdU stock solution with PBS (see Note 1);
d. Before processing immunostaining, prepare 1× Click-iT EdU reaction buffer, Alexa Fluor azide, and 10× Click-iT EdU buffer additive according to the manufacturer's protocol.

4.3 Protocol

The protocol is performed as follows:

a. Perform in vivo cerebellar electroporation in P6 mouse pups;
b. On day 1 after electroporation, anesthetize mouse pups on ice for 2–3 min (see Note 2);
c. Use the forceps to lift up the skin, and then subcutaneously inject EdU at 50 mg/kg as described (Zeng et al., 2010) with a 26-gauge needle;
d. Put mouse pups on the heating pad;
e. Return mouse pups to their mother around 1 h after injection (see Note 3);
f. For cell-cycle analysis, perform transcardial perfusion (see Note 4) at 3 h after EdU injection; for the cell-cycle exit assay, perform transcardial perfusion and collect the cerebella at 1 day after EdU injection, as the estimated cell-cycle duration of GNPs in postnatal murine cerebella is around 19 h (Fujita, 1967);
g. Perform immunostaining in cerebellar sections. For cell-cycle analysis, use the combination of antiphospho histone-H3 (pH3) and anti-GFP as primary

antibodies; for the cell-cycle exit assay, use the combination of anti-Ki67 and anti-GFP as primary antibodies (see Note 5). Complete the immunostaining using corresponding secondary antibodies;

h. Perform the EdU-labeling reaction as described in the manufacturer's protocol. We incubate the tissues in Click-iT reaction cocktail for 30 min. Protect the samples from light;

i. Finalize the immunostaining of DNA by DAPI or Hoechst labeling.

The combination of a successful in vivo cerebellar electroporation with EdU labeling enables the identification of the cell-cycle distribution of GNPs in response to the genetic manipulation in vivo. Because the cell-cycle analysis is completed within 1 day after cerebellar electroporation, we usually collect some sections for Ki67 immunostaining. This additional procedure is useful to confirm that all the electroporated GFP^+ cells are within the cell cycle. Under this circumstance, we count the percentage of cells in the G1 phase by calculating the EdU^- $pH3^-$ cells among all GFP^+ cells. We measure the percentage of GFP^+ electroporated cells in the G2/M phase by calculating the $pH3^+$ cells among all GFP^+ cells, as the expression of pH3 is noticeable in the G2 phase through the whole mitotic stage (Hendzel et al., 1997). The exposure of EdU in vivo for 3 h induces the labeling of all GNPs in the S phase and a portion of GNPs entering into the early G2/M phase (Florio et al., 2012). Therefore, we count the percentage of S-phase cells by calculating the percentage of EdU^+ cells among the GFP^+ cells, subtracting by the percentage of EdU and pH3 double-positive cells among the GFP^+ cells. Examples of EdU labeling in the electroporated cerebellum (Fig. 3A) and the representative images with summary of the formula used for cell-cycle analysis are shown in Fig. 3B and C, respectively.

Finally, as the cell-cycle exit index is defined by how many proliferative cells labeled by EdU at day 1 leave the cell cycle at day 2 after electroporation (Fig. 3D), we thus calculate the cell-cycle exit index by dividing the total number of GFP^+ $Ki67^-$ EdU^+ cells by all GFP^+ EdU^+ cells (Fig. 3E).

4.4 Notes

Note 1: We usually inject 100 μL of EdU working solution in each mouse pup at 50 mg/kg, as previously described (Zeng et al., 2010). For example, if a P6 CD-1 mouse pup is around 5 g, we mix 500 μL EdU stock solution with 500 μL PBS, producing a final EdU concentration of 2.5 mg/mL. This 1 mL of solution can be used for 10 pups.

Note 2: In order to prevent the leakage of the injected EdU solution from the skin due to the movement of mouse pups, we usually anesthetize mouse pups on ice prior to EdU injection.

Note 3: The mouse pups do not need to be returned to their mother if proceeding to a cell-cycle analysis, in which the cerebella are collected at 3 h after EdU injection.

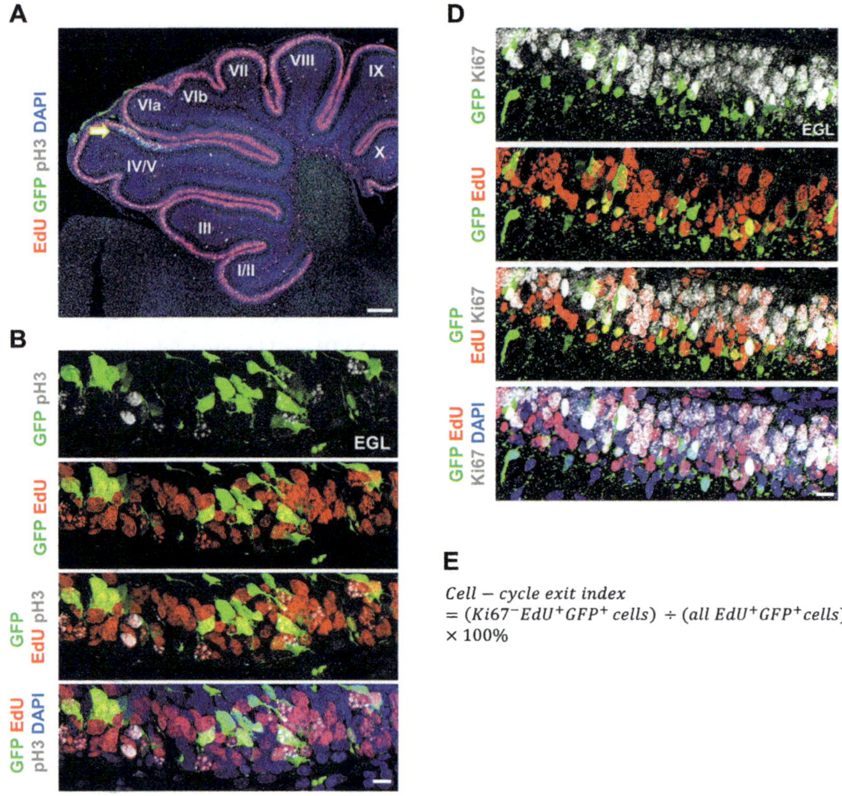

C

$$G1\ phase\ \% = (pH3^{-}EdU^{-}GFP^{+}\ cells) \div (all\ GFP^{+} cells) \times 100\%$$

$$S\ phase\ \% = (pH3^{-}EdU^{+}GFP^{+}\ cells) \div (all\ GFP^{+} cells) \times 100\%$$

$$G2\ to\ M\ phase\ \% = (pH3^{+} cells) \div (all\ GFP^{+} cells) \times 100\%$$

E

$$Cell-cycle\ exit\ index = (Ki67^{-}EdU^{+}GFP^{+}\ cells) \div (all\ EdU^{+}GFP^{+} cells) \times 100\%$$

FIG. 3

Representative images of immunostaining markers used for examining neuronal cell distribution at different cell-cycle stages after EdU incorporation. (A) A representative image showing successful in vivo cerebellar electroporation and EdU incorporation. P6 cerebella are electroporated with plasmids encoding GFP (*green*) for 1 day, following by EdU injection for 3 h. Cerebella sections are immunostained with pH3 (*gray*) and counterstained with DAPI (*blue*). EdU-labeled DNAs are detected by a Click-iT EdU Cell Proliferation Kit (*red*). The arrow indicates the distribution of electroporated cells (GFP^{+}) in the primary fissure (between lobules IV/V and VIa). Scale bar: 200 μm. (B) Magnified images with separated channels for studying the cell-cycle distribution at D1 after electroporation. P6 cerebella are treated and immunostained as described in (A). Scale bar: 10 μm. (C) Equations for counting the percentage of GFP^{+} electroporated cells in each cell-cycle phase. (D) Representative images used for studying cell-cycle exit in the developing cerebellum. P6 cerebella are electroporated with plasmids encoding GFP (*green*) for 2 days. At 24 h after electroporation, mouse pups are injected with EdU (red), and cerebella are collected after an additional 24 h. Cerebella sections are immunostained with Ki67 (*gray*) and counterstained with DAPI (*blue*). EdU-labeled DNA is detected by a Click-iT EdU Cell Proliferation Kit (*red*). Scale bar: 10 μm. (E) The equation for calculating the cell-cycle exit index.

Note 4: We anesthetized mouse pups on ice until they demonstrated no pain response to external stimuli by gently pinching their tail and toes. After exposure the thoracic cage, we create a tiny incision at the right atrium (RA) of the heart using butterfly scissors, and infuse PBS into the left ventricle (LV) of the heart until no more bloody efflux is detected in the RA. We then infuse 4% PFA (by using a half volume of PBS) into the LV for fixation.
Note 5: As the GFP signal can be quenched after the application of EdU-labeling reactions, we use anti-GFP immunostaining to detect the GFP signals of electroporated cells.

5 Concluding remarks

The methods described here provide first-hand information for understanding neuronal proliferation, differentiation, cell-cycle distribution, and cell-cycle exit by manipulating genes of interest in GNPs in vivo. By applying these EdU-labeling and cerebellar electroporation techniques, we found that knockdown of Cep120, a centriolar protein implicated in JS, impairs neuronal differentiation and cell-cycle exit in the developing cerebellum, leading to the accumulation of ectopic cells in the EGL (Chang et al., 2021). Our findings provide novel insight into the differentiation defects and ectopic neurons observed in pathological cerebellar tissues in JS patients. Moreover, a parallel analysis, by characterizing the pathological features of human JS samples with known genetic etiology, will be crucial for understanding all the aspects of JS.

Acknowledgments

We thank those the confocal imaging core facilities (IBMS, AS-CFII-108-116) of Academia Sinica and Mrs. Show-Rong Ma for confocal imaging assistance. This study was supported by grants from the Ministry of Science and Technology, Taiwan (MOST-109-2326-B001-010; MOST-110-2917-I-564-010) and Academia Sinica, Taiwan (AS-IA-109-L04).

References

Ayrault, O., Zhao, H., Zindy, F., Qu, C., Sherr, C. J., & Roussel, M. F. (2010). Atoh1 inhibits neuronal differentiation and collaborates with Gli1 to generate medulloblastoma-initiating cells. *Cancer Research*, *70*, 5618–5627.

Bashford, A. L., & Subramanian, V. (2019). Mice with a conditional deletion of Talpid3 (KIAA0586)—a model for Joubert syndrome. *The Journal of Pathology*, *248*, 396–408.

Chang, C. H., Chen, T. Y., Lu, I. L., Li, R. B., Tsai, J. J., Lin, P. Y., et al. (2021). CEP120-mediated KIAA0753 recruitment onto centrioles is required for timely neuronal differentiation and germinal zone exit in the developing cerebellum. *Genes & Development*, *35*, 1445–1460.

Chang, C. H., Zanini, M., Shirvani, H., Cheng, J. S., Yu, H., Feng, C. H., et al. (2019). Atoh1 controls primary cilia formation to allow for SHH-triggered granule neuron progenitor proliferation. *Developmental Cell*, *48*, 184–199.e185.

Chizhikov, V. V., Davenport, J., Zhang, Q., Shih, E. K., Cabello, O. A., Fuchs, J. L., et al. (2007). Cilia proteins control cerebellar morphogenesis by promoting expansion of the granule progenitor pool. *The Journal of Neuroscience: The Official Journal of the Society for Neuroscience*, *27*, 9780–9789.

Damerla, R. R., Cui, C., Gabriel, G. C., Liu, X., Craige, B., Gibbs, B. C., et al. (2015). Novel Jbts17 mutant mouse model of Joubert syndrome with cilia transition zone defects and cerebellar and other ciliopathy related anomalies. *Human Molecular Genetics*, *24*, 3994–4005.

Di Pietro, C., Marazziti, D., La Sala, G., Abbaszadeh, Z., Golini, E., Matteoni, R., et al. (2017). Primary cilia in the murine cerebellum and in mutant models of medulloblastoma. *Cellular and Molecular Neurobiology*, *37*, 145–154.

Florio, M., Leto, K., Muzio, L., Tinterri, A., Badaloni, A., Croci, L., et al. (2012). Neurogenin 2 regulates progenitor cell-cycle progression and Purkinje cell dendritogenesis in cerebellar development. *Development (Cambridge, England)*, *139*, 2308–2320.

Friede, R. L., & Boltshauser, E. (1978). Uncommon syndromes of cerebellar vermis aplasia. I: Joubert syndrome. *Developmental Medicine and Child Neurology*, *20*, 758–763.

Fujita, S. (1967). Quantitative analysis of cell proliferation and differentiation in the cortex of the postnatal mouse cerebellum. *The Journal of Cell Biology*, *32*, 277–287.

Goetz, S. C., & Anderson, K. V. (2010). The primary cilium: A signalling Centre during vertebrate development. *Nature Reviews Genetics*, *11*, 331–344.

Gusel'nikova, V. V., & Korzhevskiy, D. E. (2015). NeuN as a neuronal nuclear antigen and neuron differentiation marker. *Acta Naturae*, *7*, 42–47.

Hendzel, M. J., Wei, Y., Mancini, M. A., Van Hooser, A., Ranalli, T., Brinkley, B. R., et al. (1997). Mitosis-specific phosphorylation of histone H3 initiates primarily within pericentromeric heterochromatin during G2 and spreads in an ordered fashion coincident with mitotic chromosome condensation. *Chromosoma*, *106*, 348–360.

Juric-Sekhar, G., Adkins, J., Doherty, D., & Hevner, R. F. (2012). Joubert syndrome: Brain and spinal cord malformations in genotyped cases and implications for neurodevelopmental functions of primary cilia. *Acta Neuropathologica*, *123*, 695–709.

Kenney, A. M., & Rowitch, D. H. (2000). Sonic hedgehog promotes G(1) cyclin expression and sustained cell cycle progression in mammalian neuronal precursors. *Molecular and Cellular Biology*, *20*, 9055–9067.

Lloyd, R. V., Erickson, L. A., Jin, L., Kulig, E., Qian, X., Cheville, J. C., et al. (1999). p27kip1: A multifunctional cyclin-dependent kinase inhibitor with prognostic significance in human cancers. *The American Journal of Pathology*, *154*, 313–323.

Menezes, J. R., & Luskin, M. B. (1994). Expression of neuron-specific tubulin defines a novel population in the proliferative layers of the developing telencephalon. *The Journal of Neuroscience: The Official Journal of the Society for Neuroscience*, *14*, 5399–5416.

Miyata, T., Maeda, T., & Lee, J. E. (1999). NeuroD is required for differentiation of the granule cells in the cerebellum and hippocampus. *Genes & Development*, *13*, 1647–1652.

Miyazawa, K., Himi, T., Garcia, V., Yamagishi, H., Sato, S., & Ishizaki, Y. (2000). A role for p27/Kip1 in the control of cerebellar granule cell precursor proliferation. *The Journal of Neuroscience: The Official Journal of the Society for Neuroscience*, *20*, 5756–5763.

Nakashima, K., Umeshima, H., & Kengaku, M. (2015). Cerebellar granule cells are predominantly generated by terminal symmetric divisions of granule cell precursors. *Developmental Dynamics: An Official Publication of the American Association of Anatomists*, *244*, 748–758.

Pan, N., Jahan, I., Lee, J. E., & Fritzsch, B. (2009). Defects in the cerebella of conditional Neurod1 null mice correlate with effective Tg(Atoh1-cre) recombination and granule cell requirements for Neurod1 for differentiation. *Cell and Tissue Research*, *337*, 407–428.

Parisi, M. A., Doherty, D., Chance, P. F., & Glass, I. A. (2007). Joubert syndrome (and related disorders) (OMIM 213300). *European Journal of Human Genetics*, *15*, 511–521.

Romani, M., Micalizzi, A., & Valente, E. M. (2013). Joubert syndrome: Congenital cerebellar ataxia with the molar tooth. *The Lancet Neurology*, *12*, 894–905.

Roussel, M. F., & Hatten, M. E. (2011). Cerebellum development and medulloblastoma. *Current Topics in Developmental Biology*, *94*, 235–282.

Spassky, N., Han, Y. G., Aguilar, A., Strehl, L., Besse, L., Laclef, C., et al. (2008). Primary cilia are required for cerebellar development and Shh-dependent expansion of progenitor pool. *Developmental Biology*, *317*, 246–259.

Sun, X., & Kaufman, P. D. (2018). Ki-67: More than a proliferation marker. *Chromosoma*, *127*, 175–186.

ten Donkelaar, H. J., Hoevenaars, F., & Wesseling, P. (2000). A case of Joubert's syndrome with extensive cerebral malformations. *Clinical Neuropathology*, *19*, 85–93.

Umeshima, H., Hirano, T., & Kengaku, M. (2007). Microtubule-based nuclear movement occurs independently of centrosome positioning in migrating neurons. *Proceedings of the National Academy of Sciences of the United States of America*, *104*, 16182–16187.

van Dorp, D. B., Palan, A., Kwee, M. L., Barth, P. G., & van der Harten, J. J. (1991). Joubert syndrome: A clinical and pathological description of an affected male and a female fetus from the same sibship. *American Journal of Medical Genetics*, *40*, 100–104.

Weyer, A., & Schilling, K. (2003). Developmental and cell type-specific expression of the neuronal marker NeuN in the murine cerebellum. *Journal of Neuroscience Research*, *73*, 400–409.

Wu, C., Yang, M., Li, J., Wang, C., Cao, T., Tao, K., et al. (2014). Talpid3-binding centrosomal protein Cep120 is required for centriole duplication and proliferation of cerebellar granule neuron progenitors. *PLoS ONE*, *9*, e107943.

Yachnis, A. T., & Rorke, L. B. (1999). Neuropathology of Joubert syndrome. *Journal of Child Neurology*, *14*, 655–659, discussion 669–672.

Zeng, C., Pan, F., Jones, L. A., Lim, M. M., Griffin, E. A., Sheline, Y. I., et al. (2010). Evaluation of 5-ethynyl-2′-deoxyuridine staining as a sensitive and reliable method for studying cell proliferation in the adult nervous system. *Brain Research*, *1319*, 21–32.

9780443185861